Marine and Coastal
Information S...

Two week
loan

Please return on or before the last
date stamped below.
Charges are made for late return.

IS 239/0799

INFORMATION SERVICES PO BOX 430, CARDIFF CF10 3XT

Marine and Coastal
Geographical Information Systems

Edited by
DAWN J.WRIGHT
Oregon State University
USA

and

DARIUS J. BARTLETT
University College Cork
IRELAND

TAYLOR & FRANCIS
Founded 1798

First published 2000 by Taylor & Francis,
11 Fetter Lane, London, EC4P 4EE

Simultaneously published in the USA and Canada by
Taylor & Francis Inc., 325 Chestnut Street, Philadelphia PA 19106

Taylor & Francis is an imprint of the Taylor & Francis Group

British Library Cataloguing in Publication Data

A catalogue record for this book is available from the British Library.

ISBN 0-7484-0862-2 (hbk)
ISBN 0-7484-0870-3 (pbk)

Library of Congress Cataloguing-in-Publication Data are available

Printed and bound by Biddles Ltd., Guildford and King's Lynn

Dedication

In memory of

Millington Lockwood
1943-1999

This book is dedicated to the memory of Millington Lockwood for the outstanding role he played in the development of the field of marine and coastal geographic information systems (GIS).

The lifetime achievements of this sincere and professional public servant cannot be conveyed succinctly. Millington Lockwood was born in San Francisco, California in 1943 and grew up in the small western New York town of Springville. He earned a B.S. degree in Meteorology and Oceanography from the State University of New York Maritime College and a M.S. degree in Environmental Systems Management--Ocean Affairs from American University. Following four years as a merchant seaman he came to work at the National Oceanic and Atmospheric Administration (NOAA).

For the past 29 years he has held a variety of positions within NOAA and was most recently the technical advisor for ocean mapping and marine GIS to the Director of the Coast and Geodetic Survey. He was actively involved in U.S. national policy with various aspects of implementation of the National Spatial Data Infrastructure (NSDI) with a special interest in the users and applications of digital marine information. His extraordinary contributions to the NSDI over the past few years are the culmination of 15 years of building and nurturing interagency relationships.

He was a prolific writer, having authored over 40 papers, most of which have focused on the importance of geospatial data in the marine environment. He also was instrumental in the dissemination of information relating to geospatial data. He worked to produce a number of books and publications to support the transfer of knowledge to as broad a community as possible. Millington was a "one-stop shop" for information on coastal and ocean data. He was an active mentor and role model to many young people just beginning their professional careers. His work ethic and professionalism have influenced young and old alike. Moreover, he worked to educate everyone at all levels. This work included assisting students as a volunteer Boy Scout counselor, sailing coordinator for the visually handicapped, science fair judge, Sunday-school teacher, and an advisor to senior level policy makers within state and local government.

People who knew Millington Lockwood best would describe him variously as professional, energetic, genuine, persistent, mercurial, humorous, and altruistic. Rarely would one find a negative word or thought associated with this man. He was an unconfined thinker, yet strongly practical, and guided by heartfelt ideals. Millington had a rare and exceptional ability to draw out the best in people without drawing attention to himself. This was true of both his professional and personal life. He was a persuasive leader and team builder, undesiring of personal credit. Millington loved his work, influenced national policy and research perspectives for spatial data, and mentored a generation of scientists, technologists, and policy makers who are now advocates for his vision and teachings.

Millington lost a courageous fight against cancer on July 22, 1999. He is survived by his wife, Susan and children, Lisa and Billy.

Contents

PART III INSTITUTIONAL ISSUES

Series Introduction

Welcome

The *Research Monographs in Geographical Information Systems* series provides a publication outlet for research of the highest quality in GIS, which is longer than would normally be acceptable for publication in a journal. The series includes single- and multiple-author research monographs, often based upon PhD theses and the like, and special collections of thematic papers.

The need

We believe that there is a need, from the point of view of both readers (researchers and practitioners) and authors, for longer treatments of subjects related to GIS than are widely available currently. We feel that the value of much research is actually devalued by being broken up into separate articles for publication in journals. At the same time, we realise that many career decisions are based on publication records, and that peer review plays an important part in that process. Therefore a named editorial board supports the series, and advice is sought from them on all submissions.

Successful submissions will focus on a single theme of interest to the GIS community, and treat it in depth, giving full proofs, methodological procedures or code where appropriate to help the reader appreciate the utility of the work in the Monograph. No area of interest in GIS is excluded, although material should demonstrably advance thinking and understanding in spatial information science. Theoretical, technical and application-oriented approaches are all welcomed.

The medium

In the first instance the majority of Monographs will be in the form of a traditional text book, but, in a changing world of publishing, we actively encourage publication on CD-ROM, the placing of supporting material on web sites, or publication of programs and of data. No form of dissemination is discounted, and prospective authors are invited to suggest whatever primary form of publication and support material they think is appropriate.

The editorial board

The Monograph series is supported by an editorial board. Every monograph proposal is sent to all members of the board which includes Ralf Bill, António Câmara, Joseph Ferreira, Pip Forer, Andrew Frank, Gail Kucera, Peter van Oostrom, and Enrico Puppo. These people have been invited for their experience in the field, of monograph writing, and for their geographic and subject diversity. Members may also be involved later in the process with particular monographs.

Future submissions

Anyone who is interested in preparing a Research Monograph, should contact either of the editors. Advice on how to proceed will be available from them, and is treated on a case by case basis.

For now we hope that you find this, the sixth in the series, a worthwhile addition to your GIS bookshelf, and that you may be inspired to submit a proposal too.

Editors:

Professor Peter Fisher
Department of Geography
University of Leicester
Leicester
LE1 7RH
UK
Phone: +44 (0) 116 252 3839
Fax: +44 (0) 116 252 3854
Email: pff1@le.ac.uk

Professor Jonathan Raper
School of Informatics
City University
Northampton Square
London
UK
Phone: +44 (0) 171 477 8000
Fax: +44 (0) 171 477 8587
Email: raper@soi.city.ac.uk

Foreword

Michael F. Goodchild

This is a book about the two-thirds of the Earth's surface that is covered by salt water, and about the interface between this marine environment and the terrestrial one that occurs along the world's coastlines. As a linear feature, the coastline in principle occupies no area, yet it has enormous significance given its proximity of the vast majority of the world's population; its role in biodiversity; and its importance as the source of much marine degradation. The book is also about geographic information systems, a rather vague term attached to software that is used for handling, displaying, analysing, and modelling information about the locations of phenomena and features on the Earth's surface. Finally, the book is about geographic information science, defined as the set of issues of fundamental importance that arise in the development and application of geographic information systems.

We know a great deal about how to apply computers to information about the terrestrial surface. Our landmasses are covered with highly accurate geodetic control networks, from which the locations of features can be surveyed with high accuracy. Our satellites collect abundant information about the radiative properties of the surface, and from these remotely sensed images, we construct detailed maps of vegetation, mineral resources, and population distribution. We use aerial photographs to construct detailed maps of landforms, cities, and land ownership, and GPS to record accurately the locations of roads, railroads, and other features. Information stored in digital form can be shared between users over electronic networks, analysed to discern patterns and anomalies, used to make predictions about future patterns, and used to support complex spatial decisions by planners and politicians.

When it comes to the marine and coastal environment, however, things are much less straightforward. Although the location of the coastline remains largely fixed (but affected daily by tides), there are no fixed features in the marine environment except at on the ocean floor, and very little of a static nature that can be mapped except for ocean depth. The tradition of multiscale mapping at public expense that has dominated the production of geographic data about the terrestrial surface simply has no equivalent for the remaining two-thirds of the Earth. Even the coastline is beset with definitional problems, and there are many instances of the same coastline being mapped by different agencies using different definitions.

The terrestrial surface provides the perfect Newtonian frame, a rigid co-ordinate system that can be used to position any feature (minor problems of crustal movement and the wobbling of the Earth's rotational axis aside). In the ocean, however, the lack of a rigid frame and the mobility of many features of interest open the possibility of representations that are object-centred, and models that are Lagrangian rather than Eulerian. The marine environment suggests interesting possibilities for radically different ways of representing spatial phenomena that avoid the dependence on an absolute co-ordinate system so characteristic of GIS.

GIS is also founded on the assumption that the two horizontal dimensions are essentially equivalent, that representations are rotationally invariant. Coastlines challenge this assumption, since there are clearly major differences between what happens transverse to the shoreline, and what happens along it. Spatial resolutions can be very different, since the detail needed to represent processes and forms transverse to the shore tends to be much greater than the detail needed to represent variation along the shore. In the three-dimensional marine environment, there is comparable dissimilarity between the horizontal and vertical dimensions.

At the same time, the marine environment presents all of the problems familiar to students of geographic information science: problems of scale and accuracy, the representation of time-varying information, the persistence of objects through time, and issues of generalisation. Many of these are explored in this book in a marine and coastal setting, adding usefully to what we already know from terrestrial applications.

One of the most difficult impediments to greater use of GIS in marine and coastal applications, besides the fundamental issues of representation, has been the comparative lack of data. The sensors that could provide detailed, four-dimensional data about the dynamic marine environment generally do not exist, although enormous improvements in sensing technology have occurred in the past decade. Such data would inevitably require massive storage, and would still challenge the capacities of today's computing systems, despite the improvements in price and performance that have occurred recently. In the past, cartographers and others developed sophisticated methods of generalisation and abstraction to deal with precisely this problem, at a time when technology was unable to handle the volume of two-dimensional detail available about static, terrestrial features. Perhaps recent developments in technology are opening up a new research arena that will focus on generalisation methods for four-dimensional (dynamic, three-dimensional) data.

Marine and coastal environments also present a rich set of problems of a societal and institutional nature. Concepts of land ownership had much to do with the development of primitive methods of mapping, measurement, and geometry among early agrarian societies, and the cadaster remains the most fundamental and detailed of GIS layers. But the marine environment challenges many of these ancient notions, and several chapters in the book address applications of GIS that have to do with resolving jurisdictional disputes, conflicts over resources, and preservation of diminishing marine fish stocks.

From the perspective of geographic information science, the growing interest in marine and coastal applications is enormously fascinating. Much can be learned from efforts to apply existing software in this novel and challenging environment. Several chapters of the book deal also with designs for new software that can handle the special requirements of these applications, by modelling with dynamics, moving objects, unequal resolutions, linear systems, depth, etc., and by providing associated methods of visualisation and analysis. Fundamentally, the book is about taking a technology that evolved to deal with problems on dry land, and stimulating its further evolution to deal with a new set of applications that has many similarities, but also many substantial differences. Just as fish adapted to the terrestrial environment by evolving into amphibians, so GIS must adapt to the marine and coastal environment by evolution and adaptation. This book provides a first glimpse of what that evolution may entail, and what software systems will eventually emerge to handle the spatial information needs of marine and coastal scientists and managers.

Preface

Dawn J. Wright and Darius J. Bartlett

As befitting the electronic age that has affected us all, this book was conceived in 1997 as a result of acquaintances forged completely over the Internet. At the time, the editors did not even know what the other looked or sounded like! We had both realised, from opposite sides of the Atlantic, and from our wanderings on the seashore, the oceans and the sea bed, that marine and coastal GIS applications were finally gaining wider acceptance in our respective communities. Equally importantly, an essential "critical mass" of leading scholars in both the marine and coastal realms was at last providing the necessary leadership and inspiration to help guide and further develop these closely-related application areas; while, at the same time, rapid take-up and evolution of the technology of GIS meant a rapidly-expanding user base for the results of this research.

The time seemed ripe to produce a book, particularly with the United Nations International Year of the Ocean (1998) fast approaching. As we spread word of our project across many an ocean, we were pleased to welcome an international authorship on board, a fortuitous mix of well-established oceanographers and GIS specialists with up-and-coming newcomers, many of whom will hopefully join the next generation of leading scholars.

From the start, the intention was to produce a book which focuses on the potential and progress of GIS *research* in the marine and coastal realms. It is not meant to be an introductory text in either marine/coastal science or in GIS. We hoped to include papers on a wide range of themes, bringing not only theoretical constructs to bear but also technical innovations and empirical results arising from recent and continuing work, all from a variety of scholarly settings and countries. We believe that we have succeeded in this, as evidenced by the three major parts of the book: the first section, on *conceptual/technical* issues, covers the most pressing theoretical challenges in marine and coastal GIS; the second, *applications*, takes the reader from microscale coastal habitats to marginal basins to seafloor spreading centres; while the third section, which we refer to as *institutional* issues, includes data and archiving policies, boundary delimitation, error and accuracy, and other like concerns. The chapters alternate topically between marine and coastal emphases, with a few of them (particularly in the conceptual section) addressing both marine and coastal issues.

We hope that the completeness and relevance of the book will be such that anyone with interests in marine/coastal environments as well as in GIS applications will find it useful. We anticipate that oceanographers, marine and coastal geographers, coastal resource managers and consultants, marine technologists, government researchers, and graduate students will find it particularly valuable.

Unlike so many other edited volumes, this book is not the result of a specialist meeting or conference. Again, most of the communications between editors and authors, between authors themselves, and with the publishers, were conducted over the Internet. Those involved have, nevertheless, felt some sense of camaraderie, and a few of our authors have prevailed despite the additional challenges of personal injury, life-threatening illness, surgery, and even the birth of a new baby girl! In addition to our authors, we must thank the many organisations and individuals who have supported the idea of the book and contributed to it directly or indirectly in various ways (even though sometimes they were not aware they were doing so). In particular, we acknowledge with gratitude the support and encouragement of the Commission on Coastal Systems of the International Geographical Union, and of the Working Group on Marine Cartography of the International Cartographic Association. At an individual level, we thank René Andersen, Pieter Augustinus, Christophe Durand, Daniele Ehrlich, Gail Langran-Kucera, Nagendra Kumar, Nadia Ligdas, Gerry Maxwell, Jacques Populus, Norb Psuty, and Florence Wong. The epilogue benefited from the comments of Jim Ciarrocca and Jeannie Murday of the Environmental Systems Research Institute, and Andy Wells of Erdas. We would like to thank Luke Hacker and Tony Moore of Taylor & Francis for their great editorial assistance, and, of course, Jonathan Raper and Peter Fisher for believing in the book and accepting it into their series. And our deepest thanks go to Cindy Fowler for composing the special dedication to our friend, colleague, and contributor to this book, Millington Lockwood. In Oregon, Dawn Wright would like to thank her mother, Jeanne Wright, and her dog, Lydia, for their marvellous support and inspiration, as well as graduate students and colleagues in the Marine Resource Management Program and the Department of Geosciences at Oregon State University. Darius Bartlett acknowledges with thanks the encouragement, support and intellectual stimulation provided by friends and colleagues in the Geography Department and the Coastal Resources Centre at University College Cork.

To our knowledge this book is the first of its kind to illustrate the broad usage of GIS in deep ocean and coastal environments. Our hope is that this book will inspire others to identify further potentials and challenges in marine and coastal GIS, thereby stimulating continued research in this important application domain of geographic information science.

Contributors

THE EDITORS

Dawn J. Wright
*Department of Geosciences, Oregon State University, Corvallis, OR 97331-5506,
USA; dawn@dusk.geo.orst.edu; http://dusk.geo.orst.edu;+1-541-737-1229
(phone);+1-541-737-1200 (fax)*
Dawn first encountered GIS in the early 1990s while working on her Ph.D. at the
University of California at Santa Barbara. She became acutely aware of the
challenges of applying GIS to deep marine environments when presented with the
first such data set collected from the deepsea vehicle *Argo I*, a few years after it
was used to discover the wreck of the *Titanic*. Dawn has been an assistant
professor of Geosciences at Oregon State University since 1995. She has
completed oceanographic fieldwork in some of the most geologically active
regions of the planet, including the East Pacific Rise, the Mid-Atlantic Ridge, the
Juan de Fuca Ridge, the Tonga Trench, and volcanoes under the Japan Sea and
Indian Ocean. Her research interests include application and analytical issues in
GIS for oceanographic data, particularly data conversion, management, and
metadata; the relationships between volcanic, hydrothermal, and tectonic processes
at seafloor-spreading centres; the analysis and interpretation of data from deepsea
mapping systems; and the geography of cyberspace.

Darius J. Bartlett
*Department of Geography, University College Cork, Cork, IRELAND; djb@ucc.ie;
+353-21-902835 (phone); +353-21-271980 (fax)*
Darius first encountered GIS in the early 1980s, while at Edinburgh University,
and became aware of the challenges of applying it to the coast while working for
Bill Carter at the University of Ulster in 1987-'88. He has been lecturer in GIS at
University College Cork since 1989. For many years he was co-ordinator of the
project on Coastal GIS for the International Geographical Union's Commission on
Coastal Systems; and, with Ron Furness, he was one of the founding organisers
of the CoastGIS series of biannual conferences.

THE AUTHORS

Vittorio Barale

Marine Environment Unit, Space Applications Institute, Joint Research Centre of the European Commission, Ispra, ITALY; vittorio.barale@jrc.it; +39-0332-789274 (phone); +39-0332-789034 (fax)
His main professional interest is the applications of remote sensing techniques for the assessment of biogeochemical and physical processes in the coastal and marine environment. He first started to work in the remote-sensing field in the late 1970's at the University of Milan, where he graduated in Physics. In the 1980s, he became involved in coastal and marine issues while at the Scripps Institution of Oceanography, where he also obtained his M.S. and Ph.D. Since 1990 he has been a senior scientist with the Joint Research Centre of the European Commission. He is a member of various professional associations, in particular the Consortium for International Earth Science Information Network (CIESIN) and the Mediterranean Coastal Environment organization (MEDCOAST).

Andra Bobbitt

Cooperative Institute for Marine Resources Studies, Oregon State University, Hatfield Marine Science Center, 2115 S.E. Oregon State University Drive, Newport, OR 97365 USA; bobbitt@pmel.noaa.gov; +1-541-867-0177 (phone); +1-541-867-3907 (fax)
Andra Bobbitt is a Senior Faculty Research Assistant at Oregon State University's Cooperative Institute for Marine Resources Studies (CIMRS). She has been working through CIMRS for the Vents Program since April 1991. Ms. Bobbitt developed the GIS for the Vents Program, which began in 1992 and includes the main database and interface, sea-going system and WWW-based applications. Her other responsibilities include managing a multibeam bathymetric database and developing the WWW sites for the Vents Program. Prior to her work with Vents, Ms. Bobbitt worked at the Scripps Institution of Oceanography for marine geology and bathymetric mapping programs. She has been participating in research expeditions since 1984.

Bronwyn Cahill

Marine Institute, Irish Marine Data Centre, 80 Harcourt Street, Dublin 2 IRELAND; Bronwyn.Cahill@marine.ie; http://www.marine.ie/datacentre; +353-1-4757100 (phone); +353-1-4784899 (fax)
Bronwyn joined the Marine Institute's Data Centre in 1993 after graduating from the University of Plymouth with a degree in Ocean Science, and has been manager of the Data Centre since 1996. During this time, she has been involved in a number of multidisciplinary data management programmes, most notably within the EU Marine Science and Technology (MAST) programme, and was responsible for the design and implementation of the Marine Data Centre's quality management system accredited to the ISO9002 (EQNET) 1994 standard. Her interest in GIS grew from a need to manipulate spatial and temporal marine data and to develop innovative techniques to manage and visualise marine data.

Órla Ní Cheileachair

*Marine Institute, Irish Marine Data Centre, 80 Harcourt Street, Dublin 2
IRELAND; Orla.Ni@marine.ie; http://www.marine.ie/datacentre;
+353-1-4757100 (phone); +353-1-4784899 (fax)*
Órla joined the Marine Institute's Data Centre in 1993 with a Masters in Benthic Biology from National University of Galway. Prior to joining the Data Centre, she gained valuable experience working in the field with an environmental consultancy. At the Data Centre she has been active in the area of electronic data publishing and was the project co-ordinator for the Marine Science and Technology (MAST) supported initiative EDAP (Electronic DAta Publishing), which developed the Guideline on Electronic Publication for Marine Projects. Órla currently leads a development team for multidisciplinary marine data management, and is actively involved in the application of GIS techniques to manage and visualize marine data.

Ken Foote

*Department of Geography, University of Texas at Austin, Austin, TX 78712-1098
USA; k.foote@mail.utexas.edu; +1-512-232-1592 (phone); +1-512-471-5049 (fax)*
Ken is the Erich W. Zimmermann Regents Professor of Geography and Director of the Environmental Information Systems Laboratory at the University of Texas at Austin. He teaches GIS, computer research techniques, cultural and historical geography, and has led several Web-based instructional development projects. These include the Geographer's Craft Project, completed in 1996, to create one of the very first online textbooks in geography. He is now leading the Virtual Geography Department Project to develop a clearinghouse for instructional materials in the Worldwide Web. His recent publications include the co-edited *Re-reading Cultural Geography* (1994) and *Shadowed Ground: America's Landscapes of Violence and Tragedy* (1997) which received the Association of American Geographers J. B. Jackson Prize in 1998.

Cindy Fowler

*NOAA Coastal Services Center; 2234 South Hobson Ave., Charleston, SC 29405-
2413 USA; cfowler@csc.noaa.gov; http://www.csc.noaa.gov; +1-843-740-1249
(phone); +1-843-740-1315 (fax)*
Cindy Fowler has a Bachelor of Science degree in Geography from the University of South Carolina and a Master of Science Degree in Natural Resource Information Systems from the Ohio State University. She has over 19 years experience working with geographic information systems, remote sensing, and other forms of spatial technologies. Cindy has experience in private industry, government service and university settings supporting the fields of forestry, natural resources, cadastres, geodetic science and coastal resource management. Currently, she is a Senior Spatial Data Analyst in the U.S. Department of Commerce's (DOC) National Oceanic and Atmospheric Administration (NOAA) in Charleston, South Carolina where she combines her love of the coast with her passion for geospatial technologies. Cindy's research interests are related to coastal and marine GIS, and especially the data needed to support them. She is particularly interested in the technical and legal implications related to marine cadastral data development.

Christopher Fox

National Oceanic and Atmospheric Administration, Pacific Marine Environmental Laboratory, Newport, OR 97365 USA; fox@pmel.noaa.gov; +1-541-867-0276 (voice);+1-541-867-0356 (fax)

Dr. Chris Fox has served as a principal investigator within the Vents Program of NOAA's Pacific Marine Environmental Laboratory since June, 1985 and also holds the rank of associate professor (courtesy) at Oregon State University's College of Oceanic and Atmospheric Sciences. He leads a diversified research program in marine mapping, geophysics, and underwater acoustics and in recognition of his efforts in developing the U.S. Navy's Sound Surveillance System for environmental applications, was awarded the Department of Commerce Gold Medal in 1994. Prior to his service with NOAA, Dr. Fox worked for the U.S. Naval Oceanographic Office, where he participated in a wide variety of studies including the numerical modelling of seafloor microtopographic roughness and the development of automated cartographic mapping from multibeam sonar systems. His current interests in GIS applications include multidisciplinary seafloor investigations, marine mammal research, the development of portable GIS systems for use in oceanographic fieldwork, and the development of web-accessible tools for providing data access.

Ron Furness

Australian Hydrographic Office, Locked Bag 8801, South Coast Mail Centre NSW 2521, AUSTRALIA; rfurness@ozemail.com.au

Ron Furness is the Director of Coordination and Development at the Australian Hydrographic Office. A cartographer by training, he has worked in the field of marine charting for the greater part of his career. He has been involved in electronic charting since the early nineteen seventies and served on a number of the early IHO committees, which led to the development of ECDIS. He is a past Federal President of the Mapping Sciences Institute, Australia and chairs the International Cartographic Association.

Christopher Gold

Department of Geomatics, Laval University, Quebec City, Quebec G1K 7P4 CANADA; Christopher.Gold@scg.ulaval.ca; +1-418-656-3308 (phone); +1-418-656-7411 (fax)

Chris started out as a geologist, and during his Ph.D. became fascinated by the problems of spatial data. He has been actively involved in the manipulation and display of data in agriculture, geography, geology, water resources and forestry, among others. He strongly believes in the importance of the development of spatial algorithms and data structures in an academic setting, and is active in encouraging collaboration between computer science, especially computational geometry, and geomatics. He currently holds an Industrial Chair in Geomatics Applied to Forestry at Laval University.

Chris Goldfinger

College of Oceanic & Atmospheric Sciences, Marine Geology, Active Tectonics Group, 104 Ocean Admin Bldg, Oregon State University, Corvallis, OR 97331-5503 USA; gold@oce.orst.edu; +1-541-737-5214 (phone); +1-541-737-2064 (fax)
Chris has been using GIS in marine tectonics research since 1990, when he was impressed with the challenges of trying to integrate a variety of digital and analog marine data. He has been involved and using GIS in a range of projects from gas hydrates to great subduction earthquakes. He developed one of the early GIS based real-time towfish navigation programs for marine surveys. He has been on the research faculty at Oregon State University College of Oceanic and Atmospheric Sciences since 1995.

Michael F. Goodchild

Department of Geography, University of California, Santa Barbara, CA 93106-4060 USA; good@ncgia.ucsb.edu; +1-805-893-8049 (phone); +1-805-893-7095 (fax)
Mike Goodchild holds degrees from Cambridge University (Physics) and McMaster University (Geography) and is currently chair of the Department of Geography, UC-Santa Barbara, as well as chair of the Executive Committee of the National Center for Geographic Information and Analysis (NCGIA). His research interests include GIS, environmental modelling, geographical data modelling, spatial analysis, location theory, accuracy of spatial databases, and statistical geometry.

Gerry Hatcher

Monterey Bay Aquarium Research Institute, Moss Landing, CA 95039 USA; gerry@mbari.org; +1-831-775-1758 (phone); +1-831-775-1620 (fax)
Gerry first became interested in GIS as a graduate student in 1990 while working in the Ocean Mapping Development Group at the University of Rhode Island. While there, he completed a masters of science degree in ocean engineering with a thesis entitled *GIS as a Data Management Tool for Seafloor Mapping*. Since then he has been employed at the Monterey Bay Aquarium Research Institute (MBARI) developing GIS applications to assist oceanographic science and marine operations, and has spent many days at sea. His applications have been used on science missions in areas as diverse as Antarctica, Alaska, the Indian Ocean, and the Mariana Trench.

Donald J. Huebner

Department of Geography, University of Texas at Austin, Austin, TX 78712-1098 USA; djhuebner@mail.utexas.edu; +1-512-471-5116 (phone); +1-512-471-5049 (fax)
Don began working in GIS as a research and teaching assistant for Ken Foote at the University of Texas at Austin. In addition to developing teaching modules for the Geographer's Craft Project, Don has applied GIS to tracking rabies distribution in coyotes, and range expansion and distribution of feral hogs in Texas. Currently as a doctoral candidate, he is working on a landscape ecology project for a central New Mexico mountain range. This project is using GIS for reconstructing past, present, and future landscape conditions and will model the effects of urban expansion on this locale. Despite the brief foray into coastal GIS, his primary research interests are in landscape history and processes, particularly in the mountain west of North America.

Theresa Kennedy
Department of Earth and Environmental Engineering, Henry Krumb School of Mines, Columbia University, 500 West 120th St., New York, NY 10027 USA; tk346@columbia.edu; +1-212-854-1568 (phone); +1-212-854-7081 (fax)
Theresa became interested in the management and vizualisation of geographic information in 1992 whilst working for the British Geological Survey. Her interest took her into research at the Open University in Milton Keynes where she studied the application of GIS to managing and integrating regional geological datasets and then on to the Irish Marine Data Centre in Ireland where she became involved in all aspects of data management for coastal and oceanographic data. She is currently a Staff Research Associate with Columbia University and the New York-New Jersey Clean Ocean and Shore Trust (COAST) where she is the technical project manager for the Virtual Harbour Estuary Project, an initiative to assimilate and integrate GIS data resources for the Hudson Estuary System.

Rongxing (Ron) Li
Department of Civil and Environmental Engineering and Geodetic Science, The Ohio State University, 470 Hitchcock Hall, 2070 Neil Avenue, Columbus, OH 43210-1275 USA; li.282@osu.edu; see http//: shoreline.eng.ohio-state.edu; +1-614-292-6946 (phone); +1-614-292-2957 (fax)
Dr. Ron Li is an associate professor at the Department of Civil and Environmental Engineering and Geodetic Science of The Ohio State University. He has B.S. and M.S. in Surveying Engineering from Tongji University in Shanghai and a Ph.D. in Photogrammetry and Remote Sensing from the Technical University of Berlin. Ron has been a guest editor of three special issues on coastal and marine GIS and an associate editor of the international *Journal of Marine Geodesy*, published by Taylor & Francis. He was a GIS specialist for the Asian Development Bank (ADB) project "Institutional Strengthening of Malaysian Shoreline Management," 1994-1996. His research interests include digital mapping, spatial data structures, coastal and marine GIS, photogrammetry and remote sensing.

Millington Lockwood
National Oceanic and Atmospheric Administration, Office of the Coast Survey, 1315 East West Highway, Silver Spring, MD 20910 USA; millington.lockwood@noaa.gov; +1-301-713-2777 x171 (phone); +1-301-713-4019 (fax)
Millington Lockwood, to whom this book is dedicated, had nearly 30 years in the field of marine information systems including coastal mapping, navigation, marine geology, surveying, and marine positioning. His recent interests were in the area of digital data dissemination, formatting, and data standards as a chairman of a working group of the US Federal Geographic Data Committee (FGDC). Specific interests dealt with the data quality aspects of shoreline and bathymetric data for the U.S. Coastal and Great Lakes regions. Millington was recently guest editor of a special content volume of the *Journal of Surveying and Land Information Systems* (Vol. 58, No. 3) on coastal zone GIS.

Anne Lucas
Department of Geography, University of Bergen, Breiviksveien 40, N-5045 Bergen
NORWAY; anne.lucas@nhh.no; +47-55-959-659 (phone); +47-55-959-393 (fax)
Anne was introduced to GIS through the Canada Land Data System in 1979, when her work with Environment Canada involved the development of methodologies for identifying landscapes sensitive to acid rain. Later she experienced GIS from the business and application development side while at the IBM Bergen Scientific Center. She has been working with coastal and marine applications since then and now lectures in GIS, remote sensing and ocean studies at the University of Bergen.

Brian McAdoo
Department of Geology and Geography,Vassar College, Box 735, Poughkeepsie,
NY 12604 USA; brmcadoo@vassar.edu, +1-914-437-7703 (phone);
+1-914-437-7577 (fax)
Brian is a marine geologist who gets seasick in the bathtub. He found that by using submarine bathymetry in a GIS framework, he could explore the world's oceans from the comfort of his office. Brian received a B.S. in Geology from Duke University, a Dip. Sci. Geology from the University of Otago (New Zealand), and a Ph.D. in Geology from the University of California, Santa Cruz. Currently, he is an assistant professor at Vassar College, where he teaches coastal and marine geology in a hands-on and affordable (compared to the cost of oceanographic research) GIS framework.

Norman Maher
Monterey Bay Aquarium Research Institute, P.O. Box 628, Moss Landing, CA
95039;nmaher@mbari.org; 831-775-1714 (phone); 831-775-1620 (fax)
Norman, a Humboldt State University graduate, began using a GIS while at the U.S. Geological Survey in 1990 for mapping continental shelf sediments offshore San Francisco. For the past nine years he has been using GIS for display and analysis of multibeam bathymetry, sonar imagery, and geologic data. His current focus at MBARI is on mass wasting and canyon formation processes in the Monterey Bay region.

Geoff Meaden
Department of Geography, Canterbury Christ Church College, North Holmes
Road, Canterbury, Kent CT1 1QU, UNITED KINGDOM; g.j.meaden@cant.ac.uk
Geoff is a Senior Lecturer in the Department of Geography at Canterbury Christ Church University College in the UK. In the early 1980s his Ph.D. research made use of GIS techniques to establish the best locations for trout farming in England and Wales. From this work Geoff developed his main teaching and research interests, which are in biogeography and GIS, though he is now also the Director of a Marine Fisheries GIS Unit. The unit is currently engaged in several projects including zoning for mariculture in Sri Lanka, mapping hydrodynamics and fish relationships in the Straits of Dover and in developing a fisheries electronic log book which is integrated to a GIS. He carries out regular GIS and fisheries related assignments for the Food and Agriculture Organisation of the United Nations, for whom he has co-authored the two major works written to date on GIS applications to fisheries, and he has helped in the development of a digital fisheries atlas. He has presented numerous papers on this subject, and was the invited keynote speaker

at the 1st International Symposium on GIS in Fisheries Science, held in Seattle in March, 1999.

Hal Palmer

MRJ Technology Solutions, 10560 Arrowhead Drive, Fairfax, VA, USA 22030-7305; hpalmer@mrj.com; +1-703-277-1239 (phone); +1-703-385-4637 (fax)
Hal is a marine geologist who earned his B.S. in Geology from Oregon State University. Following Army service, he completed an M.S. and Ph.D. in Marine Geology at the University of Southern California. Although he has conducted diving and geophysical surveys around the world, his first encounter with GIS was as a consultant engaged in environmental studies for MRJ's clients in various maritime industries. He has applied GIS to studies ranging from terrain analyses for amphibious landings to pipeline and telecommunications cable route selection. Recent work involves the development of a global maritime boundaries database as a commercial product.

Lorin Pruett

MRJ Technology Solutions, 10560 Arrowhead Drive, Fairfax, VA, USA 22030-7305; MaritimeBoundaries@mrj.com or Lpruett@mrj.com; +1-703-277-1879 (phone); +1-703-385-4637 (fax); +1-703-385-0700 (switchboard)
Lorin has a B.S. in Geology and an M.A. in GIS. He has been designing, developing and maintaining large geographic database systems since the mid-1980s. Lorin started out consulting in the oil and gas industry where he was responsible for the redesign and maintenance of the Eastern Gas Devonian Shales Database for the Gas Research Institute. Since 1989 he has been at MRJ using ESRI's Arc/INFO GIS software to build global environmental databases including The Global Maritime Boundaries Database reference featured in this book. In 1995, Lorin won the Outstanding Masters Degree Project in Technology award for the Circum-Atlantic Project prototype CDROM he developed for a startup USGS/IGU project. Lorin has been editor for, contributed to, or authored numerous contract related documents.

Jonathan Raper

Department of Information Science, School of Informatics, City University, Northampton Square, London EC1V 0HB UK; raper@soi.city.ac.uk; +44-171-477-8415 (phone); +44-171-477-8584 (fax)
A geographer and geomorphologist by training, Jonathan Raper has been on the faculty of Information Science at City University since 1998. His research interests geographical information science are in the following areas: the handling and analysis of environmental information, the incorporation of multimedia data into geographical information systems and digital libraries, analysis of geographical data policy, and the philosophy of spatial and temporal representation.

Chris Roberts

Australian Hydrographic Office, Locked Bag 8801, South Coast Mail Centre NSW 2521, AUSTRALIA
Chris Roberts is a senior cartographer in the Australian Hydrographic Office. He has over 20 years experience in all aspects of manual and digital nautical chart production. Mr Roberts represents Australia on several international Working Groups concerned with the preparation and maintenance of the standards and specifications relating to ECDIS.

Andy Sherin
*Geological Survey of Canada (Atlantic), Bedford Institute of Oceanography, 1
Challenger Drive, P.O. Box 1006, Dartmouth, Nova Scotia B2Y 4A2 CANADA;
sherin@agc.bio.ns.ca; +1-902-426-7582 (phone); +1-902-426-1466 (fax)*
Andy was a late comer to GIS after being the first curator of the Geological Survey
of Canada's marine sample collection in1974 and for twenty years developing
information systems for marine geoscience in a data base management system
environment. His interest in the display and analysis of marine geoscience data led
him to explore the use of GIS tools and to the development of his first major GIS
application in 1994, a coastal information system based upon dynamic
segmentation. He has also applied GIS to marine applications including
preliminary mapping of marine geological surveys using dynamic segmentation,
3D visualisation of high-resolution seismic interpretations and the integration of
geoscience data for environmental planning. Andy is also interested in
interorganizational cooperation for data exchange to support integrated coastal
management. He has pursued this interest since 1991 by representing the
Geological Survey of Canada on the Atlantic Coastal Zone Information Steering
Committee, an intergovernmental and intersectoral organization providing a forum
for the development and coordination of a coastal zone information infrastructure
for Atlantic Canada.

Yafang Su
*Office of Academic Computing, University of California, Los Angeles, 5931 Math
Sciences Addition, Box 951557, Los Angeles, CA 90095-1557 USA;
yafang@ucla.edu; http://www.oac.ucla.edu/people/yafang_su/;
+1-310-825-7418 (phone), +1-310-206-7025 (fax)*
Yafang is currently a senior GIS analyst and consultant in the High Performance
Computing and Visualization Group of the UCLA Office of Academic Computing.
She also serves as an Editorial Board member for the GIS journal *Geo-
Information-Systeme*. Her current research interests are Internet GIS, GIS system
integration, and scientific visualisation. She was first interested in applying GIS
and scientific visualisation in oceanography during her post doctorate research in
1996-1997 at the Monterey Bay Aquarium Research Institute, California. Before
then, she was an associate professor in GIS at the State Key Lab of Resources and
Environment Information System (LREIS), Institute of Geography, Chinese
Academy of Sciences, Beijing, China. She was the first person to apply GIS to
study the investment environment in China.

Nancy von Meyer
*Fairview Industries, Blue Mounds WI 53517 USA;
nancy@fairview-industries.com; +1-608-437-6701 (phone);
+1-608-437-6702 (fax)*
Nancy became interested in GIS at the University of Wisconsin Engineering
College while working on projects related to parcel mapping and surveying for
Dane County. Through work on cadastral data standards, she became familiar with
offshore legal boundaries, and rights and interests from the Mineral Management
Service. In 1997 Nancy started a GIS and cadastral related project with the Coastal
Services Center (CSC) in Charleston, South Carolina. Working with the
knowledgeable and enthusiastic CSC staff, she is continuing to learn more about
shoreline and coastal boundary, rights, and regulation topics.

Herman Varma

Canadian Hydrographic Service, Maritime Region, Bedford Institute of
Oceanography, 1 Challenger Dr., Dartmouth, Nova Scotia B2Y 4A2 CANADA;
hvarma@helical.ns.ca; +1-902-426-5376 (phone)

Herman Varma is currently the Head of Cartographic Research at the Bedford Institute of Oceanography. He is a graduate of Dalhousie University and has been working on projects such as dolphin robot vehicles, and dense data aquisition systems such as laser infrared digital airborne radar (LIDAR), and multibeam/sweep systems. His last project resulted in the formulation of an Oracle spatial data option (SDO) product through the use of helical hyperspatial codes (HHcodes). Oracle's (SDO) is the first extension to standard relational database management system technology to support spatial data handling. Herman is currently involved in building a very, very large spatio/temporal databases for the Canadian Hydrographic Service.

Robert Ward

Australian Hydrographic Office, Locked Bag 8801, South Coast Mail Centre NSW
2521, AUSTRALIA

Commander Robert Ward is a naval officer and hydrographic surveyor who has served both afloat and ashore in a wide variety of appointments associated with navigational chart making and surveying. He has served on the teaching staff at the Royal Navy School of Hydrographic Surveying in England and was Officer in Charge of the Royal Australian Navy Hydrographic School in Sydney. Operational postings have included Officer in Charge of the RAN Detached Survey Unit, Executive Officer of the oceanographic research vessel HMAS *Cook* and command of the Australian Navy's largest and most capable survey ship, HMAS *Moresby*. Commander Ward is currently employed in the Australian Hydrographic Office dealing with national and international policy concerning the use and implementation of digital charting technology.

CHAPTER ONE

Down to the Sea in Ships: The Emergence of Marine GIS

Dawn J. Wright

1.1 INTRODUCTION

In the mid-1960s Roger Tomlinson recognised that digital computers could be used quite effectively to map out and analyze the vast quantities of information being collected by the Canada Land Inventory. The resulting statistical and cost-benefit analyses were used to develop management plans for large rural areas throughout the whole of settled Canada. One of the conclusions of the initial effort was that computerisation was going to be the best alternative for developing these management plans, in spite of the primitive computers of that time and their high costs. Roger Tomlinson called this new kind of "computerisation" the "geographic information system" and the rest, as they say, is history. Since that time geography and GIS have enjoyed an especially close relationship, extending, as Johnston (1999) notes, "well beyond the commonality of titles." Geography has been the academic "home" of much of the continuing research, development and training in GIS, as well as a means of survival for the discipline when the pressures of academic justification or public sector funding have reared up (Johnston, 1999). This chapter gives a brief review of the development of *marine* GIS, highlighting a series of "firsts" in this application domain, due mainly to the efforts of geographers (or geomatics specialists in some of the Canadian terminology), along with, or in addition to, those of oceanographers. It should be noted that this review reflects the author's own observations and value judgements; different views certainly may exist elsewhere. However, what seems clear, as is the case with the history of GIS in general (e.g., Coppock and Rhind, 1991), is that there were many initiatives concerned with different facets of marine GIS, usually occurring independently and often in ignorance of each other, and frequently originating because of the curiosity, interest, and sometimes the outright courage of certain individuals.

For the sake of discussion, the domain of marine, as opposed to coastal, GIS is here defined as the deep, open water and ice beyond human sight of the coast (i.e., the swash zone, bays, dunes, estuaries, coastal wetlands, and the like). Bartlett (1999) gives a review of the development of coastal GIS in the next chapter. The distinction is made because marine and coastal GIS, both as application domains and as user communities, have, until recent years, developed fairly independently of each other, just as traditional oceanography departments in North America have often grouped together biological, chemical, physical, and geological studies of the ocean as "oceanography science programs," while creating a separate category for coastal studies, particularly if the emphasis is on coastal resource management. It may be fair to say that marine applications of GIS have been more in the realm of basic science whereas coastal applications, due in part to the intensity of human activities, have encompassed both basic and applied science, as well as policy and management. A full review of research issues endemic to marine GIS (i.e., what sets a marine GIS apart from the traditional, land-based GIS), such as the multiple dimensionality and dynamism of marine data, the inherent fuzziness of boundaries, the great need for spatial data structures that vary their relative positions and values over time, etc. will not be mentioned here as they are already covered in full by Li and Saxena (1993), Lockwood and Li (1995) and Wright and Goodchild (1997).

1.2 GEOGRAPHERS DISCOVER WHAT OCEANOGRAPHERS ALREADY KNEW

As mentioned earlier, interest and developments in marine GIS have been due mainly to the efforts of geographers and oceanographers. The involvement of geographers in marine GIS has been especially interesting, since throughout most of the history of geography as an academic discipline the study of the oceans beyond the realm of the nearshore has escaped attention (Steinberg, in press; Wright, in press). Indeed, American geographers, have contributed little to marine research until recent decades, although the first textbook of modern marine science, written by Lt. Matthew Fontaine Maury of the U.S. Navy in 1855, was entitled *The Physical Geography of the Sea*. It was the post World War II exploitation of offshore resources, as well as the environmental movements of the 1960s arising from coastal population and industrial growth, that directed some American geographers to the open water (West, 1989; Psuty *et al.*, in press).

During the 1970s and 1980s, as the support for research into *global* Earth systems and the effects of human-induced environmental change steadily increased, geographers began to broaden their focus past traditional boundaries. Marine geography received a major boost in the 1990s with the advent and popularity of Earth System Science (ESS), an interdisciplinary initiative seeking to understand the *entire* Earth system (atmosphere, oceans, ice cover, biosphere, crust, and interior) on a global scale (Williamson, 1990; Nierenberg, 1992). Other recent factors increasing the exposure of marine geography, and ultimately marine GIS, include rising global environmental awareness and concerns, increased pollution and the endangerment of marine fish populations, a heightened understanding of the role of marine life in maintaining the global ecosystem, new opportunities for marine mineral extraction, new techniques for undertaking marine exploration, the 1994 activation of the United Nations Convention on the Law of the Sea, and the designation of 1998 as the International Year of the Ocean (Psuty *et al.*, in press).

1.3 A SERIES OF FIRSTS

1.3.1 Government and Academia

One of the precursors to marine GIS was an automated mapping effort developed in the early 1960s by oceanographers of the U.S. National Ocean Survey (NOS). The NOS had at their disposal computer resources that were prohibitively expensive to others at the time, and pioneered the production of "figure fields" and matrices of depth values for the creation of hundreds of nautical charts (Coppock and Rhind, 1991).

The 1970s and 1980s witnessed the development of sophisticated technologies for ocean data collection, resulting in an explosion of data and information to usher in the 1990s. Realizing the potential of GIS for managing, interpreting, and visualizing these data, an American oceanographer, in collaboration with a software developer from Dynamic Graphics, Inc., published in 1990 one of the first articles on the potential of marine GIS (Manley and Tallet, 1990). The article, featured in the magazine of The Oceanography Society, focused not only on the data management and display functions that GIS was well known for, but was far-sighted in its discussion of truly 3-dimensional property modelling, volumetric visualizaton, and quantitative analysis in GIS, particularly for physical and chemical oceanographic data. Two years later *Sea Technology* magazine featured its first article on marine GIS, highlighting the use of the technology for search and recovery of lost objects on the seafloor (Caswell, 1992). At around the same time, oceanographers and geographers in the U.S., Canada, the U.K., and Europe presented the first results of a variety of marine GIS applications, including: the modelling of tidal currents and winds in Canada (Keller *et al.*, 1991); a digital marine atlas in the U.K. (Robinson, 1991); the monitoring of water quality in the New York Bight (Hansen *et al.*, 1991); the first processing of Exclusive Economic Zone (EEZ) data along the U.S. west coast (Langran and Kall, 1991); the monitoring of pollution outflow and diffusion in Scandinavia (Dimmestøl and Lucas, 1992); and the investigation of natural oil seeps in the Gulf of Mexico (MacDonald *et al.*, 1992).

Serendipitously, the early 1990s also witnessed, by way of the U.S. Global Change Program, the creation of the Ridge Interdisciplinary Global Experiments (RIDGE) program, a highly successful research initiative of the U.S. National Science Foundation. This was to lead indirectly to the first graduate student theses in marine GIS, as well as later pioneering efforts by the Vents Program of U.S. National Oceanic and Atmospheric Administration. RIDGE was launched in response to the growing realisation that knowledge of the global mid-ocean ridge (seafloor-spreading centers) was fundamental to the understanding of key processes in a multitude of disciplines: marine biology, geochemistry, physical oceanography, geophysics, and geology (National Research Council, 1988). Throughout the 1990s this has prompted several major co-ordinated experiments on the seafloor, involving a multiple instrument arrays for the study of geological, physical, chemical, and biological processes within and above the seafloor (Detrick and Humphris, 1994). The resulting data range from measurements of temperature and chemistry of hydrothermal vent fluids and plumes, to microtopography of underwater volcanoes, to magnitudes and depths of earthquakes beneath the seafloor, to the biodiversity of hydrothermal vent fauna (Psuty *et al.*, in press). On the east coast of the U.S., one of the major data centers that supported the initial mapping efforts of RIDGE was the University of Rhode Island Ocean Mapping Development Center (OMDC). OMDC lent a great deal of support to the first known American graduate thesis in marine GIS (Hatcher, 1992), leading to a Master of Science degree in Ocean Engineering. The thesis described the development of a raster marine GIS, based on GRASS, for the collection of geological data from the

Narragansett Bay, and the processing and mapping of these data both at sea and on shore. At the same time, on the west coast of the U.S., the first marine GIS study to be presented at the Association of American Geographers Annual meeting (Wright *et al.*, 1992a), described the implementation of a vector GIS (Arc/INFO) for data collected from the first comprehensive, large-scale survey of the distribution of deepsea hydrothermal vents along the East Pacific Rise. Based on the success of this initial survey, as well as the discovery of an eruption at this site in 1991, RIDGE funded many subsequent research cruises to this region throughout the 1990s. The region has also gained quite a bit of notoriety in the news media. The results from the first implementations of GIS at the East Pacific Rise were incorporated later into the first known American doctoral dissertation focusing on marine GIS (Wright, 1994), which led to a joint degree in physical geography and marine geology.

1.3.2 The Commercial Sector Awakens

By the early 1990s land-based applications accounted for the lion's share of the commercial market for GIS software, dictating the development pathways for much of the industry. The commercial sector catered to the most profitable domain-specific niches in the GIS market (e.g., public utilities, transportation, hydrology, forestry, location-allocation modelling, etc.). However, as more marine practitioners in academic circles were discovering the utility of GIS, they began to make their special application needs known to commercial vendors, encouraging them to increase the functionality of their products for this new market of users. Concurrently, commercial shipping, government and military practitioners were in need of better nautical charting capabilities for safer navigation.

As early as 1987, MRJ, Inc. (now MRJ Technology Solutions) incorporated Arc/INFO into their marine analysis applications, and in the late 1980s to early 1990s marketed some of the first customised software solutions for these applications using the Arc/INFO, Genamap, and Erdas GIS packages. In 1993 they released the *Marine Data Sampler*, one of the first commercially available CD-ROM collections of global oceanographic images and data sets. These were coupled with ArcView to introduce GIS to the ocean professional, and to demonstrate the display and analysis powers of GIS when linked with available marine observations (MRJ, 1993).

In 1991, the first marine GIS posters to be presented at the Environmental Systems Research Institute (ESRI) User Conference (Carrigan *et al.*, 1992; Wright *et al.*, 1992b) were among the maps chosen by ESRI President Jack Dangermond for inclusion in the publication, *Arc/INFO Maps*, which annually highlights the multidisciplinarity of Arc/INFO users. This planted a seed for the oceanographic applications that ESRI developers would champion later on in the decade (e.g., ESRI, 1996/1997). At around the same time, Universal Systems, Ltd, in collaboration with the Canadian Hydrographic Service and the Ocean Mapping Group of the University of New Brunswick, set about developing and marketing one of the first commercial marine GIS packages in North America, called CARIS GIS, with the accompanying Hydrographic Information Processing System. Released in 1992-'93, these products were expressly designed for the processing, visualisation, and display of large quantities of bathymetric sounding data, as well as the production of high quality nautical charts. They were the precursors to the full suite of CARIS Marine Information System software now available. Intergraph joined the nautical charting market in 1993 with one of the first implementations of the Electronic Chart Display and Information Systems or ECDIS (Scott *et al.*, 1993; Ward *et al.*, 1999).

1.3.3 Into the Mainstream

High quality marine GIS abstracts, papers, and technical reports continued to appear in various conference proceedings (e.g., Bobbitt *et al.*, 1993; Drutman and Rauenzahn, 1994; Déniel, 1994; Triñanes *et al.*, 1994; Wright *et al.*, 1994). These gave further exposure to marine applications of GIS, often educating the oceanographic community about the potential of the tool as well as the science behind it. Chief among these were the report of Hamre (1993), which addressed the important issue of the oceanographic user specifications needed for the development of a sound marine GIS, and that of Lucas *et al.* (1994) in addressing the equally important issue of spatial metadata management for oceanographic applications.

As Goodchild (1992) notes, however, it is the transition from publication in conference proceedings to publication in well-respected, peer-reviewed journals that may best establish the legitimacy of a speciality in the eyes of some. Li and Saxena (1993) published one of the first of such in *Marine Geodesy*, describing some of the important differences between terrestrial and marine applications of GIS, and presenting the results of an integrated system for the exploration and development of the EEZ around the Big Island of Hawaii. Mason *et al.* (1994) published the results of an extensive marine GIS effort in the *International Journal of Geographical Information Systems* (now the *International Journal of Geographical Information Science*). The study combines time-dependent satellite with *in-situ* oceanographic data for the interpretation of mesoscale (~20 km) ocean features and the prediction of climate change. In the following year, the first peer-reviewed effort connected to the RIDGE initiative appeared in the *Journal of Geophysical Research* (Wright *et al.*, 1995). The main thrust of the paper is on geological interpretations at the crest of the East Pacific Rise, namely the abundance, width, and distribution of seafloor fissures in relation to the ages of lava flows and the distribution of hydrothermal vents. But there are also sections devoted to the processing, analysis, and mapping of these data using GIS. An extremely important effort, also appearing in 1995, was the first special issue of a peer-reviewed journal devoted entirely to marine GIS. This issue of *Marine Geodesy*, edited by Rongxing Li, included papers on a new conceptual data model for deepsea bathymetry (Li *et al.*, 1995), ocean disposal and monitoring of environmental impacts in the Farallon Islands (Hall *et al.*, 1995), detection of waste disposal sites on the seafloor (Chavez, Jr. and Karl, 1995), and a new spatial data structure for integrating marine GIS with spatial simulation (Gold and Condal, 1995). Two additional papers appeared in *Marine Geodesy* in 1997 (Goldfinger *et al.*, 1997; Wright *et al.*, 1997), which brings the narrative of this chapter up to the inception of *Marine and Coastal Geographical Information Systems*.

The chapters in this book that are devoted in whole or in part to what was defined earlier as the *marine* realm, illustrate the present "state-of-the-art" from a purely conceptual and institutional standpoint (Cahill *et al.*, 1999; Gold, 1999; Li, 1999; Lucas, 1999; Sherin, 1999; Varma, 1999), as well as via applications for a wide range of locations throughout the world's oceans (Figure 1.1). Happily for the growth of marine GIS, there are many more studies that could be mentioned, if only space allowed.

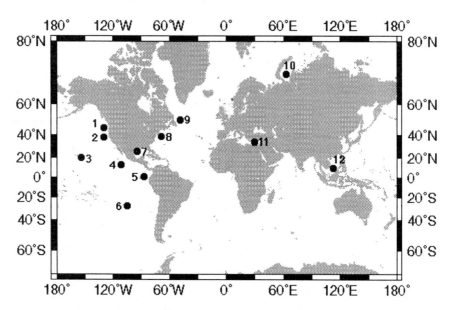

Figure 1.1 Locations of marine GIS study areas described in the chapters of this book: (1) Oregon, USA continental margin and Juan de Fuca Ridge (Goldfinger, 1999; Fox and Bobbitt, 1999; McAdoo, 1999); (2) California, USA continental margin (Hatcher and Maher, 1999; McAdoo, 1999; Su, 1999); (3) Hawaii (Li, 1999); (4) northern East Pacific Rise (Wright, 1999); (5) territorial seas of Central and South America (Palmer and Pruett, 1999); (6) southern East Pacific Rise (Wright, 1999); (7) Gulf of Mexico continental margin (McAdoo, 1999); (8) northeastern USA continental slope (McAdoo, 1999); (9) Flemish Cap, offshore Newfoundland, Canada (Sherin, 1999); (10) Kara Sea, offshore Siberia, Russia (Lucas, 1999); (11) marginal and enclosed basins of the Mediterranean (Barale, 1999; Palmer and Pruett, 1999); (12) South China Sea (Palmer and Pruett, 1999).

1.4 CONCLUSION

Suffice it to say that marine GIS has certainly "arrived" as a well-established application domain. The initial impetus for developing a marine speciality in GIS was the need to automate the production of nautical charts and to more efficiently manage the prodigious amounts of data that are now capable of being collected at sea. With the understanding that ocean research is very costly (~$10,000-$25,000 per day is typical), yet deemed extremely important by large funding agencies (due in large part to the recognised importance of ESS), marine GIS is primed to make even more important contributions to both ocean science and geographic information science. The speciality has triumphed by successfully adapting to a technology designed primarily for land-based applications and structured in a 2-dimensional framework that does not match the ocean environment. By explicitly recognising and attempting to overcome the limitations of GIS, marine geographers and oceanographers have succeeded in improving its fundamental toolbox, while extending the methodological framework for its applications. This is not to say that serious problems and challenges no longer exist. Bartlett and Wright (1999) in the Epilogue discuss these. But the overall result has been a progression from applications merely for collection and display of data to complex simulation, modelling, and the development of new oceanographic research methods and concepts. This, coupled with recent capabilities for "seeing" the ocean environment in unprecedented detail (e.g., Goldfinger, 1999; Wright, in press) hold tremendous promise for our ability to achieve an even better understanding of and level of protection for the marine environment.

1.5 REFERENCES

Barale, V., 1999, Integrated geographical and environmental remotely-sensed data on marginal and enclosed basins: The Mediterranean case, in this volume, Chapter 13.

Bartlett, D. J., 1999, Working on the frontiers of science: Applying GIS to the coastal zone, in this volume, Chapter 2.

Bartlett, D. J. and Wright, D. J., 1999, Epilogue, in this volume, Chapter 23.

Bobbitt, A., Lau, T.-K. and Fox, C.G., 1993, Integrating multidisciplinary data sets from the Juan de Fuca Ridge using geographic information systems. *EOS, Transactions of the American Geophysical Union*, **74**, p. 88.

Cahill, B., Kennedy, T. and Ní Cheileachair, O., Managing marine and coastal data sources: A national oceanographic data centre perspective on GIS, in this volume, Chapter 19.

Carrigan, B., Holland, B., Cox, A., Miller, D. and Landsman, E., 1992, Interlocking habitats of Monterey Bay. In *ARC/INFO Maps 1991*, edited by Dangermond, J. (Redlands, California: Environmental Systems Research Institute), p. 47.

Caswell, D.A., 1992, GIS: The "big picture" in underwater search operations. *Sea Technology*, **33**, pp. 40-47.

Chavez, P.S., Jr. and Karl, H.A., 1995, Detection of barrels and waste disposal sites on the seafloor using spatial variability analysis on sidescan sonar and bathymetry images. *Marine Geodesy*, **18**, pp. 197-211.

Coppock, J.T. and Rhind, D.W., 1991, The history of GIS. In *Geographical Information Systems: Principles and Applications*, 1, edited by Maguire, D.J., Goodchild, M.F. and Rhind, D.W. (New York: John Wiley and Sons), pp. 21-43.

Déniel, J.-L., 1994, Electronic navigational chart data. In *Oceans 94* (Brest, France: IEEE), pp. 541-544.

Detrick, R.S. and Humphris, S.E., 1994, Exploration of global oceanic ridge system unfolds. *EOS, Transactions, American Geophysical Union*, **75**, pp. 325-326.

Dimmestøl, T. and Lucas, A., 1992, Integrating GIS with ocean models to simulate and visualize spills. In *4th Scandinavian Research Conference on GIS* (Helsinki, Finland), pp. 1-17.

Drutman, C. and Rauenzahn, K.A., 1994, Marine geophysics modeling with geographic information systems. In *Oceans 94* (Brest, France: IEEE), pp. 528-531.

Environmental Systems Research Institute, 1996/1997, Making waves in oceanographic research and development. *ArcNews*, **18**, p. 31.

Fox, C. G. and Bobbitt, A. M., 1999, NOAA Vents Program marine GIS: Integration, analysis and distribution of multidisciplinary oceanographic data, in this volume, Chapter 12.

Gold, C.M. and Condal, A.R., 1995, A spatial data structure integrating GIS and simulation in a marine environment. *Marine Geodesy*, **18**, pp. 213-228.

Gold, C.M., 1999, An algorithmic approach to marine GIS, in this volume, Chapter 4.

Goldfinger, C., 1999, Active tectonics: Data acquisition and analysis with marine GIS, in this volume, Chapter 18.

Goldfinger, C., McNeill, L., Kulm, L. and Yeats, R., 1997, Case study of GIS data integration and visualization in marine tectonics: The Cascadia subduction zone. *Marine Geodesy*, **20**, pp. 267-289.

Goodchild, M.F., 1992, Geographical information science. *International Journal of Geographical Information Systems*, **6**, pp. 31-45.

Hall, R.K., Ota, A.Y. and Hashimoto, J.Y., 1995, Geographical information systems (GIS) to manage oceanographic data for site designation and site monitoring. *Marine Geodesy*, **18**, pp. 161-171.

Hamre, T., 1993, *User requirement specification for a marine information system*, Technical Report 74 (Nansen Environmental and Remote Sensing Center).

Hansen, W., Goldsmith, V., Clarke, K. and Bokuniewicz, H., 1991, Development of a hierarchical, variable scale marine geographic information system to monitor water quality in the New York Bight. In *GIS/LIS '91 Proceedings* (Atlanta, Georgia: ACSM-ASPRS-URISA-AM/FM), pp. 730-739.

Hatcher, G. 1992, *A Geographic Information System as a Data Management Tool for Seafloor Mapping* (Master's Thesis, University of Rhode Island, Narragansett, Rhode Island).

Hatcher, G. and Maher, N., 1999, Real-time GIS for marine applications, in this volume, Chapter 10.

Johnston, R.J., 1999, Geography and GIS. In *Geographical Information Systems. Volume 1: Principles and Technical Issues*, 1, edited by Longley, P.A., Goodchild, M.F., Maguire, D.J. and Rhind, D.W. (New York: John Wiley & Sons), pp. 39-47.

Keller, C.P., Gowan, R.F. and Dolling, A., 1991, Marine spatio-temporal GIS. In *The Canadian Conference on GIS '91 Proceedings* (Ottawa), pp. 345-358.

Langran, G. and Kall, D.J., 1991, Processing EEZ Data in a marine geographic information system. In *1991 EEZ Symposium on Mapping and Research*,

Portland, Oregon, edited by Lockwood, M. and McGregor, B.A. (Portland, Oregon: U.S. Geological Survey Circular 1092), pp. 127-129.

Li, R., 1999, Data models for marine and coastal geographic information systems, in this volume, Chapter 3.

Li, R. and Saxena, N.K., 1993, Development of an integrated marine geographic information system. *Marine Geodesy*, **16**, pp. 293-307.

Li, R., Qian, L. and Blais, J.A.R., 1995, A hypergraph-based conceptual model for bathymetric and related data management. *Marine Geodesy*, **18**, pp. 173-182.

Lockwood, M., and R. Li. 1995, Marine geographic information systems: What sets them apart? *Marine Geodesy* , **18**, pp. 157-159.

Lucas, A., 1999, Representation of variability in marine environmental data, in this volume, Chapter 5.

Lucas, A., Abbedissen, M.B. and Budgell, W.P., 1994, A spatial metadata management system for ocean applications: Requirements analysis. In *ISPRS Working Group II/2 Workshop on the Requirements for Integrated GIS* (New Orleans, Louisiana), pp. 1-13.

MacDonald, I.R., Best, S.E. and Lee, C.S., 1992, Biogeochemical processes at natural oil seeps in the Gulf of Mexico: Field-trials of a small-area benthic imaging system (SABIS). In *First Thematic Conference on Remote Sensing for Marine and Coastal Environments* (New Orleans, Louisiana), pp. 1-7.

McAdoo, B., 1999, Mapping submarine landslides, in this volume, Chapter 14.

Manley, T.O. and Tallet, J.A., 1990, Volumetric visualization: An effective use of GIS technology in the field of oceanography. *Oceanography*, **3**, pp. 23-29.

Mason, D.C., O'Conaill, M.A. and Bell, S.B.M., 1994, Handling four-dimensional geo-referenced data in environmental GIS. *International Journal of Geographical Information Systems*, **8**, pp. 191-215.

MRJ, Inc., 1993, *Marine Data Sampler* (Oakton, Virginia: MRJ, Inc.).

National Research Council, 1988, *The Mid-Ocean Ridge: A Dynamic Global System* (Washington, D. C.: National Academy Press).

Nierenberg, W.A., 1992, *Encyclopedia of Earth System Science* (San Diego, California: Academic Press).

Palmer, H. and Pruett, L., 1999, GIS applications to maritime boundary delimitation, in this volume, Chapter 21.

Psuty, N.P., Steinberg, P.E. and Wright, D.J., in press, Coastal and marine geography. In *Geography in America at the Dawn of the 21st Century*, edited by Gaile, G.L. and Willmott, C.J. (New York: Oxford University Press).

Robinson, G.R., 1991, The UK digital Marine Atlas Project: An evolutionary approach towards a Marine Information System. *International Hydrographic Review*, **68**, pp. 39-51.

Scott, D.J., Miller, P.D., Loewenstein, F.C. and Langran, G., 1993, ECDIS: A shipboard navigation system. *Microstation Manager*, **12**, pp. 40-43.

Sherin, A., 1999, Linear reference data models and dynamic segmentation: Application to coastal and marine data, in this volume, Chapter 7.

Steinberg, P., in press, Navigating to multiple horizons: Toward a geography of ocean-space. *The Professional Geographer*.

Su, Y., 1999, A user-friendly marine GIS for multi-dimensional visualisation, in this volume, Chapter 16.

Triñanes, J.A., Cotos, J.M., Tobar, A. and Arias, J., 1994, A geographic information system for operational use in pelagic fisheries-FIS. In *Oceans 94*, Brest, France (Brest, France: IEEE), pp. 532-535.

Ward, R., C. Roberts, and R. Furness, 1999, Electronic chart display and information systems (ECDIS): State-of-the-art in nautical charting, in this volume, Chapter 11.

Williamson, P., 1994, Integrating Earth system science. *Ambio*, **23**, p. 3.

Varma, H., 1999, Applying spatio/temporal concepts to correlative data analysis, in this volume, Chapter 6.

Wright, D.J., 1994, *From Pattern to Process on the Deep Ocean Floor: A Geographic Information System Approach* (Ph.D., University of California, Santa Barbara, California).

Wright, D.J., 1999, Spatial reasoning for marine geology and geophysics, in this volume, Chapter 8.

Wright, D.J., in press, Getting to the bottom of it: Tools, techniques, and discoveries of deep ocean geography. *The Professional Geographer*.

Wright, D.J. and Goodchild, M.F., 1997, Data from the deep: Implications for the GIS community. *International Journal of Geographical Information Science*, **11**, pp. 523-528.

Wright, D.J., Haymon, R.M. and Woods Hole Oceanographic Institution Imaging Group, 1992, Analysis of ocean floor data using GIS techniques. In *Association of American Geographers Annual Meeting* (San Diego, California: Association of American Geographers), p. 262.

Wright, D.J., Haymon, R.M. and Woods Hole Oceanographic Institution Imaging Group, 1992b, Applications of ARC/INFO to the studies of the ocean floor. In *ARC/INFO Maps 1991*, edited by Dangermond, J. (Redlands, California: Environmental Systems Research Institute), pp. 28-29.

Wright, D.J., Haymon, R.M., Macdonald, K.C. and Goodchild, M.F., 1994, The power of geographic information systems (GIS) for oceanography: Implications for spatio-temporal modelling of mid-ocean ridge evolution. In *Proceedings of The Oceanography Society Pacific Basin Meeting* (Honolulu, Hawaii: The Oceanography Society), p. 66.

Wright, D.J., Fox, C.G. and Bobbitt, A.M., 1997, A scientific information model for deepsea mapping and sampline. *Marine Geodesy*, **20**, pp. 367-379.

Wright, D.J., Haymon, R.M. and Fornari, D.J., 1995, Crustal fissuring and its relationship to magmatic and hydrothermal processes on the East Pacific Rise crest (9°12'-54'N). *Journal of Geophysical Research*, **100**, pp. 6097-6120.

CHAPTER TWO

Working on the Frontiers of Science: Applying GIS to the Coastal Zone

Darius J. Bartlett

2.1 INTRODUCTION

Applying GIS to the coast is an activity in which a number of frontiers and boundaries need to be considered. The first of these relates to the nature of the coastal system itself. The coast represents one of the most important boundary zones on planet Earth, marking the dynamic, three-way interface between land, sea and the overlying atmosphere. These three very distinct but interlinked environments, two fluid and one comparatively enduring, meet at the ever-changing line of the shore. Most people will intuitively recognise the existence of the shoreline, although its precise location is less easy to determine; while defining the landward, seaward and indeed the vertical extents of the coastal zone is even more problematic. Carter (1988) defines the coastal zone as being "that space in which terrestrial environments influence marine (or lacustrine) environments and vice versa." According to such a definition, in many parts of the world entire countries and districts are entirely contained within the coastal zone (examples include the Netherlands, much of Bangladesh, the Nile Delta, the Atlantic and Gulf coasts of the United States, as well as innumerable Pacific islands).

Secondly, the coast can be seen as a cultural and conceptual frontier, representing a transition zone between the known and the comparatively unknown. Perhaps somewhat surprisingly in view of humanity's long association with the sea, there is very great variability in both the quantity and the quality of coastal data and knowledge available to the scientist or the decision-maker (see, for example, Smith and Piggott, 1987). Indeed, it is often popularly suggested that we know more about the surfaces of our near planetary neighbours than we do about our coastal waters, the continental shelves and the deep ocean floor.

Thirdly, the line of the shore is often used as an administrative and jurisdictional division. As well as frequently marking the division between national (land) and international (water) space (Carter, 1988), the coastal zone is also itself usually subject to many different levels of government, judicial responsibility, and management. In Ireland, for example, administrative responsibility for the Irish coastal zone is currently divided among 15 County Councils, 2 Corporations, 6 County Boroughs, 14 Urban District Councils, 7 Town Commissioners and 2 Port authorities (National Coastal Erosion Committee, 1992; Institute of Public Administration, 1997); while numerous central governmental departments, non-governmental organisations, state- and semi-state bodies, development agencies, defence and security forces, tourism concerns, and other bodies also have interests and activities that impinge on the shore. Some of these organisations focus on the terrestrial side of the divide, while others are concerned with the offshore and deep-sea environment. Very few have interests, or management structures, that encompass both elements; and in many cases, each body or organisation will have its own, unique, aerial units, geographic boundaries and limits of responsibility. Similar divisions of responsibility and authority are commonplace in most coastal states. Given the diversity of interest groups, stakeholders, managerial authorities and administrative structures that converge at the shore, conflicts are almost inevitable between and among coastal users, managers, developers and the wider public, as well as between human society and the natural environment.

Finally, many of the current limits of GIS science and technology are also encountered at the coast. A decade ago, Chrisman (1987) suggested that "some of the current success [in GIS] is achieved by exploiting the easy part of the problems. The tough issues, temporarily swept under the rug, will re-emerge, perhaps to discredit the whole process." Coastal GIS remains to this day a "tough issue." While applying GIS technology has undoubtedly contributed to improved knowledge and shoreline management practices, it has managed this mostly by "exploiting the easy part of the problem". Indeed, in 1993, it has even been claimed that many coastal implementations have achieved their success *in spite of*, and not *because of* the fundamental character of current GIS (Bartlett, 1993a). Specific problems, many raised and discussed further in subsequent chapters of this book, include the quest for effective conceptual and data models of coastal objects and phenomena; representing and modelling the three-dimensional nature of the coastal system within a GIS framework; and handling the temporal, dynamic properties of shoreline and coastal processes.

2.2 WHY APPLY GIS TO THE COAST?

Human societies have always enjoyed a close relationship with the coast, although the nature of this relationship has not always been harmonious. Traditional uses of coastal space have included trade and conquest, migration and defence and, in some cases, a focus for cultural and spiritual identity (Bartlett and Carter, 1990; Carter, 1988). Today, at least 40% of the world's human population now lives on or near the coast, and the proportion of coastal dwellers is increasing at a very much faster rate than that of the overall population (Carter, 1988).

The coastal zone also provides access to physical and other resources. For many countries of the world the oceans, and especially the continental margins, provide an important primary source of protein. It is estimated that more than 99% of the world's catch of marine fish species are caught within 320km of land, and more than 50% of the total biological production of the ocean takes place in the coastal zone (Holt and Segenstam, 1982, quoted in Salm and Clark, 1989; see also Meaden, 1999). Minerals and vital hydrocarbons also come from many of our coastal waters (Cooke and Doornkamp, 1990; Jefferies-Harris and Selwood, 1991) while we also use the offshore zone as a convenient disposal ground for sewage and for domestic and industrial wastes, including toxic and/or radioactive materials. As well as attracting industry, the coastal fringes of our landmasses are also becoming increasingly sought after for leisure and tourism developments, both organised and informal. All of these activities may contribute significantly to regional, national and international economic performance (Bartlett, 1993a).

Finally, because of its dynamic nature, the coastal system is one of the most hazardous locations to live in. Minimising the human and economic consequences of flooding, erosion and other impacts is a major consideration for many coastal societies, particularly if recent warnings about possible near-future changes in global climate (thereby possibly increasing storm frequency and impact) and rises in world sea-level are proved correct (e.g., Titus, 1987; Carter, 1988, 1990; Devoy, 1992)

Multiple use of coastal space, the implications of coastal processes on human society, and the fragility of the marine environment and its coastal fringe, all requires that rational, integrated and sustainable management strategies be developed. Traditional coastal management, developed over many centuries, has mostly been reactive, predominantly localised, and largely based around civil engineering works designed to protect specific bits of shoreline against specific real or perceived threats. It is only comparatively recently that the rise in environmentalism world-wide has led to a change towards a new ethos in coastal management, one based on longer-term planning, more regional scales of investigation, and greater use of so-called "soft engineering" methods of shoreline protection (Carter, 1988; Bartlett and Carter, 1990). The terms "Coastal Zone Management" (CZM) or, occasionally, "Integrated Coastal Zone Management" (ICZM) are frequently used to refer to this latter type of coastal management. The primary objective of CZM may be stated as being to "devise a framework within which Man may live harmoniously with nature" (Carter, 1988).

While development and implementation of integrated coastal management policies is now established and internationally recognised ideal, the tools and methodologies for achieving such goals are still under development. It is clear, however, that for any management of the shore to be effective, it is necessary for the policies to be based on informed decision-making. This in turn requires ready access to appropriate, reliable and timely data and information, in suitable form for the task at hand. Since much of this information and data is likely to have a spatial component, GIS have obvious relevance to this task, and have the potential to contribute to coastal management in a number of ways. These include:

- the ability to handle much larger databases and to integrate and synthesize data from a much wider range of relevant criteria than might be achieved by manual methods. This in turn means that more balanced and co-ordinated coastal management strategies may be developed for considerably longer lengths of shoreline, spanning administrative divisions and even national borders where required;

- encouragement for the development and use of standards for coastal data definition, collection and storage, which promotes compatibility of data and processing techniques between projects and departments, as well as ensuring consistency of approach at any one site over time (Kennedy-Smith, 1986; AGI, 1996; Buchanan, 1997; ENVALDAT, 1997; Bartlett et al., 1998; Lockwood and Fowler, 1999);

- the use of a shared database (especially if access is provided via a data network) also facilitates the updating of records, and the provision of a common set of data to the many different departments or offices that might typically be involved in management of a single stretch of coast. A shared database implies reduction or elimination of duplicated records, and thus the potential for significant economic savings as well as improved operational efficiency;

• as well as providing efficient data storage and retrieval facilities, GIS also offers the ability to model, test and compare alternative management scenarios, before a proposed strategy is imposed on the real-world (e.g., Lee *et al.*, 1992; Ligdas, 1996). Computer technology allows the consideration of much more complex simulations; their application to very much larger databases; and also enables compression of temporal and spatial scales to more manageable dimensions (Langran, 1990; Varma, 1999).

2.3 A BRIEF HISTORICAL PERSPECTIVE ON THE APPLICATION OF GIS TO THE COAST

Although a few pioneering articles relating to the application of GIS to coastal issues do occur in academic journals and other official publications, most of the early literature is restricted to government reports (e.g., Carter, 1976) conference presentations (e.g., Eberhart and Dolan, 1980) and the "grey" literature (e.g., Eberhart, 1980). Much of this early literature was primarily concerned with remote sensing of coastal waters, and focused on the acquisition of data from satellite imagery, with GIS being a largely-implied later part of the data processing operation (e.g., Diop, 1981, 1982; Caixing, 1986).

Analysis of this literature suggests that interest in applying GIS to the coast first emerged in the early- and mid-1970s, at a time when GIS itself was still very new technology. Much of the earliest references are aspirational, rather than describing work in progress: in 1972, for example, Ellis wrote of the need for a coastal zone information system which "relates data, information, predictive techniques, environmental interactions, methods of analysis and applications into one system of procedures, tools and instructions for use by planners" (Ellis, 1972). Similar desiderata were identified by Schneidewind (1972), in a review of the information needs for coastal development planning. It was some years, however, before integrated geo-data processing systems for coastal science or management appeared.

The first, pioneering applications of GIS to the coast mostly focused on the ability of computer systems to store and retrieve data (Bartlett, 1993a). Most only used spatial concepts such as absolute location or relative position in a limited way, and tended also to have somewhat limited graphics (including hard copy generating) capabilities. Typical examples of such systems included the US Marine Resources Council's Management Information System (MIMS) for Long Island (Ellis, 1972); the Resources and Management Shoreline (RAMS) database developed for handling data relating to development permits in Chesapeake Bay (Eberhart, 1980; Eberhart and Dolan, 1980); the TRIP database developed, by the Tourism and Recreation Research Unit at Edinburgh University for the Countryside Commission for Scotland and the Scottish Tourist Board, as an aid to tourism planning, (TRRU, 1977); and the Canada Coastal Information System (CIS) developed by the Geological Survey of Canada (Fricker and Forbes, 1988) (Bartlett, 1993a). At the same time, considerable progress was being made elsewhere, in the development of techniques for numerical modelling of coastal processes by computer (see, for example, Komar, 1983). To this day, however, coastal process modelling largely remains a separate branch of computing, and most such modelling is at best only loosely coupled to GIS *per se*.

In part, these early developments of coastal GIS were frequently constrained by the hardware and software then available, and also by generally low levels of awareness shown by potential users, regarding the functionality and capabilities of GIS (Green, 1987). It should be remembered also that commercial GIS packages only became relatively available – especially in Europe – in the early 1980s. These tended to be very expensive when compared to present-day costs; they were relatively much more complex to install and maintain; and, with very few exceptions, they ran exclusively on mainframe or higher-end workstation computers. PC-based GIS were restricted to comparatively simple raster systems.

One illustration of the influence these restrictions frequently had on the development and application of coastal GIS, and the typical steps that had to be taken to overcome these, may be drawn from the experiences of the present author who, in the summer of 1987, was engaged as research assistant at the University of Ulster, Coleraine, working under the direction of the late Professor (then Dr.) Bill Carter. The objective of the project was to undertake on a detailed survey of the coast of Northern Ireland (Carter and Bartlett, 1988a, 1988b, 1988c), in order to assess the extent and causes of coastal erosion, and suggest management options for dealing with the problem. During the course of the investigation, the researchers had to collect, integrate and process a wide diversity of data, relating to the geometry of the shore (beach extent, embayment curvature and dimensions, cliff slope, etc.), composition (solid geology, sediment characteristics, vegetation, made ground and armoured shorelines, etc.), natural processes (meteorology, sediment transport, wave energy parameters, etc.), human occupancy and activities (including beach sand extraction and other resource exploitation, civil engineering structures, etc.), cultural heritage (archaeological sites, etc.), and many other parameters. As well as collecting current data, the investigation required an historical time-series of coastal changes to be built up, so that trends and predictions could be assessed. Although extremely varied in terms of type and resolution, it was recognised that most of the data to be collected for the research were spatial, that is they could be referenced to one or more locations, and geography provided the key to their integration and analysis. It was therefore decided to computerise the collection, analysis and management of data associated with the study, and thereby explore the potential and limitations of GIS to assist in research of this type.

Like many other comparatively early applications of GIS to the coast, including those referred to above, the system developed for the Erosion Survey also focused on data storage and retrieval. A literature review conducted at the start of the project, to establish the then state-of-the-art, revealed just how new and comparatively undocumented the technique of coastal GIS was, with very few published role models available to give guidance. It was therefore necessary to develop our own approaches and techniques as the project developed.

After careful review of the project data-handling requirements, the available human and financial resources, and the technologies at our disposal, a loosely-coupled hybrid approach was adopted: at the core of the system was a database, created on a desk-top personal computer (a then top-of-the-range 80383-powered machine running MS-DOS) using the Oracle relational database management system (Bartlett, 1988a). Since the purpose of the research was to assist in shoreline erosion risk assessment and management, the primary geographical focus was the line of the shore, and it was decided to conceptualise and model the coastal system as a simple line object. The coastline of County Antrim, extending from the mouth of the River Bann in the north to Larne Harbour in the south, was therefore digitised as a single line, and the co-ordinates of the resulting vertices were entered into an Oracle database as a "long-thin" database table comprising three fields: a unique sequential identifier number; the x-co-ordinate ; and the y-co-ordinate. Each vertex was entered as a separate record in the database. Thematic data relating to coastal attributes (presence and type of built structure, wave climate observations, shoreline geology, land use, etc.) were then entered into separate database tables. At each along-shore change in attribute, the start and end vertices for that particular segment were logged alongside the attribute properties in the thematic tables. A series of SQL macros were developed for proximity searching, in order to match and snap these start- and end-point vertices to the nearest vertices on the digitised line, thus linking the geometric and the thematic databases. Further macros and *ad hoc* querying, as appropriate, allowed some, admittedly limited, investigation of spatial interrelationships between variables of interest.

Graphic output and visualisation of results was similarly limited, largely due to constraints imposed by the technologies then available. Map generation for the project was achieved by selecting and retrieving the vertices, relating to the shoreline segment(s) of interest, from the database, exporting these data from Oracle, and transferring these via the internal campus computer network to the university mainframe computer. Once transferred, a series of macros imported the data to the GIMMS mapping package running on the mainframe, which enabled the production of suitable plots. This process also enabled graphs, scattergrams and bar charts to be created from statistical data likewise abstracted and exported from the database to GIMMS (Bartlett, 1988b).

Although primitive by today's standards, the system as described above did enable the objectives of the Coast Survey to be achieved (Carter and Bartlett, 1988a, 1988b, 1988c, 1990), and gave valuable insight into the many issues that confront the development of GIS for coastal zone problems. As the chapters in the present volume testify, resolution of several of these problems remains, to this day, a high priority for most coastal GIS research agendas. In particular, in its reliance (admittedly cumbersome in the way it was implemented) on an attempted segmentation of the shore, whereby the attribute and the geometric data were separated, and where results of *ad hoc* database searches of the attribute data were linked to the corresponding sections of the coast, the approach adopted in the Northern Ireland survey anticipated similar, much more recent and more efficient, approaches to dynamic segmentation of the shore (Bartlett, 1993b; McCall, 1995; Sherin, 1999).

2.4   APPLYING PROPRIETARY GIS TO THE COAST

Since the late 1980s, in particular, there has been an upsurge in the application of proprietary GIS packages to the coast. Initially this tended to be based on software running on mainframe computers but, increasingly, workstations and desktop/ personal machines are being used, as the performance of these increases and a greater range and amount of suitable software appears on the market. Many GIS vendors and products are now involved, and several examples of such applications are provided in the chapters of this book. At the same time, the application to typical coastal problems of most current proprietary GIS is still limited and, as is indicated in the Epilogue, the drive to extend GIS functionality further into the marine and shore environments is a concern of many, if not most, major system developers. For the end-user, the application of proprietary software has both costs and benefits. The advantages include: ' *Privately owned* .

- freedom from the need to write, develop, maintain in-house software;
- improved access to data, since a growing amount of data is being collected and made available using standard interchange formats, many of which are based on the data formats of particular vendors;
- easier vertical (within a particular project) and horizontal (between projects) integration of data sets and applications;
- easier implementation of data standards for data quality and exchange;
- the user-base and pool of expertise of proprietary software is invariably going to be more extensive than that for systems developed in-house. This in turn can lead to reduced need for system-specific training, improved learning curves for new personnel, and greater access to assistance and advice in case of difficulty.

Against this, the disadvantages of applying proprietary GIS packages to the coast also have to be considered. In particular, most commercial GIS were developed for land-based applications, and are built around cartographic metaphors, data models and fundamental paradigms optimised for conditions found in terrestrial (including on-shore social and economic) environments. These paradigms and models frequently are poorly suited to coastal data, where boundaries between key variables are less easy to define; where a much greater range of spatial scales and resolutions have to be considered; where there is a greater need to work in three spatial dimensions; and where the temporal dimension is also fundamental to many analyses. Notwithstanding the many advances in technology, coastal zone GIS remains comparatively immature and, as suggested earlier, is situated firmly at the cutting edge of geoinformation science and technology. "

2.5   THE CHALLENGE OF APPLYING GIS TO THE COAST

2.5.1 Conceptual Models of the Shore

Problems

One of the most important lessons to be learned from collective experience in the· field of coastal GIS, both published and unpublished, is the importance of rigorous data modelling before work starts on implementing a GIS database. Any GIS database is a model of reality, and it is important to establish a thorough and rigorous conceptual model (Peuquet, 1984), in order to successfully identify and represent all those features of the real world considered relevant to the study. Unfortunately, where the coast is the focus of study, this modelling is made particularly difficult, partly due to the multiplicity of professions, stake holders and other groups with interests at the shore; and partly due to the inherently fuzzy and indeterminate nature of so many coastal entities and phenomena.

We have already seen that there is a multiplicity of different sectors, public and private, with interests at the coast. It is essential to establish the user-base for whom the coastal zone GIS is intended, and the view of the coast and its component elements that is relevant to this constituency, since this operational context will have a direct bearing on the more technologically-oriented questions of hardware and software selection, and the overall architecture of the system. It will also define the information products expected from the system, and hence the types of data, and the processes performed on these, that are required in order to produce the desired output.

One useful approach to deriving this conceptual model, where time and resources permit, is for the database designer to pay one or more field visits to the coast in the company of those scientists, administrators or other professionals for whom the system is being developed. During this field visit, the prospective end-users are invited to describe, in purely verbal terms, using the language and terminology of their discipline, the coastal zone that they are seeing. By examining the elements, attributes and relationships between these that they identify, the data modeller can thereby derive a first-order impression of what is important and what is of lesser concern to these users within the coastal system. While the results of this modelling will be *ad hoc* and unstructured, they do provide a useful basis for subsequent more formal entity-relationship modelling based on this provisional schema.

2.5.2 Data Availability and Access

Effective decision-making on the coast requires the decision-maker to have a genuine understanding of the morphological, biological and human-oriented processes likely to be encountered within the coastal system. This level of understanding will only be obtained if accurate, timely and appropriate information is available for consultation (French, 1991); while information may be recognised as the end-product derived from the processing of data. An organised, planned and coherent coastal database should therefore be seen as a *sine qua non* of good coastal management. Unfortunately, in practice the availability of data can rarely be assumed. Smith and Piggott (1987, p. 14) note that "the collection and management of a coastal database calls for quite high levels of expenditure and long time periods. Often the coastal manager's greatest problem is to convince his political masters [sic] and his rate-payers that the work is worth doing at all."

Data relating to the coast may be obtained through either or both of two separate routes. Firstly, data may be obtained by direct field survey and measurement, by processing of satellite or airborne imagery, from shipboard instruments (sonar, etc.), or by means of automatic data loggers (automatic wave recorders or tide gauges for example) (Terwindt, 1992). Unfortunately, data collected by such means may be expensive, time-consuming, may require specialist equipment or survey skills, and in some cases may be precluded on logistical grounds or when a long time-series of data are required.

As an alternative to direct capture, the data to be used may be secondary or derived (Maguire, 1989). This latter category includes data from published or archived materials, including maps and charts, aerial photographs, air-borne or satellite remotely sensed imagery, and observations relating to climate or sediments. These data may typically be recorded in the form of published or unpublished literature. Like directly captured data, these secondary data also have both strengths and weaknesses. On the positive side, secondary data may frequently be cheaper and more accessible than primary data. However, unless the data are accompanied by good documentation and copious metadata, they may be of uncertain reliability or provenance, thus rendering the user vulnerable to unwitting propagation of errors and uncertainties contained in the original data sets.

In practice, and perhaps somewhat surprisingly in view of humanity's long association with the sea, there is very great variability in both the quantity and the quality of coastal data; while, even if the data do exist, strategic and/or commercial considerations may frequently make access to these data difficult. The continuing importance of data capture to both coastal and marine GIS is further discussed in several chapters in the present volume.

2.5.3 Data Modelling Issues

The path from data to information to understanding applies equally to both conventional and automated methods of coastal data management. Data are the raw materials upon which any attempts at meaningful and effective coastal management must therefore depend. It is important that the data not only exist, but also that they are accessible in a suitably structured form (Peuquet, 1994).

In the early days of coastal GIS, limitations in computer hardware and software imposed significant constraints on the choice of data model available to the system developer. Raster systems, while comparatively simple to program and easy to implement, did carry important overheads in terms of data volumes requiring processing and the resulting trade-offs that had to be made between high-resolution but equally high-volume databases on the one hand versus lesser volumes of data but at the expense of loss of spatial definition (these issues are clearly less critical in the late 1990s, as a result of recent dramatic improvements in computer processing power, parallel advances in bulk data storage technologies, and the data retrieval efficiency of modern software). Meanwhile, vector systems tended to be more complex, and were poorly suited to representing the dynamics and fuzzy boundaries of many objects and phenomena at the coast.

Since the early 1990s, however, much research has been devoted to extending the repertoire of data models available for coastal (and other) GIS applications: they include the dynamic segmentation of line objects (Bartlett, 1993a; McCall, 1995; Sherin, 1999); Voronoi and other tessellations of the plane (Gold, 1992, 1999; Gold and Condal, 1995); octrees and other structures for three-dimensional data visualisation and modelling (Li, 1999; Raper, 1999); and the coupling of cellular automata with spatial data models (Burrough and McDonnell, 1998). While many of these techniques remain largely experimental, results obtained and reported to date indicate that they hold considerable potential for advancing the utility of GIS within coastal and marine environments.

2.5.4 Institutional Issues

As with any new scientific methodology or emergent technology, successful take-up and implementation of GIS is as dependent on awareness and other human factors as it is on purely technical issues. This is as true in the case of coastal applications of the technology as it is as elsewhere.

In order to smooth the learning curve for coastal managers new to GIS, and also to accelerate the development of new and coastal-specific techniques and methods for applying these technologies, in the late 1980s the Commission on the Coastal Environment (since re-named to the Commission on Coastal Systems) of the International Geographical Union adopted a project to investigate and publicise the use of GIS for coastal management. A number of products arose out of this initiative, including publication of a position paper on the then state-of-the-art of coastal zone GIS (Bartlett, 1993a) and of annotated bibliographies (Bartlett, 1993d; Fell *et al.*, 1997); establishment of an on-line discussion forum for students and practitioners in the field of coastal and marine GIS, which is currently subscribed to by over 500 members on five continents[1]; and organisation of a number of conferences and workshops. Of particular note regarding the last of these, in 1995 the IGU Commission joined forces with the Working Group on Marine Cartography of the International Cartographic Association, to organise the First International Symposium on GIS and Computer Mapping for Coastal Zone Management (CoastGIS '95). This meeting, which was held in Cork, Ireland, in February 1995 (Furness, 1995) attracted some hundred or so delegates. CoastGIS has since become a regular, biennial event, with a growing attendance level, reflecting the increased interest in, and acknowledged importance of, the subject: the second meeting took place in Aberdeen, Scotland (Green, 1997); and a third meeting in the series is due to convene in Brest, France, in September 1999.

---

[1] The SEA-GIS list: to subscribe, send a request to LISTSERV@LISTSERV.HEA.IE with the request "subscribe SEA-GIS <your name>" in the body of the message (not in the subject header field).

2.6 CONCLUSION

The application of GIS to the coastal zone has come a long way since the first, tentative, steps were taken in the late 1970s and early 1980s. Some of the difficulties in coastal GIS have been at least partially resolved, through advances in hardware and software; while much current research focuses on attempts to overcome the remainder. The various chapters of this book contain many examples showing how the technologies are being successfully applied, while others demonstrate how and where other, as yet unresolved, issues are being addressed. Despite these initiatives and developments, coastal (and marine) applications of GIS are still comparatively immature compared to many other application areas.

The chapters of this book that follow illustrate some of the more significant recent advances and applications in the field. They also point to several of the remaining problem areas and current technical or conceptual hurdles yet to be overcome, demonstrating that there is still abundant scope for further research, and for both existing and new professionals to make their own contributions to this important area of geographic information science.

2.7 REFERENCES

AGI, 1996, Guidelines for Geographic Information Content and Quality: For Those Who Use, Hold or Need to Acquire Geographic Information. *AGI Publication number 1/96* (London: Association for Geographic Information).

Bartlett, D.J., 1988a, GIS applications for regional studies: The Antrim Coast erosion survey. In *Merlewood Research Papers* (Cumbria, England: Institute for Terrestrial Ecology, Merlewood).

Bartlett, D.J., 1988b, The Use of GIMMS in the Antrim Coast erosion survey. In *Proceedings, Third Annual GIMMS Users' Conference and Workshop*, England, April 20-21, 1988 (Oxford Polytechnic).

Bartlett, D.J., 1990, Spatial data structures and coastal information systems. In *Proceedings EGIS'90, First European Conference on Geographical Information Systems,* Amsterdam, The Netherlands, April 10-134, 1990 (J. Harts, H.F.L. Ottens, H.J. Scholten and D.A. Ondaatje eds), **1**, pp. 30 - 39.

Bartlett, D.J., 1993a, GIS and the Coastal Zone: Past, Present and Future. *AGI Publication number 3/94* (London: Association for Geographic Information).

Bartlett, D.J., 1993b, Space, time, chaos and coastal GIS. In *Proceedings, Sixteenth International Cartographic Congress*, Cologne, Germany, May 3-9, 1993.

Bartlett, D.J., 1993c, Coastal zone applications of GIS: Overview. *In* St Martin, K. (Ed), *Explorations in Geographic Information Systems Technology Volume 3: Applications in Coastal Zone Research and Management* (Worcester, Massachusetts: Clark Labs for Cartographic Technology and Analysis).

Bartlett, D.J., 1993d, The Design and Application of Geographical Information Systems for the Coastal Zone: An Annotated Bibliography. *NCGIA Technical Report 93-9* (Santa Barbara, California: National Center for Geographic Information and Analysis).

Bartlett, D.J., and R.W.G. Carter, 1990, Seascape ecology: The landscape ecology of the coastal zone. *Ekologia* (CSFR), **10**, pp. 43-53.

Bartlett, D.J., Buchanan, C, Joyce, D and Tobin, D., 1998, Avoiding the GIGO syndrome: Quality control in coastal GIS databases. In *International Conference on Education and Training in Integrated Coastal Management: The Mediterranean Prospect,* Genoa, Italy, May 25-29, 1998.

Burrough, P.A. and McDonnell, R., 1998, *Principles of Geographical Information Systems* (Oxford: Oxford University Press).

Buchanan, C., 1997, National and International initiatives versus regional applications: A regional user's perspective on data quality standards. *IOC/IODE Ocean Data Symposium: Summary of Proceedings,* Dublin, October 1997.

Caixing, Y., 1986, The application of remote sensing in the investigation of the resources and environment in the coastal zone of China. *The Application of Remote Sensing Techniques to Coastal Zone Management and Environmental Monitoring.* Ohaka, Bangladesh, November 18 – 26 1986, pp. 185-197.

Carter, R.W.G., 1988, *Coastal Environments* (London: Academic Press).

Carter, R.W.G., 1990, *The Iimpact on Ireland of Change in Mean Sea Level.* Programme of Expert Studies on Climate Change, Number 2 (Dublin: Department of the Environment).

Carter, R.W.G. and D.J. Bartlett, 1988a, *Coast Erosion and Management in the Antrim Coast and Glens and the Causeway Coast Areas of Outstanding Natural Beauty* (University of Ulster, Department of Environmental Studies, for the Department of the Environment).

Carter, R.W.G. and D.J. Bartlett, 1988b, *Causeway Coast AONB: Coast Erosion and Management* (University of Ulster, Department of Environmental Studies, for the Department of the Environment).

Carter, R.W.G. and D.J. Bartlett, 1988c, *Ballycastle Beach: Evolution and Management Issues (*University of Ulster, Department of Environmental Studies, for the Department of the Environment).

Carter, R.W.G. and D.J. Bartlett, 1990, Coast erosion in northeast Ireland, Part 1, Sand beaches, dunes and river mouths. *Irish Geography,* **23**, pp. 1-16.

Carter, V., 1976, Computer mapping of coastal wetlands. *USGS Professional paper 929.*

Chrisman, N.R., 1987, Design of information systems based on social and cultural goals. *Photogrammetric Engineering and Remote Sensing,* **53**, pp. 1367-1370.

Cooke, R.U. and Doornkamp, J.C. 1990, Coastal Environments. *In* Cooke, R.U. and Doornkamp, J.C., (Eds) *Geomorphology in Environmental Management* (Oxford: Clarendon Press), pp. 269-302

Devoy, R.J.N., 1992, Questions of Coastal Protection and Human Response to Sea-Level Rise in Ireland and Britain. *Irish Geography ,* **25**, pp. 1-22.

Diop, E.S., 1981, Une methode de determination de d'étude des differents taxons-paysages des milieux estuariens: example de la Casamance et du Saloum. *Rapport de l'UNESCO sur les Sciences de la Mer,* **117**.

Diop, E.S., 1982, L'imagerie Landsat et la cartographie des formations quaternaires des zone lagunaires et estuariennes de l'Afrique de l'Ouest. Méthodologie et résultants obtenus. *In* Lassere, P. and Postma, H., (Eds), *Actes du symposium international sur les lagunes* (Special volume of *Oceanologica Acta: revue européene d'oceanologie) ,* pp. 95-99.

Eberhart, R.C., 1980, Pressures on the edge of Chesapeake Bay: 1973-1979. *Environmental Programs Report CPE-7905 (*Maryland, USA: Johns Hopkins University).

Eberhart, R.C. and Dolan, T.J., 1980, Chesapeake Bay Development Pressures: RAMS database analysis. *Proceedings, Coastal Zone '80.*

Ellis, R.H., 1972, Coastal Zone Management System: a combination of tools. *In* Marine Technology Society (Editors), *Tools for Coastal Zone Management.* (Washington, D.C.: Marine Technology Society).

ENVALDAT, 1997, *Establishing the Value of Environmental Data Products. A guide to present practice in coastal zone management.* Version 3, Final Draft for Dundee Workshop. (Unpublished working paper produced by HR Wallingford Ltd. and project partners, as part of the ENVALDAT-Customer Valuation of Environmental Data project, funded by the European Commission)

Fell, B., Gavelek, R., Bartlett, D. and Miller, A.H., 1997, *Coastal Management. A Bibliography of Geographic Information System Applications* (Charleton, S.C.: NOAA Coastal Services Center). (Also available from the Internet: http://www.csc.noaa.gov/gisbib)

French, D.P., 1991, Quantitative mapping of coastal habitats and natural resource distributions: application to Narragansett Bay, Rhode Island. *Coastal Zone '91 Proceedings,* **1**, pp. 407-415.

Fricker, A. and Forbes, D.L., 1988, A system for coastal description and classification. *Coastal Management,* **16**, pp. 111-137.

Furness, R. (Ed.), 1995, *Proceedings, CoastGIS'95, the First International Symposium on GIS and Computer Mapping for Coastal Zone Management.* Cork, Ireland, February 3-5, 1995.

Gold, C., 1992, An object -based dynamic spatial model, and its application in the development of a user-friendly digitising system. *Proceedings, 5th International Symposium on Spatial Data Handling,* August 3-7, 1992, Charleston, South Carolina, pp. 495-504.

Gold, C.M., 1999, An algorithmic approach to marine GIS, in this volume, Chapter 4.

Gold, C. and Condal, A., 1995, A spatial data structure integrating GIS and simulation in a marine environment. *Marine Geodesy ,* **18**, pp. 213-228.

Green, D.A., 1997, *Proceedings, CoastGIS'97, the Second International Symposium on GIS and Computer Mapping for Coastal Zone Management,* Aberdeen, Scotland, 1995.

Green, N.P.A., 1987, Teach yourself geographical information systems. The design, creation and use of demonstrators and tutors. *International Journal of Geographical Information Systems,* **1**, pp. 279-290.

Institute of Public Administration, 1997, *Administration Yearbook and Diary 1998* (Dublin, Ireland: Institute of Public Administration).

Jefferies-Harris, T. and Selwood, J., 1991, Management of Marine Sand and Gravel: A seabed information system. *Land and Mineral Surveying,* **9**, pp. 6-8.

Kennedy-Smith, G.M., 1986, Data Quality: A Management Philosophy. *Proceedings, AutoCarto London,* **1**, pp. 381-401

Komar, P.D., 1983, Computer models of shoreline changes. *In* Komar, P.D. (Ed) *CRC Handbook of Coastal Processes and Erosion* (Boca Raton, Florida: CRC Press Inc).

Langran, G., 1990, Tracing temporal information in an automated nautical charting system. *Cartography and Geographic Information Systems*,**17**, pp. 291-299.

Lee, J., Park, R.A. and Mausel, P., 1992, Application of geoprocessing and simulation modeling to estimate impacts of sea level rise on the northwest coast of Florida. *Photogrammetric Engineering and Remote Sensing.*

Li, R., 1999, Data models for marine and coastal geographic information systems, in this volume, Chapter 3.

Ligdas, C.N.A., 1996, The study of coastal processes in the north east of England using a GIS. *In* Tausik, J and Mitchell, J. (Eds). *Partnership in the Coastal Zone* (European Coastal Association for Science and Technology/Samara Publications), pp. 515 – 524.

Lockwood, M. and Fowler, C., 1999, Significance of coastal and marine geographic information systems within the context of the United States national geospatial data policies, in this volume, Chapter 20.

McCall, S., 1995, The application of dynamic segmentation in the development of a coastal geographic information system. *Proceedings, CoastGIS'95, the First International Symposium on GIS and Computer Mapping for Coastal Zone Management* (Cork, Ireland), pp. 305-312.

Maguire, D., 1989, *Computers in Geography* (Harlow, Essex: Longman Scientific and Technical).

Meaden, G., 1999, Applications of GIS to fisheries management, in this volume, Chapter 15.

National Coastal Erosion Committee, 1992, *Coastal Management: A Case for Action*. Irish County and City Engineers' Association, National Coastal Erosion Committee, July 1992.

Peuquet, D., 1984, A conceptual framework and comparison of spatial data models. *Cartographica* , **21**, pp. 66-113.

Raper, J., 1999, 2.5- and 3-D GIS for coastal geomorphology, in this volume, Chapter 9.

Salm, R.V. and Clark, J.R., 1984, *Marine and Coastal Protected Areas: A Guide for Planners and Managers* (Gland, Switzerland: International Union for Conservation of Nature and Natural Resources), 302 pp.

Schneidewind, N.F., 1972, Information systems and data requirements: coastal development planning. *In* Brahtz, J.F.P. (Ed), *Coastal Zone Management.*

Sherin, A., 1999, Linear reference data models and dynamic segmentation: Application to coastal and marine data, in this volume, Chapter 7.

Smith, A.W.S. and Piggott, T.L., 1987, In search of a coastal management database. *Shore and Beach,* **55**, pp. 13-20.

Terwindt, J.H.J., 1992, *Users Report on Instruments for Coastal Research.*(International Geographical Union Commission on the Coastal Environment. Working Group Coastal Instrumentation), 31 pp.

Titus, J.G. (Ed), 1987, *Greenhouse Effect, Sea Level Rise and Coastal Wetlands.* United States EPA Publication EPA-230-05-86-013.

Varma, H., 1999, Applying spatio/temporal concepts to correlative data analysis, in this volume, Chapter 6.

CHAPTER THREE

Data Models for Marine and Coastal Geographic Information Systems

Rongxing Li

3.1 INTRODUCTION

Modelling the ever-changing marine and coastal environment requires spatial and temporal modelling techniques. It deals with traditional spatial information such as coastal terrain data, bathymetric data, and coastal land cover data. However, other less unconventional aspects of spatial data relating to marine and coastal environments make modelling these a very challenging issue. An example of such unconventional data might come from a shoreline that is dynamic because of water level changes, coastal erosion, and other natural and human activities. A time series data set related to this dynamic environment is often collected at one spatial location, while containing a vast volume of observations (wind, current, wave and others). Furthermore, in modelling the deep-ocean environment there is a lack of basic landmarks or infrastructure features for use as references, compared to those existing in land-based systems (transportation networks, hydrographic features, and various levels of administrative or other boundaries). In some cases, bathymetric data may be the only data available for establishing a deep-ocean database; while a further difficulty, is how to associate attributes to the data in order to facilitate spatial queries.

Considering the complexity of the marine and coastal environment, an appropriate data model is crucial to the success of the entire information system: this is the focus of the current chapter. First, some general considerations are presented about data models. Second, some data models used in the marine and coastal environment are introduced. Finally, brief conclusions are given.

3.2 GENERAL CONSIDERATIONS

3.2.1 Data Model Design

Design of a data model consists of four phases, namely, external design, conceptual design, logical design, and internal design (Laurini and Thompson, 1992; Li *et al.*, 1998). *External design* is an initial stage. The real world is simplified according to application requirements, because information from the real world is too rich to be entirely included in a database. The special data needs of existing and potential users should be addressed and accommodated. The result of the external design is usually a data dictionary describing spatial and attribute data to be included. In the *conceptual design* phase, a model for organising the data is constructed. Based on the external design result, spatial objects are defined as entities, and attributes are associated with the entities. Associations/relations are used to describe relationships between the entities. To implement the relationships, cardinality numbers are assigned to each relation, which check the consistency of the model. The data model designed at this point is independent of a specific database management system and independent of computers to be used. The result of the conceptual design is usually termed an entity-relation (ER) model. The ER model is converted to tables in the *logical design* phase, if it is decided that relational databases are to be generated. Other schema of databases may require different styles for the presentation of the data; for example, object-oriented GIS software may employ a flat-file approach instead, while network and hierarchical DBMS have been rarely used for GIS applications. Usually, each entity or association/relation is converted to a separate table. However, some tables may be combined into one, or one table may be split into a few tables in order to conform to consistency requirements of standard spatial relational databases. In *interior design*, functions and capabilities of hardware and software have to be considered. Interior design provides a basis for data model implementation.

In addition, georeferencing of all data sets involved is an essential issue in light of the fact that many data sets describe dynamic phenomena of the marine and coastal environment. For instance, consider the construction of a coastal terrain model (CTM), which covers a strip on both sides of the shoreline, by combining data from a digital terrain model (DTM) and a digital bathymetric model. Since the DTM gives heights of grid points above a reference ellipsoid, and the bathymetric data may present depths of grid points from a water level datum, a simple merge of the data sets would not produce reasonable results. Careful examination of the water level data used will lead to a better understanding of both data sets. As the next step, the depths of the bathymetric data need to be converted to elevations relative to the ellipsoid, considering the water-level datum and differences between the ellipsoid being used and the geoid. Decisions to use multiple referencing systems or a unified reference system in a model should be made based on individual cases.

2.2.2 Review of Object Representations

Object representations describe geometric characteristics of spatial features and provide associations with attributes. They are very important for implementation of logical design results. The most popular representations in commercial software packages are vector and/or raster based. A vector-based representation describes two-dimensional objects by a set of points, lines and polygons that are topologically combined. Attributes are associated with the geometric representation through, for example, relational tables. On the other hand, a raster representation depicts two-dimensional objects by assigning geometric (boundary) and/or attribute (class) values to regularly distributed pixels. To describe higher-dimensional objects, usually of a volumetric and/or temporal nature, an extension from the vector and raster representations is necessary.

Figure 3.1 illustrates two groups of three-dimensional (3-D) representations, namely surface-based and volume-based representations. Surface-based geometric representations depict geometric characteristics of objects by micro-surface cells or surface primitives. Four popular and important surface-based representations are grids, shape models, facet models, and boundary representations.

- A *grid* or raster is the most popular numerical description of object (including terrain) surfaces in GIS and digital mapping. Surface points are defined in a grid on the XY plane. For every grid point there is a corresponding elevation. If the original data are a set of irregularly distributed surface points, an interpolation is necessary to build a grid. This interpolation procedure is also called gridding. A surface description with a grid is very easy to understand. Based on this grid, a contour map or a 3-D mesh can easily be generated. Some operations used in image processing, such as filters, can also be applied to grids. However, a multi-valued surface for which there is more than one Z value for each grid point may not be represented. This is why a grid is also called a 2.5 dimensional representation.

- A *shape model* describes an object surface by derivatives (slopes) of surface points. Instead of Z values, slopes in X and Y directions are provided for every grid point. A normal vector can be defined at each grid point. Thus, the relative relationship between grid points, namely the shape of the surface, is known. However, its absolute position and orientation in the co-ordinate system remain undetermined. If one or more Z values of grid points are available, Z values of all other grid points can be calculated using the slope information. The storage requirement of a shape model is proportional to the dimension of the grid, which is usually twice that of a grid of the same dimension. An important application of shape models is reconstruction of surface models by shape from shading techniques (Li, 1993).

- A *facet model* approximates an object surface by surface cells, which may be of varying shapes. For example, a triangle facet model has its surface cells defined as triangles of various sizes, and is called a triangulated irregular network (TIN). In areas where the surface varies very slowly, triangles of large sizes can be used to represent the surface; while in areas with rapid surface changes finer triangles can approximate surface variations more precisely. The storage space needed for a triangle facet model is proportional to the number of triangles used, which depends on surface characteristics and the required resolution of the resulting model. Because of their stable structure and simplicity for processing, triangle facet models are widely used in generation of DTMs, visualisation of 3D object surfaces, ray tracing and spatial database conversions (Li, 1991).

- A *boundary representation* (B-rep) has a hierarchical data structure in which an object surface is usually composed of four categories of predefined primitives: point, edge, face and volume elements. An element of a B-rep consists of geometric data, an identification code of element categories, and its relationship to other elements. Free-form surfaces, which cannot be represented by simple analytical mathematical functions, may be described by B-spline functions. The storage requirement needed for a B-rep is approximately proportional to the number of primitives used. Grids, shape models and facet models are suited for describing object surfaces with irregular shapes. However, B-reps give exactly the surface geometry of objects with regular shapes.

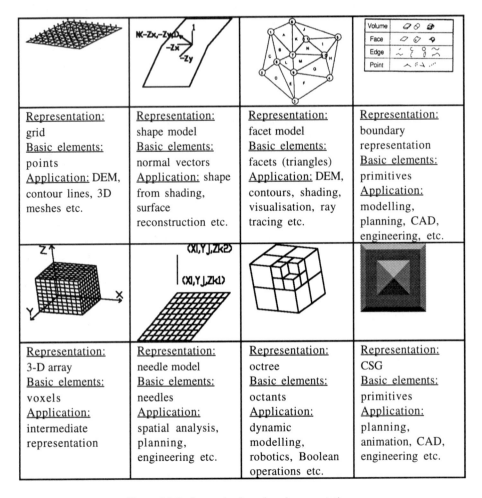

Figure 3.1 Surface and volume based representations.

Volume-based geometric representations (Figure 3.1) describe the interior of objects by using solid information, instead of surface information. Therefore, they are also called solid models. Volumetric information is necessary, for example, for operations in ocean mining, water column modelling, coastal planning, and interference detection for automated navigation. Four volume-based representations are described.

• A *3-D array* has its elements of either zero or one, where zero means the background and one indicates the occupation of the element by objects. Therefore, it is also called a 3-D binary array. Only one bit of memory space is needed to store an element. Suppose that an object is scanned in the 3-D binary array whose elements are initialised with zero. This scanning procedure results in a 3-D binary array with its elements of value 1 representing solid information of the object. The higher the scanning resolution, the greater the dimension of the array. Thus, the storage requirement is proportional to the number of elements, that is, the dimension of the 3-D binary array. Because of its large storage requirement, a 3-D binary array is rarely used in computational practices for large data sets.

- A *needle model* uses run-length encoding to reduce the storage requirement. If a 3-D binary array is defined in a co-ordinate system, for every grid point on the XY plane, array elements along the Z direction are checked. Only those segments in the Z direction with continuous elements of value 1 are considered and called run lengths. For each vertical segment, also called a needle, only its starting Z co-ordinate and run length are required. Needle models are often used to represent solid information, for example, multi-layers of water-column-related information or geological subsurfaces.

- An *octree* is a more compact and efficient volume-based representation (Samet, 1990; Li, 1991). An original octant (similar to a 3-D binary array) is defined, which is the smallest cube containing the object. At the first level, this original octant is divided into 8 suboctants by halving the original octant in three directions. Each suboctant is then checked to determine whether it is occupied by the object. The suboctants are classified into three categories: (1) F=full (occupied by the object); (2) E=empty (no object element in the suboctant); and (3) P=partial (partially occupied). P-octants are further subdivided into eight suboctants at the next level, which are again classified. The partition procedure continues until all suboctants are either F- or E-octants. Those final octants are presented by their octree codes, i.e. listed digits at different levels. Storage space requirement for an octree is relatively low in comparison to other volume-based representations. It increases when the required octree resolution is high (i.e., the size of the smallest suboctants decreases). Because of their hierarchical data structure, octree representations are very efficient in spatial analysis, Boolean operations, and database management.

- *Constructive solid geometry (CSG)* represents an object by a combination of predefined primitives such as cubes, cylinders, cones and complex primitives. Relationships between the primitives include geometric transformations and Boolean operations. Usually, CSG has a tree structure where leaf nodes correspond to Boolean set operations (Mortenson, 1985). Storage space of a CSG representation increases as the number of primitives increases. Since leaves of a CSG tree store only geometric parameters of primitives and their relationships, a CSG representation usually generates a very compact spatial database. Volume-based geometric representations of 3D binary arrays, needle models and octrees are capable of representing objects with irregular shapes. Conversely, CSG representations are well suited for describing regularly shaped objects.

The above-extended surface- and volume-based representations offer a wide range of choices of geometric representations for modelling objects in marine and coastal GIS. Grid and TIN models have been used to describe coastal terrain topography and bathymetry (e.g., Wright *et al.*, 1998). Shape models can be used for densifying bathymetric data from sonar imagery. Needle models and octrees are efficient for modelling volumetric objects such as sediments on or underneath the seafloor and suspending objects in water columns. Finally, protection structures and other artificial objects can be represented by B-rep and CSG.

3.3 SOME SPECIFIC MARINE AND COASTAL DATA MODELS

3.3.1 Shoreline

A shoreline in a navigation chart produced by NOAA's National Geodetic Survey is defined as a line intersected by coastal land and the mean lower-low water (MLLW) level that is the synoptic average over a 19.2-year lunar/solar cycle. In practice the shoreline is usually now derived from aerial photographs that are taken and matched with tidal records (Slama *et al.*, 1980). In a digital procedure, the shoreline may be derived from an intersection between a coastal terrain model (CTM) and a water surface model corresponding to the MLLW.

Considering the definition and nature of the CTM and the water surface, the shoreline is not a static line. Thus, the shoreline representation should be flexible to accommodate any future changes, which may include geometric changes caused by physical conditions such as erosion (Li *et al.*, 1998), and attribute changes along the shoreline such as land use changes, protection status changes and others. A shoreline can be defined based on line segments of a polygon layer, for example, L_2, L_5 and L_8 in Figure 3.2a. Attributes associated with the shoreline segments, such as erosion status, can also be stored with the segments. However, instead of constant attribute information along a segment (for instance, L_5), there may be three different values corresponding to three subsegments. Usually, we would have to break the segment into three subsegments in order to assign the three attributes. In fact, the attribute values may change frequently so that the subsegments would have to be remodified. An extreme case would be that the segment is assigned with multiple attributes such as erosion status, protection type, land use, and others. The need for attribute changes will make the database management almost impossible.

A technique called dynamic segmentation, originally designed for transportation applications, has been applied in shoreline management (Li *et al.*, 1998) and in analysing potential impacts of sea level rise (Bartlett *et al.*, 1997; Sherin, 1999). The version of dynamic segmentation implemented in commercial GIS packages such as Arc/INFO of the Environmental Systems Research Institute (Environmental Systems Research Institute, 1992) allows the shoreline to be defined as a curved line (Figure 3.2b). A point along the shoreline can be defined by a distance, also called measure, between the start of the shoreline to the point concerned. In such a way, a subsegment, for example, the first subsegment of L_5, is defined by a pair of measures (from M_{51} and to M_{52}) and the attribute information is recorded independently of the line geometry (Bartlett *et al.*, 1997). This model is especially effective for assigning multiple classes of attributes to a shoreline, by adding more columns in an event table where each class occupies two columns (from and to) to depict its subsegments. It is advantageous that changes of attribute

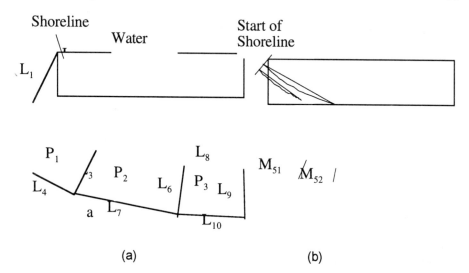

Figure 3.2 Shoreline: (a) Shared with a polygon layer
and (b) dynamic segmentation.

status on the same shoreline do not require dissolving segments or breaking the shoreline. It is achieved by simply changing the measures. The above characteristics, in addition to others, make dynamic segmentation an increasingly popular model in coastal applications.

3.3.2 Seafloor

Bathymetric data that describe seafloor topography are usually in the form of a grid or TIN. In the marine environment, where seamounts mid-ocean ridges, trenches and abyssal seafloor are dominant features, difficulties in managing bathymetric and related data include, among others: (1) association of attribute data; (2) multiple resolutions; and (3) large volume of data sets. A hypergraph-based bathymetric data model was proposed by Li *et al.* (1995), which sets the convenience for data description and efficient data management and rendering as top priorities.

Figure 3.3a shows the bathymetry of the Cook Seamount area off the Big Island of Hawaii. Since the data are available on a grid, it is natural and easy to display and render the data as a 3-D mesh (Figure 3.3b). However, bathymetric features such as seamount peaks, ridges and valleys are not explicitly represented. These features are the most significant characteristics in this area. According to the specific application, various attributes may be associated with them. Otherwise, the attributes may have to be associated with each grid point in order to be fully integrated.

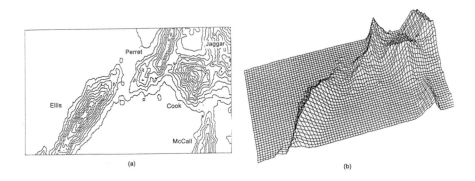

(a)

(b)

Figure 3.3 Bathymetry of the Cook Seamount area off the Big Island of
Hawaii: (a) 2-D contour lines, and (b) a 3-D mesh model.

A hypergraph structure can provide a graph-based description of the
bathymetry, and can organise these features hierarchically. Furthermore, attributes
are explicitly associated with the features (Figure 3.4). Generally, objects with
common attributes are grouped together to form a substructure. Raw data including
track lines, bathymetric data (grid), and other field survey data make a substructure
accessible through the root. General information for an area such as area name,
boundary of the area (usually defined by longitude and latitude), map projections,
datum, and other details, is also placed at the same level for easy access. Geometric
description is organised hierarchically according to seamount peaks. This
substructure contains common attributes such as seamount names, positions of
peaks, contour lines of the peaks, etc. Each substructure corresponds to one table.
To describe interrelationships between objects and substructures, additional tables
might be established. In such a case, redundancy is often introduced. Therefore,
interrelationships should be reduced to a minimum.

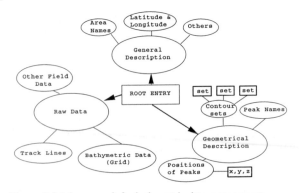

Figure 3.4 A hypergraph for bathymetric data management.

status on the same shoreline do not require dissolving segments or breaking the shoreline. It is achieved by simply changing the measures. The above characteristics, in addition to others, make dynamic segmentation an increasingly popular model in coastal applications.

Figure 3.5 gives a detailed explanation of the substructure of geometrical description in Figure 3.4. At the first level, the seamounts in the area, named Ellis, Perret, Cook, Jaggar, and McCall, are defined as objects. Attributes associated with the seamounts including seamount names, co-ordinates of peaks, major mining materials, etc. are linked to each object. The object "base" is created to represent the seafloor bottom, for example, with a depth range from the maximum depth of the seafloor in the area to a certain depth that is below all defined seamounts in the area. Contour sets describing the geometry of seamounts are at the second level. For each seamount, the contour lines are organised in a top-down fashion and linked to corresponding seamounts through a relation. There are no overlaps of contours between the seamounts and the base. However, there might be contour lines traversing more than one seamount. In this case, one contour line may have relations with more than one seamount. Detailed geometric characteristics of contours are given at the third level where co-ordinates of vertices are stored and related to corresponding contours.

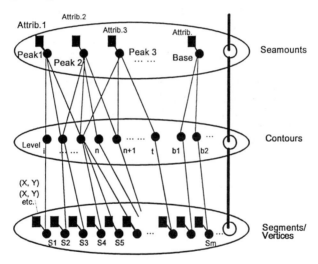

Figure 3.5 Substructure of geometric description in Figure 3.4.

This data structure has multiresolution features for graphics rendering. Horizontally, vertices with 2-D co-ordinates construct the surface of a 3-D object at fixed heights. In the vertical direction, the surface is sampled by contours. Therefore, the resolution of the whole model is dependent on the density of contour vertices and the contour intervals. Thus, the vertical sampling can be analysed well using Nyquist sampling theory, while the density of contour vertices varies according to bathymetry. In visualising the seafloor, observers tend to pay much more attention to the foreground than to the background. Thus, the rendering quality of distant seamounts may not need to be as high as for closer ones, although they should not be neglected. Moreover, since the display screen has a limited resolution, it is not reasonable to render far objects in full detail since they may actually be represented by only a few pixels on the screen. If objects are rendered in a multiresolution scheme, in which far objects are in lower resolutions and close ones in higher resolutions, the quality of the whole scene should not be much reduced. Meanwhile, the rendering speed is expected to be improved. The hypergraph model is well suited to this implementation, because the data are organised hierarchically according to resolution. However, the rendering software needs to be improved to more effectively handle the multiresolution model. Figure 3.6 illustrates a rendering scene where Ellis Seamount (from Figure 3.3a) is in full resolution, while other seamounts are in half resolution. The same shading/colouring scheme is used for both high- and low-resolution objects. Thus, the whole screen remains harmonic and the rendering speed increases.

Figure 3.6 Multiresolution graphics rendering of Cook Seamount bathymetric data.

3.3.3 Time Series

Time series data for the coastal and marine sectors may include logs or processed data of wave and wind observations, sedimentary and current measurements from underwater devices, and other hydrographic and oceanographic observations. They often come with a measured location indicating where the data were collected and a large volume of observations. Although an average of wind strength, water level, or temperature would be sufficient for many applications, measurements at certain times or a distribution of measurements would provide much more detailed information or correlation with other spatial or non spatial data. Information derived from the data using spatial statistic analysis may make the data more useful.

Two ways to handle time series data are as simplified or full-scale data sets. The first approach performs data pre-processing and only stores important statistic derivatives of a time series data set, such as means, standard deviations, histograms, etc. Metadata are also made available to describe the original data set. Information in the metadata may include when, where and how the data were acquired, data format, accuracy, and others. The time series data information is linked to a GIS through the location information. On the GIS interface, the location is represented as a point symbol. An event table lists all locations and file names of the corresponding meta data sets, and statistic derivatives. Thus, by clicking on a point symbol on the interface, descriptive information and statistic derivatives of the time series data can be queried. The original time series data may be requested if necessary. The advantages of this approach are its low requirement of space and computational power, and its simplicity and efficiency in database management.

The full-scale data approach stores the original time series data in a GIS. The event table links directly to the time series data although metadata and statistic derivatives may also be included. Because of the nature of the time series data, an integration of a statistic software package in the GIS environment may be necessary. Such a statistic package supplies analysis tools unavailable in GIS. Results from both GIS and the statistics package can be seamlessly combined to support higher level operations. A typical example would be simulation and prediction of shorelines if an additional hydrological modelling system were integrated.

3.4 CONCLUSION

The importance of data models for marine and coastal GIS has been recognised by researchers and developers since the early days of GIS applications for these environments. Although we are able to handle marine and coastal spatial data, we still cannot do so efficiently (e.g., Wright and Goodchild, 1997). This is especially true when dealing with temporal data, integration of GIS with simulation and modelling systems, and large interdisciplinary coastal information infrastructures. Future research efforts should be directed toward, among others, a fuller understanding of the temporal nature of marine and coastal spatial data, the influence of uncertainty of spatial data on marine and coastal modelling, and the integrity of coastal and marine spatial data from interdisciplinary data sources.

3.5 REFERENCES

Bartlett, D., Devoy, R., McCall, S. and O'Connor, I., 1997, A dynamically segmented linear data model of the coast. *Marine Geodesy*, **20**, pp.137-151.
Berge, C., 1973, *Graphs and Hypergraphs* (New York: North-Holland Publishing Co.).
Environmental Systems Research Institute, 1992, *ARC/INFO User's Guide 6.0: Dynamic Segmentation* (Redlands, California: Environmental Systems Research Institute).

Fritsch, D., 1990, Towards three-dimensional data structures in geographic information systems. In *European Conference on GIS*, pp.335-345.

Laurini, R. and Thompson D., 1992, *Fundamentals of Spatial Information Systems*, (London: Academic Press Ltd).

Li, R., Cho, W.K., Ramcharan, E.K, Kjerfve, B. and Willis, D.H., 1998, A coastal GIS for shoreline monitoring and management: Case study in Malaysia. *Surveying and Land Information Systems,* **58**, pp.157-166.

Li, R., Qian, L., and Blais, J.A.R., 1995, A hypergraph-based conceptual model for bathymetric and related data management. *Marine Geodesy*, **18**, pp.172-182.

Li., R., 1993, Shape from shading: A method of integration of sonar images and bathymetric data for ocean mapping. *Marine Geodesy*, **15**, pp.115-127.

Li, R. and Saxena, N.K., 1993, Development of an integrated marine geographic information system. *Marine Geodesy*, **16**, pp.293-307.

Li, R., 1991, An algorithm for building octree from boundary representation. In *Intelligent Design and Manufacturing for Prototyping*, edited by Bagchi, A. and Beaman, J.J. (PED-Vol. 50, ASME), pp.13-23.

Lockwood, M. and Hill, G.W., 1988, Developing a 10-Year EEZ seafloor mapping and research program. *Marine Geodesy*, **12**, pp.167-175.

Mortenson, M.E., 1985, *Geometric Modelling* (New York: John Wiley & Sons).

Samet, H., 1990, *The Design and Analysis of Spatial Data Structures* (Reading, Massachusetts: Addison-Wesley Publishing Company, Inc.).

Sichritzis, T. and Lochovsky, F.H., 1982, *Data Models* (Englewood Cliffs, N.J.: Prentice-Hall, Inc.).

Sherin, A., Linear reference data models and dynamic segmentation: Application to coastal and marine data, in this volume, Chapter 7.

Slama, C.C., Theurer, C. and Henriksen, S.W., 1980, *Manual of Photogrammetry* (Falls Church, Virginia: American Society of Photogrammetry).

Turner, A. K., editor, 1990, *Three Dimensional Modelling with Geoscientific Information* (Dortrecht: Kluwer).

Wright, D.J. and Goodchild, M.F., 1997, Data from the deep: Implications for the GIS community. *International Journal of Geographical Information Science*, **11**, pp. 523-528.

Wright, D.J., Wood, R. and Sylvander, B., 1998, ArcGMT: A suite of tools for conversion between Arc/INFO and Generic Mapping Tools (GMT). *Computers and Geosciences*, **24**, pp. 737-744.

CHAPTER FOUR

An Algorithmic Approach to Marine GIS

Christopher Gold

4.1 INTRODUCTION

The concept of a "marine GIS" is in some ways an oxymoron. A GIS almost presupposes that "geography" equals "land". The classical GIS structure is based on manual cartography, with registered transparent overlays of different themes, and the manual extraction of relevant combinations, an approach that is much less useful in a full marine environment, although still relevant in coastal zone management. Thus polygon overlay, polygon reclassification and buffer zone generation are the central operations. These are based on some form of underlying topological structure defining polygon connectivity. Other modules, such as network management or terrain modelling, are handled by other structures.

In a marine GIS, however, many of these concepts are open to question. Where are the polygons, the topographic models or the networks? On land, most of the features are presumed to be stable, with only occasional updates. At sea, how many features are stable, and how many continuously moving? In other words, is a "marine GIS" a GIS at all, within the usual definitions? Clearly, in the long term, it is not, but in that case, one needs to define more fully which operations need to be performed, and then to examine the relevant structures.

Two properties immediately come to mind when thinking about the sea, that it is three-dimensional, and that it is continuously changing. The definition of a three-dimensional spatial marine model is of great interest, but probably beyond our current capabilities in this paper. The discussion will therefore be restricted to two dimensions, although this is even less plausible than it is on land. Nevertheless, if 2-D concepts may be expanded to 3-D then this will be a bonus.

This chapter will concentrate on 2-D models of either the water surface or of the seafloor (bathymetry). Even these are fundamentally different from a land-based GIS: many things are continuously changing their position. Any spatial structure must be able to handle routine positional update, or else we will not be able to implement any meaningful spatial model. A spatial model is here defined as the selection of the appropriate properties of real-world space that is to be simulated in the computer. This has been discussed in detail in Gold (1996), and will be summarised briefly here.

One of the properties of real-world space is that it appears to be continuous. This is clear in the case of "field" data, such as temperature or (terrestrial) elevation. It is also true for "object" data, such as houses, as this is really dealing with an abrupt density change between the surrounding air and the solid house. Is a cloud an object or a field? Treated as a less abrupt density change it is best represented as a field. The same is true of many marine features: currents, plankton fields, polynyas, etc.

Nevertheless, in the human management of information, a classification of these changes is performed in field value, to provide identifying names. Within a computer system, the "space" must be broken up into appropriate pieces for storage, classification and management purposes. However, it is important to remember that a classification *is* being performed, and space is being subdivided, with all its over-simplification and potential for error. Even when working with straightforward objects such as houses, there is an underlying spatial classification of "house" and "not house", and notions of connectivity, proximity and adjacency.

On the basis of these thoughts, which become less esoteric and more meaningful when working with something as non-solid as the sea, it becomes clear that a computer simulation of space must be able to handle field-type data, must be able to embed solid objects within this space, and must be able to handle local movement (either of water or of solid objects such as ships) within a spatial data structure. As the overall objectives of a marine GIS have still to be defined, a preliminary requirement is the ability to simulate the movement of water or objects, and to be able to monitor the system state at any moment in time. Simulation is here defined as either the use of some mathematical algorithm, as for water flow, or a series of updates of ship position, or other observations. The ability to combine field and object is required, in order to preserve such spatial information as connectivity, proximity and adjacency, and to prohibit the existence of more than one object in one location at the same time.

Since the focus is on a form of spatial partition, it is natural to look first at the grid or raster model. This partitions space into a tessellation based on some superimposed co-ordinate system. It has the particular advantage that adjacent cells are readily detected, and many algorithms have been developed to handle questions of proximity and connectivity, among other questions. Its limitations are due to the fact that it *is* purely a partition of space, and is not directly related to any embedded objects or samples of field values. Where objects are concerned, it is difficult to manipulate. In addition, it is hard to generate for the globe with satisfactory results, it has a fixed spatial resolution that must be defined in advance, and it may have very high storage costs.

A second spatial partition is the "vector" polygon structure, as used in most commercial GISs. After the real-world space has been classified, this works fairly well. However, it was defined for static polygonal structures, not for discrete objects or data points, and is not easy to update locally. It is hard, for example, to imagine a polygon-based system to manage the navigation of a ship between various moving obstacles. It has the attraction of being a tessellation, as is the raster, can be represented as a graph structure inside the computer (a big advantage), but does not readily adapt to many unconnected island structures, or to networks or other planar graphs. Its biggest handicap lies in its construction from digitised arcs, which is typically slow and error-prone.

Other spatial partitions are also used for specialised purposes. For example, 2D-trees, quadtrees of various types and R-trees are spatial indexing schemes for the rapid storage and retrieval of spatially-located objects, and triangulated irregular networks (TINs) are used for the modelling of terrain from irregularly distributed elevation values. These and other variants are discussed in Worboys (1995).

It should be noted that all of these structures are concerned only with the spatial component of our information, and not with its attributes, although this has been touched on briefly in the previous consideration of clouds and houses. Much work has been done on this aspect by many other researchers . The focus of this chapter is only on changes in values of some property or properties over space. This has been discussed further in Gold (1996), and here it is simply assumed that some classification may be made between spatial locations, to distinguish them. For simplicity, this will usually be referred to as the "colour" of the object.

At this stage, it is now appropriate to specify a spatial model and an associated spatial data structure that satisfies many of the proposed requirements. The spatial model, the Voronoï diagram, has been discovered many times in different applications over the last century or so (Déscartes, Dirichlet, Voronoï, and Thiessen, among others), and more recently has had many versions and algorithms developed within the discipline of computational geometry (Aurenhammer 1991; Okabe *et al.*, 1992). The task here is to describe it as an appropriate topological data structure for many GIS applications, especially in dynamic contexts such as the marine environment.

4.2 THE SIMPLE VORONOÏ DIAGRAM: A STATIC DATA STRUCTURE

The discussion will be restricted initially to the simple nearest-point Voronoï diagram (VD), sometimes known as the Thiessen diagram, in the Euclidean plane, although many extensions exist, and are described in the citations above. The VD may readily be embedded in the surface of the sphere, simplifying the boundary conditions in the plane. The diagram may be defined as the partitioning of the plane such that, for any set of distinct data points, the cell associated with a particular data point contains all spatial locations closer to that point than to any other. This definition may be extended beyond data points, to line segments and other, more complex objects such as houses or road segments if needed. The definitions and algorithms which follow may be extended to two or more dimensions as required, permitting the extension of this approach to a fully three-dimensional marine GIS. The simple Voronoï/Thiessen diagram may be found in various GIS packages, usually for determining the regions closest to certain resources, such as fire stations or shopping centres.

A simple VD is shown in Figure 4.1, along with the associated dual Delaunay triangulation (DT). This triangulation, often used in terrain modelling, has a triangle edge connecting each pair of objects whose Voronoï cells are adjacent. The DT consists of triangles whose circumcircles are empty. If four or more points are cocircular, the interior triangle edges may be arbitrarily selected. It has the advantage of the raster model in that it is a tessellation of the plane, and hence neighbouring points are well defined. It has the advantage of the vector polygon model in that only one object is associated with each cell.

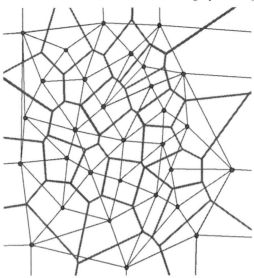

Figure 4.1 Voronoï diagram and Delaunay triangulation.

It will be shown later how to use these point objects to generate ones that are more complex. Per item, the VD requires more storage than the raster model, although fewer cells are needed to represent most spatial features. Unlike the vector polygon model, the VD is readily updated locally. This means that if a point is added, only a small number of neighbouring cells need to be modified. Thus updating is very fast. Initially, a description will be provided of a simple implementation of this algorithm and an appropriate data structure. Next, a variety of applications will be shown that become surprisingly simple with this approach.

The easiest way to start is with a large triangle on the plane, big enough to hold all the data. Alternatively, one may use a tetrahedron or octahedron on the sphere. The simplest data structure element is a triangle, with pointers to each of the three vertices and to each of the adjacent triangles (Gold *et al.*, 1977). A "walk" through the triangulation from any starting triangle easily gives the triangle containing the point to be added (Figure 4.2a). This triangle is split in three (Figure 4.2b), and all the exterior edges are tested to see if the three new triangles have empty circumcircles (Figure 4.2c). If they do not, the edge forming the diagonal between them is switched, connecting the other two vertices (Figure 4.2d), and the edges of the new triangles are tested. On the average, there are only a few switches (Gold *et al.*, 1977). This is sufficient to create the triangulation, and dual Voronoï diagram, as shown in Figure 4.1, with reasonable efficiency. One modification is to replace the triangle data structure with the quad-edge structure (Guibas and Stolfi, 1985) or as modified by Gold (1998) as the "quad-arc" structure, for more complex cases to be described later.

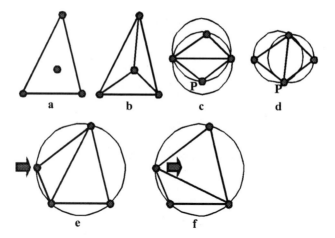

Figure 4.2 Point insertion and diagonal switching.

Here the topological element is no longer the triangle but the edge, and it may be used to store any connected graph on the orientable manifold. Only a few lines of code are needed for the implementation, as shown in Gold (1998).

4.3 EXTRACTING BOUNDARIES FROM DATA

While well known and easy to implement, the simple VD is sufficient for a variety of applications when combined with the "visibility ordering" of Gold and Maydell (1978) or Gold and Cormack (1987), which gives a method for traversing a triangulation so that each triangle is visited once only. Based on this, each vertex of the input data may be given a colour, and then the edges of the VD between vertices of different colours may be extracted, discarding the rest. This is described in Gold *et al.* (1996). Figure 4.3 gives the example of a possible geological map, where each data point identifies the rock type of an outcrop.

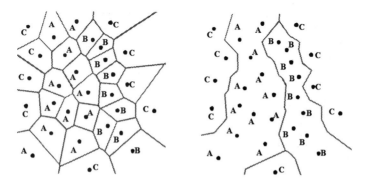

Figure 4.3 Extraction of geological boundaries.

Discarding the unwanted edges gives a first approximation to the geological boundaries (Figure 4.3). Gold *et al.* (1996) used this approach for forest map input by digitizing around the interior of each polygon in turn, and labelling each point with the polygon number. This results in a rapid approximation to the polygon map, with significant data input savings. Arcs were then exported to a conventional GIS. Gold (1997) extended this to scanned maps, where "fringe" points were generated around the black pixels on the scanned image (Figure 4.4a). These were then coloured using a "flood-fill" algorithm for each polygon interior, and then processed as before. Figure 4.4b shows the VD/DT of the points from Figure 4.4a, and Figure 4.4c shows the extracted polygon boundaries. Gold (1998) extended this again by using the quad-arc structure mentioned previously, permitting the preservation of the topological structure within small PC-based programs. This structure is capable of maintaining any connected planar graph, unlike the previous triangle structure. Thus for a static VD (or for a snapshot of a dynamic VD at some specified time) contours can be generated separating differing classes of data, as identified by the vertex colour. This is applicable to the marine GIS as well as to the terrestrial applications given.

A concept closely associated with the VD is the medial axis transform (MAT), or "skeleton". This is a subset of the VD, and is usually applied to polygons, each point on the MAT is equidistant to at least two edges of the polygon. The resulting graph may be used to classify the polygon shape in character recognition, for example (see Blum, 1967, 1973 and Blum and Nagel, 1978). In all of Figure 4.4, the discrete topology is really the skeleton of the linework, identified as separating vertices of differing colour.

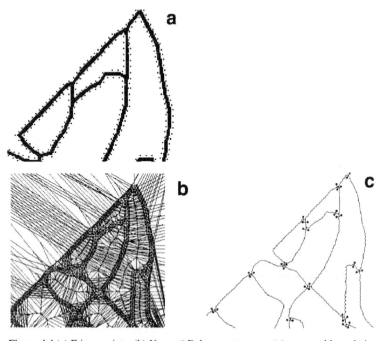

Figure 4.4 (a) Fringe points; (b) Voronoï-Delaunay structure; (c) extracted boundaries.

One problem, however, is that it is not always easy to colour the black boundary pixels so that edges may be distinguished. If there are no closed polygons, in a river network for example, all input points to the VD will have the same colour, and no skeleton will be produced. Amenta *et al.* (1998) showed that if a curve or boundary is sufficiently well sampled then the boundary may be extracted from the unordered points as the set of DT edges whose circumcircles do not contain part of the skeleton. Sufficiently well sampled was shown to be a function of the distance from the boundary to the MAT. It is usually easily achievable except at sharp corners. Gold (1999) and Gold and Snoeyink (in press) show that the extraction process may be reduced to a simple local test on each quad-edge, which is assigned to be either part of the crust (DT edge) or the skeleton (VD edge). Of interest to us is that the discrete skeleton, as an approximation to the MAT, is also extracted. A useful terrestrial example is the extraction of approximate watershed boundaries from a scanned hydrological network: the MAT gives the watershed between watercourses (Figure 4.5).

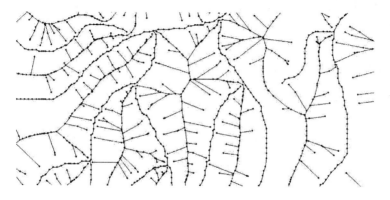

Figure 4.5 Watershed estimation from hydrology network.

While irregular in location and shape, the simple VD has many of the advantages of a raster that were previously described. This becomes valuable once we consider flow simulation, an application of interest in a marine GIS. In a raster model the flow domain, perhaps an aquifer or watershed in a terrestrial setting, is approximated by a relatively coarse grid. Each cell is given initial conditions (land elevation and water level), and the slope calculated between each pair of adjacent cells (approximated by their centroids). For each time step, flow is estimated between cell pairs, the water levels are updated, and the cycle repeated, either to a steady state, or for the duration of the simulation. The VD may also be used in the same manner: the governing equations take a little more derivation, but this is facilitated by the cell boundaries being perpendicular bisectors between the generating points (see Narasihman and Witherspoon, 1976). This is particularly efficient because the VD is more easily adapted to the form of the flow domain and boundary conditions (constant head or impermeable boundaries), requiring fewer cells than the grid.

The same advantage holds true for simple mapping, as in the geological map previously mentioned. A simple terrain model may be generated where the data point attribute is elevation rather than rock type. It is appropriate for the above flow modelling, and forms a useful technique whenever the closest data value is the safest elevation estimate, rather than some more complex function. It has proved very useful for regional volume estimates in coal resource evaluation, although the surface generated (a "prism map") has abrupt elevation changes.

A more common approach to terrain (or seafloor) modelling is to use the DT of the same data, rather than the VD, as this gives surface continuity, even though slopes and higher order derivatives are discontinuous. Attempts to generate "smooth" surfaces across triangle boundaries have not been entirely successful. However, the adaptability of being able to use data of any distribution is extremely valuable, especially with dense data such as along ships' tracks, with airborne observations and even with contours digitised from existing charts. In this last case, the crust/skeleton algorithm is of value in the generation of a correct triangulation (Figure 4.6a), as it enables the detection of gullies associated with indentations in individual contours. Figure 4.6b shows the same map generalised by retracting the minor "hair" branches on the skeleton. This smoothes minor perturbations in the "crust".

Figure 4.6 (a) Skeleton extracted from contours (top map). (b) Simplified skeleton (bottom map).

A big advantage of the DT, as opposed to any other possible triangulation, is the guarantee that the same triangulation will be produced independent of data input order, other than for cocircular points. It is also locally updateable. An improved interpolation approach using "area-stealing" will be described below. Apart from polygons and terrain models, the third data structure used in conventional GIS is to manage networks (roads, rivers) for various kinds of analysis, especially network flow. As the above structures are planar graphs, this application may be implemented with the DT, flagging the relevant edges accordingly. It is even more appropriate to use the quad-arc structure extracted from the original VD, as full connectivity is preserved, and bi-directional flow may be handled by the ordering structure imposed by the basic quad-edge topology. Thus, we have shown that, even with the simple incremental VD algorithm, most topological structures used in a traditional GIS may be maintained using the VD/quad-arc approach.

4.4 POINT DELETION: A DYNAMIC DATA STRUCTURE

In computer science terms, a data structure is "dynamic" if items may be added and deleted individually without having to rebuild a major portion of the whole thing, thus permitting rapid updating of the structure. A delete operation must now be defined in order to add to a simple incremental procedure in order to have a dynamic VD/DT data structure. While mentioned briefly in the literature as "the inverse of the incremental insertion algorithm", a detailed description of the algorithm, recently implemented, has yet to be found. Briefly, it goes as follows. Locate a triangle having the point to be deleted, by performing the "walk" operation as mentioned above. For each pair of the triangles surrounding the point to be deleted, examine the three remaining vertices. Could they be a valid Delaunay triangle in the absence of the central point? If so, switch their common edge, so that the central of the three vertices is no longer a neighbour of the point to be deleted. If the three vertices form a valid Delaunay triangle then there must be a valid circumcircle (i.e., the vertices forming the triangle are in anticlockwise order, and this circumcircle must be empty, with none of the other neighbours of the point to be deleted falling within this circle). This process is repeated with the reduced set of neighbours until only three remain. These three triangles are then reduced to one by removing the central point. As each of the triangles "cut off" is Delaunay, the resulting triangulation is also Delaunay. This is equivalent to removing the point to be deleted as the first step, and then retriangulating the remaining polygon. However, the process just described is simpler, and consistent with the point addition algorithm. In this case, Figure 4.2d reverts to Figure 4.2c for each vertex switched out. Figure 3,2b reverts to Figure 4.2a when the central point is removed.

Given the ability to add and delete points at will, there are various extensions that can be made to the previous applications, most obviously the ability to modify the data set interactively. This is particularly useful in the rapid digitizing application, where initial errors need to be corrected once they are detected. Single points or complex objects can be "moved" by deleting their composing points, and re-inserting them at the new locations. However, the biggest advantage comes when we wish to perform interpolation on our data set, and the flat triangular plates of the terrain model are inadequate.

Interpolation is the ability to extract estimates of the value of a curve or surface at intermediate locations between the data points. The simple "proximal query" or prism map described above is a form of interpolation, as is the triangulated terrain model. Where some form of "smooth" surface is desired, another approach must be taken. A good summary is given in Lam (1983). The most common approach is to take some form of weighted average of a set of neighbouring points to the location where the estimate is desired; the biggest problem is how to select these neighbours. If a "counting circle" is used, the results depend directly on its radius. The trick is to select a set of neighbours so that the contribution of a particular data point decreases to zero before it is eliminated from the set of neighbours, if not, then there is a discontinuity in the surface as a result of that point's sudden removal from the set of neighbours. Details are found in Gold (1989), where an alternative is proposed as described below. A mathematical definition of a similar solution is found in Sibson (1981).

The VD gives an intuitive definition of the neighbours of any data point, so to find the neighbours of some arbitrary query, its location should be inserted into the VD, and its immediate neighbours determined. This new point has "stolen" some of the area of each of these neighbours. That is why they are neighbours, having a boundary in common (Figure 4.7). If these stolen areas, normalised so they sum to unity, are used as weights for the elevation values of the data points, then we have a surface with certain properties. The surface is continuous, and it is "smooth" everywhere except at the data points. Its derivatives may be calculated (Anton *et al.*, 1998). A simple function may be used to make it smooth at the data points, and to conform to any slope data that may exist. Most importantly, it conforms to every data point, no matter what the distribution. Given the previous discussion, the algorithm is simple: insert the query point; find the neighbours and calculate the area of their Voronoï cells, remove the query point, recalculate the Voronoï area of the neighbours, and calculate the weighted average based on the differences in the areas.

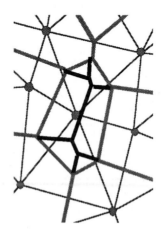

Figure 4.7 Areas stolen by point insertion.

Therefore, the "object" topological model defined by the VD may be used to derive "field" properties. The VD combines the properties of both object and field views of space, and hence of both "raster" and "vector" data structures. In a marine GIS based on the VD, bathymetric models may readily be generated, globally or locally. Of particular interest is the case of a boat navigating through a channel. Depths and slopes may be recalculated in real time, and the course adjusted accordingly. Recent depth soundings may be added to or removed from the data set, and used as required.

4.5 POINT MOVEMENT: A KINETIC DATA STRUCTURE

If a point represents the boat in the interpolation example above, it is natural to think of *moving* the boat, rather than merely deleting and re-inserting it. This has been done in the work of Roos (1990), Guibas *et al.* (1991), and Gold (1991) for two dimensions. If one point in the DT is perturbed slightly, then its Voronoï boundaries change slightly also. However, if the displacement is large enough that the circumcircle of a neighbouring triangle is entered (Figure 4.2e) then the diagonal is switched (Figure 4.2f). On leaving the circumcircle of three adjacent neighbours of the moving point the reverse operation takes place. If one can detect when the boat moves into or out of these circumcircles, then one can maintain the VD/DT topological structure. This clearly allows one of the basics of navigation, collision detection, as the examination of the boat's immediate neighbours at any point in time will give those most likely to be hit if the current course is continued. These neighbouring points could be the outlines of islands, wrecks or other obstacles. One could also perform channel navigation with the moving boat, considering the neighbouring points to be depth soundings, and using the interpolation techniques described above.

Yet another application comes to mind. Previously described was a water flow simulation by the finite difference method, where the cells (originally grids, but then the VD) expressed fixed portions of space through which the water flowed. This is the Eulerian model, but the Lagrangian model is also attractive, where "particles" (centres of mass) are free to move, along with their associated cells. "Free-Lagrange" methods have been developed using the VD, but without the dynamic data structure component (Fritts *et al.*, 1985). The dynamic VD is obviously attractive for this, and tidal modelling is being attempted using this approach. Global forces act on each particle (data point), which represents a fixed mass of water. Local forces act through the immediate Voronoï neighbours, and the result is a velocity vector associated with each point. Normally with the free-Lagrange method, a fixed time step is used, as with finite-difference or finite-element methods. This can cause problems, especially where turbulent flow is possible, as one fast-moving particle may collide with another, or else overshoot it, destroying the mesh structure. This problem may be eliminating by removing the fixed time step, and replacing it with a priority-queue of the topological events described above. That is, each moving point has its next topological event, when a triangle switch must take place, and the time of this is dependent on the particle's velocity. By storing this list of events, and processing them in temporal sequence (updating the list after

each event), the particle flow may be simulated without the usual mesh maintenance problems. Marine applications of this include simulation of tides, currents
and pollution plumes. The method is directly extendable to three dimensions. Indeed, all the point insertion, deletion and movement algorithms previously described are also feasible in three dimensions, as is the crust and skeleton extraction
process.

4.6 THE LINE SEGMENT VORONOÏ DIAGRAM

The above kinetic model may be extended towards more traditional GIS operations
by the addition of one simple observation: a curve is the locus of a moving point.
Thus, if one thinks of the moving point described above as the cartographer's
"pen", then one may use it to draw a traditional map composed of line segments or
polygons. The two-dimensional algorithm for moving points needs to be modified
slightly. The triangles may now have either points or line segments at their vertices, instead of just points. As a result, a more elaborate circumcircle calculation is
needed: instead of a circle touching the three points forming the triangle vertices,
one must calculate a circle that is tangent to any line-segment vertices as well. The
work of Anton and Gold (1997) shows how this may be done in an iterative fashion. Indeed, it may be used with any objects at the vertices of the triangle. In the
case of line segments, it is also necessary to know the *side* of the line segment that
is connected to the triangle. This may be the done with the half-line data structure
of Yang and Gold (1996) or the equivalent. When the "pen" is placed at the start of
a line, a point is inserted at that location, and as the "pen" is moved a point is split
off from this point and moved towards the destination. This is just point creation
and movement, as described above. However, the connection between the parent
point and the child (moving) point is itself an object: the desired line segment,
which is distinct from its endpoints. As the moving point progresses, the adjacency
relationships between the moving point and its current neighbours are transferred to
the trailing line: the line segment has Voronoï boundaries with all the old objects
passed by the moving point. At its destination the moving point may merely stop,
or may be merged with an object (point or line segment) at that location, allowing
the connection of line segments to form polygons or other map objects.

Indeed, the point may cross various intermediate line segments, generating
intersections as it goes. Line deletion is performed by performing the reverse operations to insertion. Figure 4.8 shows a simple map constructed in this fashion.

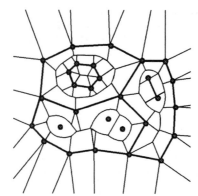

Figure 4.8 Simple line-segment Voronoï diagram.

Gold and Condal (1995) describe this as a "boat" approach to map construction: the pen is embedded in the map data structure, and interacts directly with its neighbouring objects, permitting "intelligent" operation, such as automatic polygon closure, snapping to other objects, and the detection and processing of intermediate intersections. Traditional map generation systems use the "plane" approach, otherwise known as spaghetti digitizing, where there is no interaction between the line drawing process and the underlying map (if any). The interactive approach holds many potential advantages for intelligent operations in addition to the ones described above.

In particular, the boat (or robot) model allows for a large variety of navigation or simulation applications. The "boat" may represent a real boat, and the detection of adjacent obstacles (because their Voronoï cells touch) can allow for evasive action in a simulation. The addition of some "long-distance intelligence", or goal seeking, would permit the development of automated navigation systems. It should be remembered that it is not only the boat that may be moved, but, corresponding to the marine reality, obstacles, markers, etc. may appear, disappear or move over time within the dynamic Voronoï structure. This brings us closer to the goal of a freely modifiable marine "GIS", although few of the operations correspond to those of a terrestrial GIS. Gold (1991) has discussed the use of the Voronoï method for the standard GIS operations of buffer zone, polygon-merge and overlay. Buffer zones are particularly easy, as they consist only of lines or circular arcs drawn at the appropriate distance within the Voronoï cells of line segments or points respectively. Polygon merging may be performed by the local deletion of the common boundary line segments. Polygon overlay can be performed by drawing the first map onto the second, snapping to the second map if the pen is closer than some specified tolerance.

Perhaps the most interesting observation from Gold and Condal (1995) is that exactly the same techniques are used for two-dimensional navigation (an object view of the world) as for interpolation of bathymetry (a field view). As shown in Figure 4.9, boat "A" is able to navigate towards the harbour mouth while evading the island, by examining its Voronoï neighbours. At the same time boat "B" is able to navigate to the deepest channel by using the "area-stealing" interpolation method on its adjacent depth soundings, which are preserved in a different layer than the surface obstacles. Finally, when boats A and B approach too closely to each other, this will be detected by them becoming Voronoï neighbours. This provides a straightforward model for integrating the main components of a marine GIS.

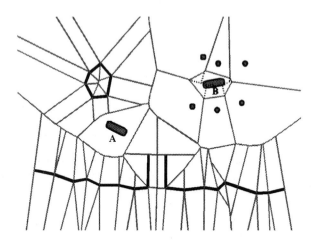

Figure 4.9 Marine navigation using the dynamic Voronoï diagram.

4.7 CONCLUSION

The algorithmic basics of a potential marine GIS have been discussed, based on the VD spatial model, and associated data structures. While the examples have been restricted to two dimensions, most of the methods are extendable to three. Because the fundamental operations of such a GIS are still not well defined, flexibility in the underlying operations is of paramount importance at this time, so as not to restrict the user's abilities by a too hasty specification of the needs of such a system: a more exploratory approach is required. However, this philosophy is just as applicable to a terrestrial GIS. As stated in the introduction, current systems are based on concepts of manual cartography, whereas today's systems are not as restricted. The algorithms described here have been implemented, are in use in two dimensions in the author's research, and may be readily reproduced, as they have in many cases been chosen for their simplicity. What perhaps is most important is the concept: a simple set of operations may be sufficiently flexible to allow for a wide variety of applications. A greater openness to new spatial models and to the expansion of the set of basic operations, especially in the light of the increasing need for dynamic systems and simulation, can only be a good thing. The exploration of the needs of this new "floating world", the examination of potential spatial models and the definition of an appropriate marine GIS, can only be a benefit to all aspects of GIS and spatial analysis.

4.8 ACKNOWLEDGEMENTS

The author would like to thank Mr. Weiping Yang, Mr. Mir Mostafavi and Mr. David Thibault for assistance with the figures.

4.9 REFERENCES

Amenta, N., Bern, M. and Eppstein, D., 1998, The crust and the beta-skeleton: combinatorial curve reconstruction. *Graphical Models and Image Processing,* **60**, pp. 125-135.

Anton, F. and Gold, C.M., 1997, An iterative algorithm for the determination of Voronoï vertices in polygonal and non-polygonal domains. In *Proceedings of the Ninth Canadian Conference on Computational Geometry*, Kingston, Ontario, pp. 257-262.

Anton, F., Gold, C.M. and Mioc, D., 1998, Local coordinates and interpolation in a Voronoï diagram for a set of points and line segments. In *Proceedings of the Second Voronoï Conference on Analytic Number Theory and Space Tilings*, Kiev, Ukraine, pp. 9-12.

Aurenhammer, F., 1991, Voronoï diagrams: A survey of a fundamental geometric data structure. *ACM Computing Surveys,* **23**, pp. 345-405.

Blum, H., 1967, A transformation for extracting new descriptors of shape. In *Models for the Perception of Speech and Visual Form*, edited by Whaten-Dunn, W. (Cambridge, Mass.: MIT Press), pp. 153-171.

Blum, H., 1973, Biological shape and visual science (Part 1). *Journal of Theoretical Biology*, **38**, pp. 205-287.

Blum, H. and Nagel, R.N., 1978, Shape description using weighted symmetric axis features. *Pattern Recognition*, **10**, pp. 167-180.

Fritts, M.J., Crowley, W.P. and Trease, H.E., 1985, *The Free-Lagrange Method. Lecture Notes in Physics*, **238** (Berlin: Springer-Verlag).

Gold, C.M., 1989, Surface interpolation, spatial adjacency and GIS In *Three Dimensional Applications in Geographic Information Systems*, edited by Raper, J. (London: Taylor & Francis), pp. 21-35.

Gold, C.M., 1991, Problems with handling spatial data: The Voronoï approach. *CISM Journal*, **45**, pp. 65-80.

Gold, C.M., 1996, An event-driven approach to spatio-temporal mapping. *Geomatica*, **50**, pp. 415-424.

Gold, C.M., 1997, Simple topology generation from scanned maps. In *Proceedings of Auto-Carto 13*, Seattle, Washington (ACM/ASPRS), pp. 337-346.

Gold, C.M., 1998, The Quad-Arc data structure. In *Proceedings of the Eighth International Symposium on Spatial Data Handling*, Vancouver, BC, pp. 713-724.

Gold, C.M., 1999, Crust and anti-crust: a one-step boundary and skeleton extraction algorithm. In *Proceedings of the ACM Conference on Computational Geometry*, Miami, Florida (ACM).

Gold, C.M., Charters, T.D. and Ramsden, J., 1977, Automated contour mapping using triangular element data structures and an interpolant over each triangular domain. In *Computer Graphics* **11**, *Proceedings of Sigraph '77*, pp. 170-175.

Gold, C.M. and Condal, A.R., 1995, A spatial data structure integrating GIS and simulation in a marine environment. *Marine Geodesy*, **18**, pp. 213-228.

Gold, C.M. and Cormack, S., 1987, Spatially ordered networks and topographic reconstructions. *International Journal of Geographical Information Systems*, **1**, pp. 137-148.

Gold, C.M. and Maydell, U.M., 1978, Triangulation and spatial ordering in computer cartography. In *Proceedings of the Canadian Cartographic Association Third Annual Meeting*, Vancouver, BC, pp. 69-81.

Gold, C.M., Nantel, J. and Yang, W., 1996, Outside-in: an alternative approach to forest map digitizing. *International Journal of Geographical Information Systems*, **10**, pp. 291-310.

Gold, C.M. and Snoeyink, J., in press, A one-step crust and skeleton extraction algorithm. *Algorithmica*.

Guibas, L., Mitchell, J.S.B. and Roos, T., 1991, Voronoï diagrams of moving points in the plane. In *Proceedings, 17th International Workshop on Graph Theoretic Concepts in Computer Science: Lecture Notes in Computer Science*, **570** (Berlin: Springer-Verlag), pp. 113-125.

Guibas, L. and Stolfi, J., 1985, Primitives for the manipulation of general subdivisions and the computation of Voronoï diagrams. *Transactions on Graphics*, **4**, pp. 74-124.

Lam, N. S-N., 1983, Spatial interpolation methods: a review. *American Cartographer*, **10**, pp. 129-149.

Narasihman, T.N. and Witherspoon, P.A., 1976, An integrated finite-difference method for analyzing fluid flow in porous media. *Water Resources Research*, **12**, pp. 57-64.

Okabe, A., Boots, B. and Sugihara, K., 1992, *Spatial Tessellations: Concepts and Applications of Voronoï Diagrams* (Chichester: John Wiley and Sons).

Roos, T., 1990, Voronoï diagrams over dynamic scenes. In *Proceedings of the Second Canadian Conference on Computational Geometry*, Ottawa, pp. 209-214.

Sibson, R., 1981, A brief description of natural neighbour interpolation. In *Interpreting Multivariate Data*, edited by Barnett, V. (New York: John Wiley & Sons), pp. 21-36.

Worboys, M.F., 1995, *GIS: A Computing Perspective* (London: Taylor & Francis).

Yang, W. and Gold, C.M., 1996, Managing spatial objects with the VMO-Tree. In *Proceedings of the Seventh International Symposium on Spatial Data Handling*, Delft, The Netherlands, pp. 711-726, 11B15-11B30.

CHAPTER FIVE

Representation of Variability in Marine Environmental Data

Anne Lucas

5.1 INTRODUCTION

That environmental data are complex needs little elaboration for many users of geographic information systems (GIS). It is well known that they are four-dimensional, showing variation in location, and depth, as well as changes through time (Goodchild, 1992; Li and Saxena, 1993; Kemp, 1997). They often exhibit indeterminate boundaries (Burrough, 1996) and are usually sampled, rather than exhaustive (Kemp, 1993). Within a GIS they are accessed, analysed, displayed and integrated. These data may be combined with other data for planning, management or decision support or used as input for environmental models. But whatever their role, the implementation of environmental data in GIS is often, at best, a compromise. When environmental data are brought into GIS, it is assumed that the representation of that data characterises the nature of the phenomena of interest. However, there is no single "best" representation that is suitable for all possible applications (Burrough, 1996). The way in which data are modelled, aggregated or generalised for use in GIS needs to be matched to the purpose it will serve.

Issues investigated in the context of environmental management tend to require a view of whole ecological systems and/or a regional perspective (Wi and Johnson, 1995). This compounds the problem of finding suitable ways of representing environmental phenomena. Data sets from diverse sources, each with their own inherent sampling structure and characteristics, as well as those measuring different phenomena or the same phenomena in the different ways, need to be brought together into a common framework. Previously archived data may need to be included in addition to data sets that may become available at a future time.

Management of marine and coastal environments has two complicating factors that are especially difficult to handle. Compared to land-based systems, marine systems tend to be sampled sparsely and infrequently and there is a reluctance to throw away potentially useful data without careful evaluation. Further, since the ocean is continuously moving, there can be significant changes in environmental conditions over very short time scales. In addition, other processes occur over longer time scales (Bartlett, 1993). Merging data sets requires more than harmonising attribute values or spatial distributions; the natural variability of the ocean in the temporal domain must also be reconciled across data sets.

This paper presents a current environmental management problem to illustrate some of the issues associated with the representation of marine phenomena in GIS. A selection of available data sources are examined as to their suitability *vis-à-vis* management needs and their ability to capture the variability of ocean phenomena. Finally, a process-based approach for aggregating data that tries to capture the spatial and temporal variability of ocean phenomena is offered.

5.2 STUDY AREA

5.2.1 Environmental Concerns

In comparison with most other areas of the world, the Arctic remains a clean environment (Arctic Monitoring and Assessment Program, 1997). However, there is concern that certain ecosystems and human populations may be under threat from pollutants, including persistent organic pollutants (POPs), heavy metals, and radioactive contamination from sources at lower latitudes. In addition, the high concentration of radioactive sources in northwestern Russia is a potential threat for the release of radionuclides into the aquatic environment. Located within the Kara and eastern Barents Seas are solid and liquid nuclear waste dumpsites (Pavlov and Pfirman, 1995). Novaya Zemlya is a former site for nuclear testing above ground, below ground and under water, and spent nuclear fuel from marine reactors have been dumped in the vicinity of the East Novaya Zemlya trough (Warden *et al.*, 1997). Evidence suggests that these sources have already leaked (Salbu *et al.*, 1997). Discharge water from the major rivers in the same area may potentially contain contaminants from the industrial areas within the watershed, although not all studies confirm this (Pfirman *et al.*, 1995). A new threat comes in the form of oil and gas development and the potential for re-suspension of contaminant-laden sediments together with spills and leaks.

The fate of water carrying potential contaminants is unknown, but there is concern that ice and water may transport these and other constituents beyond the confines of the Siberian coastal shelf, spreading throughout the Arctic basin and eventually through the rest of the worlds' oceans. These and related issues are under investigation by a number of international research organisations (Arctic Monitoring and Assessment Program, 1997, 1998).

5.2.2 The Kara Sea: Physical Description

The Kara Sea is a shallow coastal delta adjacent to the Arctic Basin along the north coast of Siberia, Russia (Figure 5.1). To the east and southeast, the islands of Severnaya Zemlya and the Taimyr Peninsula bound it. To the west, the islands of Novaya Zemlya nearly enclose the sea. The northern boundary of the Kara Sea is open to the Arctic Basin and the Barents Sea. At the eastern extreme there is a small opening to the Laptev Sea: the Vo'lkitsky Strait. And to the west, is an opening to the Barents Sea through Kara Gate. These passages are approximately

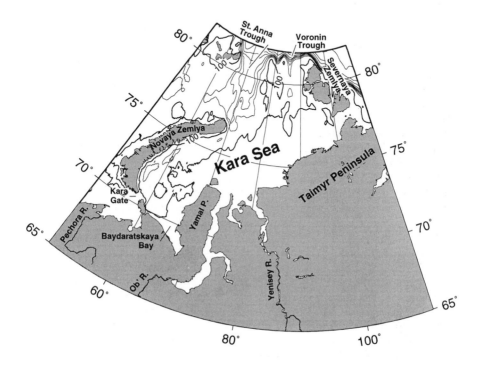

Figure 5.1 Location map for the Kara Sea with bathymetry.

60 and 45 km wide, respectively (Johnson *et al.*, 1997). The sea is generally shallow (64% of the sea is less than 100 m) with an average depth of 111 m (Pavlov *et al.*, 1994). Two troughs with depths exceeding 450_m are found at the northern boundary to the Barents Sea, these being the Voronin and St. Anna troughs; a third trough (the East Novaya Zemlya Trough) is found to the east of Novaya Zemlya with a depth of more than 400 m.

Four major rivers empty into the Kara Sea, discharging 1000 to 1400 km^3 of freshwater annually (Johnson *et al.*, 1997). This represents more than one third of the annual freshwater input into the Arctic Basin (Hanzlick and Aagard, 1980). The two major rivers, the Ob'and Yenisei, drain a land area of more than 2.58 x 10^6 km^2 and have an annual discharge of 35 x 10^3 m^3/s, more than half of which occurs during the May to July period (Pavlov *et al.*, 1994). Exchange processes with the warmer, saline North Atlantic water occur in the north through the St. Anna and

Voronin troughs. In the south, exchange occurs through Kara Gate with an inflow through the south portion and outflow through the northern section (Johnson *et al.*, 1997).

The residence time for the brackish surface waters of the Siberian shelf is approximately 3.5 years +/- 5.5 years, but within this time period, ice formed within the sea may be transported across the Arctic Ocean to the Fram Strait and into the Greenland Sea before melting. While most of the ice formed within the Kara Sea remains within the sea, especially shore-fast ice, there are pathways for potentially contaminated sea ice leaving the Kara into the surrounding seas (Pfirman *et al.*, 1997). However, water below the surface layer containing pollutants may remain in the Kara Sea for decades, while bottom waters in the outer reaches have a residence time on the order of 50 to 100 years (Schlosser *et al.*, 1995).

The ice-free sea season is short, usually from July to October. Shore-fast ice is present along the shore and into the Ob' and Yenisei estuaries through July and in some years into August (Pfirman *et al.*, 1995). One well-known feature of the Kara Sea is the polynya located near the mouth of the Ob' River and to the west of the Yamal Peninsula; this remains open through the winter and is a site for ice production (Buzov, 1991). The Novaya Zemlya ice massif found east of Novaya Zemlya is a persistent feature until later summer, in some cases remaining through the summer until the following winter (Pavlov and Pfirman, 1995). Ice formation begins in the northeast during late September and by mid-October has developed into a more or less solid cover in the Ob' and Yenisei regions (Pavlov *et al.*, 1994).

5.2.3 Problem Definition

While there is much that remains to be investigated in this environment, the current research examines the redistribution of river-borne freshwater within the Kara Sea. Variability in conditions through the ice-free season are anticipated, with lower salinity in the estuaries and near the river mouth early in the season, and increasing salinity through to freeze-up. The aim of this investigation is to examine available data sets and develop a framework for the integration of different data while capturing the spatial and temporal variability described in the data.

5.3 THE NATURE OF OCEAN DATA

Sources of ocean data are varied, and GIS users need to have more than a passing knowledge of their collection method, motivation and structure to effectively use them for environmental management (e.g., Wright and Goodchild, 1997). Any single data source can provide a number of different potential attributes. While it is redundant to state that the data should provide measures for the attribute of interest, in practice this may not always be the case, especially if using historical data. Some characteristics may not be observable under all conditions; others may not be directly observable from the raw data without additional processing or combination with other data sets. Other features may require interpretation or inference to generate the desired field; in some cases, surrogate variables must be substituted.

The specific features which are to be extracted are determined through the process of data modelling as described by Goodchild (1992) and Burrough (1996), among others. The selection of features to be included in the GIS will differ for each implementation depending upon the kinds of questions to which answers are being sought. To determine the fate of fresh river water, or map its location through time and space, data that record water salinity is the best tracer. Once more is known about the chemical composition of the river water itself, perhaps chemical constituents could be used to track the redistribution of source water.

In addition to selecting the correct attribute, the data must also appropriately represent the condition of that attribute. Variation in the phenomena is continuous while the sampling is discrete. Therefore, it is important to reconcile the sampling frequency or timing with the variation in the attribute. For example, in order to compare water levels through time, either sampling must be done at the same stage of the tide, or one of the data sets must be adjusted (Zujar *et al.*, 1997). At the very least, the user must have an understanding of the relationship between the sampling frequency and the natural variation in the attribute in order to compare multiple data sets.

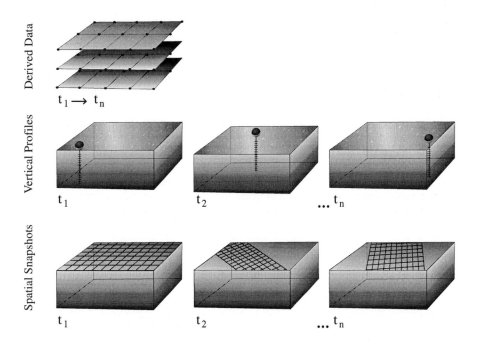

Figure 5.2 Schematic showing the three typical spatial and temporal structures of ocean data.

To simplify and reveal the complexity of possible sampling strategies, data can be characterised in terms of its spatial and temporal characteristics. Hamre (1995) uses this as a basis for the development of an exhaustive spatio-temporal data model for marine information systems. In the present context, data are described as having one or many of each of x, y (location), z (depth) and t (time) components and data from sources typically used in connection with marine applications fall roughly into three categories. Figure 5.2 illustrates the_fundamental differences in the underlying spatial nature of the data. Each of the three categories of data types will be examined in detail using data from the Kara Sea.

5.3.1 Derived Data

Derived data include climatologies generated from composite data sets, observations that have been interpolated onto a grid and results from numerical models. These data have in common the advantage of complete coverage in the x, y and usually, z dimensions; time is usually treated as a series of slices (i.e. many x, many y, many z, many t). The Levitus 94 world ocean atlas is one example of a climatology that provides global coverage of salinity, temperature and other ocean variables (Levitus *et al.,* 1994). These are averaged horizontally on a one-degree cell, vertically 10-100 m, and temporally by month, season, and year. Easy access through Internet and the variety of standard formats makes them particularly attractive for inclusion in marine GIS databases, especially given the difficulties associated with other data, as will be discussed. They may be used as input parameters for numerical models, either external to the GIS or built-in. However, if the spatial and temporal bins used for these data are too coarse for the purpose at hand, then their usefulness is limited.

A similar criticism can be made for gridded data, with the additional constraint that suitable metadata need to accompany the gridded files to be effective. Results from numerical models should not be confused with data, although they can be a valuable source of information. While the intention is to reflect reality, models only mirror reality to the point that the mathematics captures the dominant processes.

5.3.1.1 Implications for Management

The Levitus 94 data for surface salinity and surface temperature for the summer season and individual months were too coarse spatially and temporally to capture the freshwater outflow event. Gridded data were purchased from Russian sources for the years 1981-1984. CTD measurements were interpolated onto a 50-km grid at standard depths with a decorrelation scale in the objective analysis of 200 km (Figures 5.3 and 5.4). Actual locations of observations are available only in paper map format; the specific dates for the sampling are not known, and data are aggregated for each season. While there are similarities in the pattern of summer salinity surface values in the four years of data, 1981 shows exceptional distribution of freshwater through most of the bay. Subsurface data need to be examined to determine any subsurface expression of this freshwater feature.

Preliminary results from a series of numerical simulations have been produced using the MICOM (Miami Isopycnic Co-ordinate Model, Bleck *et al.*, 1992). Methods to input numerical model results are described in Lucas (1996). As input parameters, three different wind and atmospheric pressure regimes, as defined by Pavlov and Pfirman (1995), as well as real conditions from 1994 were used. Results for estimated salinity and temperature were extracted from the 180-day simulation and output on either 5.5 km or 22 km grids, with 3 variable thickness layers every 30 days (Figure 5.5). By sequencing through the model results, the evolution of the freshwater event can be traced in detail both as a surface expression and at depth. After calibration and validation the model can be used to predict the nature of events under specific combinations of conditions. These model results can also be visually compared to satellite images in terms of general characteristics (note the distinctive anvil-shaped vortex pairs at the leading edge of the front), but quantitative analyses are difficult.

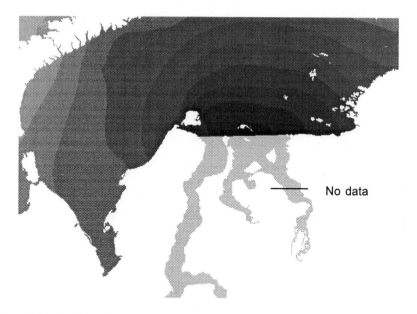

Figure 5.3 Gridded data from Levitus 94 world ocean atlas, interpolated. Salinity values range from 14 (dark) to 34 (light) psu (practical salinity units).

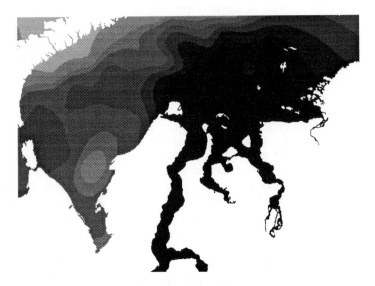

Figure 5.4 Gridded data for surface salinity 1981 obtained from Russian sources. Salinity value range dark to light: 0 to 34 psu.

Figure 5.5 Interpolated surface salinity from MICOM model output; originally on 5.5km grid for day 120 of a 180 day simulation (from W.P. Budgell, unpublished data).

5.3.2 Vertical Profiles

Vertical profiles are those data that provide information only at specific locations; there is no information about conditions either between data points, around the data collection sites, or in time between sampling repeats. *In situ* observations such as those derived from a CTD profiler taken at intervals from a stationary ship are one example (Figure 5.6a). They measure conductivity, temperature, and depth through the water column (many z) at one x, one y. During a ship-based survey there are usually there are a series of such sample sites (one t); a moored instrument, such as a current meter, will record the same information through a sampling period (many t) at a single x, y, z location.

Instruments such as CTD profilers are a rich data source; in the depth dimension, data are collected nearly continuously. The attribute of interest, salinity, must first be calculated from temperature and conductivity values using one of many instrument-specific calibration routines. While it is technically possible to store all the raw and processed data within a database, the data will likely have to be grouped prior to GIS-based analysis. If speed, disk space and user confidence are sufficient this could be performed on the fly; otherwise, given the scale and nature of the management problem, some sort of prior binning of the data through depth is appropriate. Suitable strategies for generalising the data include either averaging data across a range of depths or taking a single value at significant depths, according to a regular interval, or specific intervals that correspond to, for example, some biological process (e.g. surface layer, below thermocline).

5.3.2.1 Implications for Management

In terms of management or monitoring, an individual CTD data set at a single location is of limited use; it is a single realisation of conditions and without additional information it is unclear how far spatially and temporally this information could be extrapolated. A series of stations along a ship track is a typical product from a research cruise; data are available in digital and analogue format from the three joint Russian-Norwegian cruises which were part of the KAREX Project 1993-1995 (Endresen, 1995; Høkedal, 1995; Nygaard, 1995). To interpret the data, point data need to be processed. Numeric tags or coloured symbols are awkward to interpret visually and there are few tools in GIS to analyse point data (Wi and Johnson, 1995). Conversion to polygon or field is necessary. In Figure 5.5, the data have been interpolated to create just one realisation of a continuous field of surface salinity conditions using a weighted inverse algorithm. Routines such as kriging that might be more suited to sparse data were not available in the GIS software. For visualisation, it would have been more appropriate to limit the interpolation process to the area defined by the outermost stations; but gaps with "do data" in raster layers can be problematic.

There is an additional concern, however, and that is the presumed synopticity of the data. In the case of the KAREX93 cruise (Figure 5.6a), the 145 sites were sampled over a 4 week period (September 14 – October 16, 1993) (Høkedal, 1995). For KAREX94 (Figure 5.5b), the cruise lasted more than 8 weeks (August 15 to October 11, 1994) (Nygaard, 1995), which is nearly three-quarters of the ice-free season. All three expeditions missed the freshwater event by more than a month.

Therefore at best, data represent conditions at some (variable) time after the redistribution of freshwater has begun. The difference in spatial patterns between Figure 5.6a and 5.6b, are so extreme that they cannot be explained simply in terms of difference in timing or interpolation procedures. Interannual variability (not just seasonal variability) had been indicated based on Russian literature, although data to support this was previously unavailable.

Such a series of CTD records from a single season is insufficient to develop baseline conditions. Repeated sampling in subsequent years at the same stations can be used to develop long-term monitoring networks and eventually, establish baseline conditions. But given the observed variability between these two years, a long sampling period is indicated.

5.3.3 Spatial Snapshots

Spatial snapshots are defined as data that present a synoptic view of environmental conditions (many x, many y, one z, one t). (One can consider the scan time to be sufficiently small that the data are, in practice, synoptic.) Remotely sensed images are typical examples of this data type, providing coverage over spatial extents ranging from a few 100s of meters (low level air photos) to 1000s of km (satellite sensors). For a detailed discussion of the use of remotely sensed images in ocean applications, the reader is referred to Robinson (1995), Stewart (1985) and Victorov (1996). With a single image, there is no information about conditions before or after the image itself; a series of such data sets can provide much information about the evolution of conditions.

Limiting factors in the use of remotely sensed images include the spatial resolution of individual pixels and the spectral resolution of the sensor. The former determines the smallest feature that can be discerned on the earth's surface, the latter the environmental condition that is observable. Further discussion of spectral resolution is beyond the scope of this paper, except to say that at present, salinity is not directly observable. While salinity can be measured directly through in situ sampling, it can only be inferred through, for example, comparisons with sea surface temperature (SST), ideally made in connection with ship-based observations for calibration (Seigel *et al.*, 1996).

Timeliness of satellite imagery is usually beyond the control of the investigator, although some recent missions (e.g., SAR, SeaWiFs) give the investigator more control. This will be useful for short-term management tasks in the event of a catastrophe. Long-term data sets (since 1972) are available, which can be used for year-to-year comparisons or change detection (Lee *et al.*, 1998). Changes in mission parameters, bandwidths, resolutions during the different programs can be problematic, but are not insurmountable. Extended data sets have also been used to generate climatologies for both land and sea surfaces, producing weekly or monthly averages (Yang *et al.*, 1996).

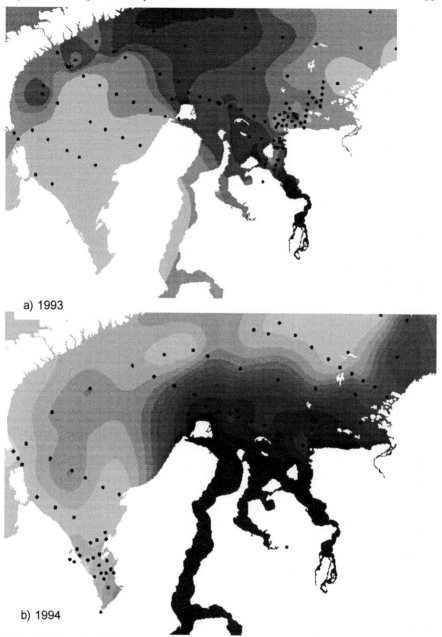

a) 1993

b) 1994

Figure 5.6 Interpolated observations from (a) 1993 and (b) 1994 *in situ* sampling. Sample stations are shown. Note the incorrectly recorded land-locked stations. Data range from dark to light: 0 to 34 psu.

5.3.3.1 Implications for Management

A series of AVHRR (Advanced Very High Resolution Radiometer) images for the Kara Sea in ice-free season June 1 to October 31, 1994 were acquired. This included 190 HRPT (High Resolution Picture Transmission) images with a spatial resolution of 5.1 km, and 40 GAC (Global Area Coverage) images with a nominal resolution of 4 km. Both sets were processed to derive SST, then geocorrected and resampled to the same grid orientation, prior to import to the GIS.

SST is a surrogate for salinity. Although the *in situ* sampling occurred within the same period as the satellite image acquisition, in most cases, cloud obscured the station and calibration of the SST with the *in situ* data could not be carried out. Relative temperature can be used as an indicator of freshwater, but it must be recognised that high skin temperatures due to latent heating, especially in shallow and protected bays, can occur and cannot always be distinguished from the warmer freshwater areas.

If an individual image provides an overview of conditions at a single point in time, then it should be possible to construct a view of the evolution of conditions through time with the 230 individual images. In practice, unless they are viewed in sequence in a movie format, there is simply too much information to process without a structured approach.

Selecting only a few key images that correspond to critical events is one way to reduce the information content. But without analysing all the images first, it would be difficult to know which images captured specific events.

Clouds severely limit the generation of SST values, even where the sea surface is visible through the cloud. Contamination by low-lying mists was a problem, although features such as polynyas, and ice distribution and concentration could be interpreted visually (Anselme, 1998). Although the images were screened for cloud cover prior to acquisition, all images showed some cloud or water vapour (Figure 5.7). Rather than rejecting images that failed to meet some arbitrary minimum cloud free parameter and risk losing valuable information, all cloud-free patches were included in the SST analysis. There are several ways in which the gaps caused by clouds can be filled to create complete coverages. A simple composite averaging of cloud-free cells loses the advantage of synopticity but if the time period over which images are averaged is sufficiently short then the composite image could be assumed synoptic. This averaged image (Figure 5.8), from the ice-free period in 1994, gives a general impression of conditions; warmest surface water in the Ob' and Yenisei Gulfs and Pechora Bay; coldest water east of Novaya Zemlya and in the northern section. Monthly averages for the Kara Sea clearly show the seasonal changes, but this method of averaging does not address the issue of interannual variability.

Statistical interpolation in time has been used with NOAA/NASA Pathfinder data to generate 10-day averages (Walker and Wilkin, 1998). While a similar approach might be suitable for identifying periods over which data may be averaged in this environment, the persistent cloud cover and the short ice-free season in the Kara Sea may limit its application.

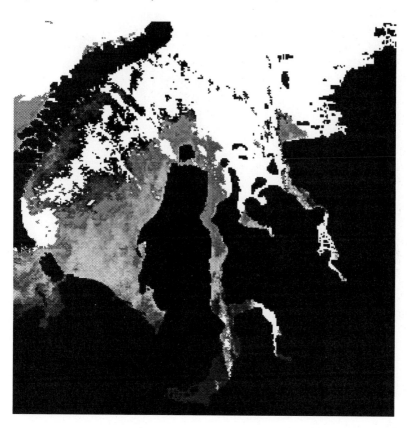

Figure 5.7 Sea surface temperature for August 14, 1994 from NOAA-12 AVHRR GAC data. Clouds (in white) cover almost half the sea area. Warm temperatures (12°C) are dark; cold temperatures (-0.4°C) are light. Remnant cloud and cloud-contaminated pixels can be seen in the otherwise, warm gulf.

Figure 5.8 Thirty-three GACs were geo-referenced and averaged together to produce this composite SST for the period July 15-September 30, 1994. Temperature range: 13.0°C to –0.4°C.

5.3.4 Evaluation

No single data source provides a complete view of the nature of ocean phenomena. Derived data give an overview of "typical" conditions, and even though they lack enough resolution in time or space to be of use in solving specific management problems, they can certainly be used where no other data are available. Vertical profiles provide detail in time and space, but only at specific locations; an overview is difficult to obtain without a dense network of stations and ordinary interpolation procedures are lacking. Satellite imagery gives synoptic coverage, but only of surface conditions and then, only of a surrogate condition.

In the case of the Kara Sea, there is a complicating factor. Data sets from different years reveal different spatial patterns in the redistribution of freshwater. This difference cannot be explained in terms of the differences in data sources, and must be due to the interannual variability. The question remains: how should the seasonal and interannual variability be handled?

Combining data to produce "integrated" data sets is difficult to implement in practice. Each data source has a different structure; most GIS require that data be converted to a common structure prior to integration. That is, point samples be interpolated to produce fields, raster data be resampled to a common grid orientation and cell resolution (Kemp, 1997). While these data structure conversions can easily be handled in GIS, the underlying problem is that the true nature of ocean phenomena is not perfectly known. Any change in the spatial and temporal structure of the data will be at best, an estimate of conditions beyond the measured sample locations. Simple averaging of data will lose or smear information. An approach is needed that will take into account the natural variation in the system (i.e. does not assume each year is the same).Rather than risk data transformation that may undermine the integrity of the measured data, this investigation seeks an alternate approach based on the temporal and spatial scales of the processes that are responsible for the patterns observed.

5.4 A PROCESS-BASED APPROACH TO DATA INTEGRATION

There are essentially two contradictory problems that need to be solved. On the one hand, the nature of the spatial and temporal variability that is evident in the different data sets must be captured. But since data are sparse, a way must be found to use all of the information that is available without changing its properties. However, if the data are simply used in their original state, GIS analytical capabilities are limited and data management becomes problematic, even though the variability is captured. Binning the like-structured data (so as to not transform the data) over generic time-space scales will produce data sets with characteristics similar to the derived fields. The issue now becomes one of determining the appropriate time-space scales for data aggregation. Relevant processes are used to determine these scales.

To develop this approach, 10 years of data are being evaluated. The initial model presented here is based on analysed data from the 1994 summer season for which there were the greatest number of data types available. And of all the data sources examined, satellite imagery is the only one that offers a synoptic view with both a frequent repeat and a long historical sequence; it forms the basis of the methodology.

Further, in this example the issue of capturing natural variation is complicated by the duality of the temporal variability. Seasonal variations were anticipated, but the data reviewed thus far also exhibit considerable interannual variability. These two temporal problems will be considered in sequence.

5.4.1 Capturing the Seasonal Variability

Rather than focusing on individual elements in the data set or the specific measurements, processes relevant to the redistribution of freshwater are identified. The characteristic time-space scales of these processes (Figure 5.9) are used to determine suitable time periods over which data may be aggregated. Based on this figure, it appears that a time scale of 10 days should capture the temporal variability of the shelf and mesoscale processes, including some of the ice dynamics. A second time scale of 1 hour is needed to capture estuarine fronts and pollution dynamics. Spatial scales are similarly bimodal, with a range of 1-10 km capturing the shelf and mesoscale structures and 10 m for the estuarine features.

To test this procedure, NOAA AVHRR data at two spatial resolutions, 4 km GAC and 5.1 km HRPT, were analysed. Features indicative of the above processes were interpreted at both scales and data were temporally aggregated using a simple composite averaging technique over a range of intervals from 2 hours to 30 days. These analyses confirmed the following. A 14-day temporal aggregation using the 4 km data is sufficient to resolve the upwelling phenomena off the Yamal Peninsula in mid-July, the two-way flow through Kara Gate that persists through most of the open water season, and the release of fresh (warm) water through the Ob' Gulf during and after ice break-up. Two-way flow in the Ob' Gulf is observed for a short period in late July. The presence of the ice massif near Novaya Zemlya is inferred from the very cold water throughout the summer season. In addition, the gradually shrinking position of the ice edge is determined from June through late July, based on the ragged edge bounding the open water. These processes are also observable with shorter aggregation periods, but with more than a 14-day period, the two-way flow and upwelling are indistinct.

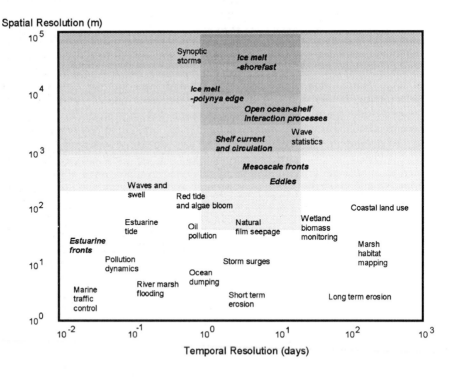

Figure 5.9 Spatial and temporal scales of ocean processes (after Dickey, 1991; Haugan and Drange, 1995; and Robin, 1995).

Figure 5.10 illustrates the effect of changing spatial resolutions on individual images. More detailed spatial structure in sea surface temperature is observed, and there is also better separation between ice and cloud throughout. However, within the Ob' Gulf, there is increased confusion as sediment-laden discharge waters are misclassified as very warm water.

On all the high resolution images, the mixing zone between the ice edge and the Yamal Peninsula clearly shows characteristic eddies and whorl patterns. The same area at 4 km is quite indistinct. These characteristic patterns are not spatially resolved at 4 km; their appearance is somewhat mottled. A similar problem arises with the same feature when the 5.1 km data are aggregated. In this case, the temporal resolution is too coarse to resolve the variability. However, once the underlying cause of this mottled pattern is recognised, the more detailed data are not necessary to determine its presence or extent.

For the current purpose of determining the distribution of the freshwater through the open water season, the __-day aggregation period and 4 km data are sufficient.

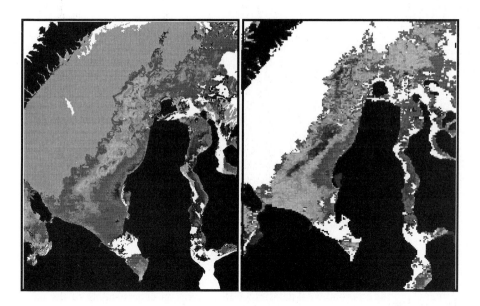

Figure 5.10 (a) 5.1 km NOAA AVHRR HRPT data from July 17, 1994 11:07. (b) 4 km NOAA AVHRR GAC data from July 17, 1994 11:04. White areas are clouds, grey are ice.

5.4.2 Identifying Interannual Variability

Having established the suitability of a 14-day aggregation period with the lower resolution data, it is possible to construct a sequence of 10 images that cover the summer period June 1 to October 31. The sequence captures the variability of the key processes in a limited number of data sets, yet include all available images in the analysis. By constructing a similar sequence for additional years, a comparison between years can be made and a suite of scenarios can be developed.

In any given year, the seasonal pattern should follow in the same sequence assuming interannual variability is not driven by external forces (e.g., meteorological forcing). What varies from year to year, will be the timing of the sequences. The hydrograph shows slight variation in the timing of the peak discharge. However, ice break-up in the spring and freeze-up in the fall can be variable in either direction. If the break-up and freeze-up are established as the "end points" in the sequence, then each seasonal cycle should fit within these limits. As shown in Figure 5.11, several scenarios can be foreseen. Where spring break-up is later than usual, the start of the freshwater discharge under the ice may be masked by the presence of ice. When a late spring is coupled with a late discharge, the progression of the sequence through the season may not advance as far. With an early freeze-up, a higher lateral rate of freezing may be observed due to the higher freshwater content of the upper layers. Other scenarios will be determined from the data as it is analysed.

5.4.3 Data Integration

The sequencing and suite of scenarios described above provides a framework for integration of other data (Figures 5.11 and 5.12). Individual vertical profiles can be coupled with the matching aggregated image for a more appropriate comparison.

The CTD provides details of conditions at depth which are unseen by the satellite. Not only are comparisons between profile and satellite temperatures possible, but this approach also opens up possibilities for better interpretation of the CTD data.

Throughout the Kara Sea, CTD profile characteristics are highly variable depending upon from which part of the sea they are sampled. Since these profiles are undersampled, there is little additional information to aid analysis of individual sites. However, the spatial representativeness of the CTD data could be derived from the satellite imagery. For example, the spatial correlation scales used for kriging could be derived from the classified SST data. This will be investigated in the next stage of the analysis

5.5 CONCLUSION

Bringing marine environmental data into GIS requires an understanding of the data sources and their characteristics as well as the processes under investigation. The natural spatial and temporal variability of the ocean needs to be acknowledged and modelled explicitly. Specifically, separating the seasonal from the interannual variability is necessary to construct a framework for data integration. Further, the time-space scales of key environmental processes are used as the criteria for data aggregation. It is believed that a process-based focus will provide a better basis for interpretation of environmental data, particularly in the four-dimensional environment of the coastal ocean.

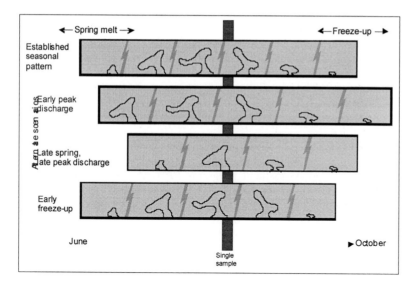

Figure 5.11 The seasonal sequence of recognizable processes or stages is established initially from analysis of satellite data for 1994. Alternate scenarios can be offered to explain at least some of the observed interannual variability. A single sample at a specific date from any one year may exhibit any range of characteristics as shown above; this makes year-to-year comparisons difficult.

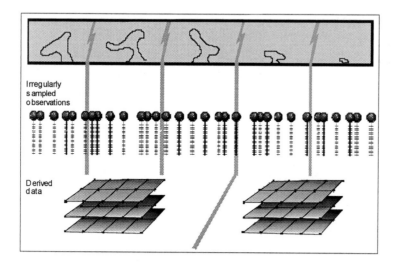

Figure 5.12 Once the sequence of processes for a specific year is determined, other data types (e.g. *in situ* observations, derived fields) can be integrated on a data element by data element basis. Rather than using absolute dates as a means for year-to-year comparisons, like stages in the seasonal cycles can be compared.

5.6 ACKNOWLEDGEMENTS

This work was supported in part by the Norwegian Research Council Project Number 120438/730. The author wishes to thank the Norwegian Polar Institute for access to data and W.P. Budgell at the Institute of Marine Research, Bergen for fruitful discussion. Kjell Helge Sjøstrøm, Department of Geography, University of Bergen, drafted Figure 5.2.

5.7 REFERENCES

Arctic Monitoring and Assessment Program, 1997, *Arctic Pollution Issues: A State of the Arctic Environment Report* (Oslo: AMAP).

Arctic Monitoring and Assessment Program, 1998, *AMAP Assessment Report: Arctic Pollution Issues* (Oslo: AMAP).

Anselme, B., 1998, Sea ice fields and atmospheric phenomena in Eurasiatic arctic seas as seen from the NOAA-12 satellite. *International Journal of Remote Sensing*, **19**, pp. 307-316.

Bartlett, D., 1993, Space, time, chaos and coastal GIS. In *Proceedings 16th International Cartography Conference*, pp. 539-555.

Bleck, R., Rooth, C., Hu, D., Smith, L.T., 1992, Salinity driven thermocline transients in a wind- and thermohaline-forced isopycnic coordinate model of the North Atlantic. *Journal of Physical Oceanography*, **22**, pp. 1486-1505.

Burrough, P.A., 1996, Natural objects with indeterminate boundaries. In *Geographic Objects with Indeterminate Boundaries*, edited by Burrough, P.A. and Frank, A.U. (London: Taylor & Francis), pp. 3-28.

Buzov, A., 1991, *Natural Factors and Their Influence on Transit Sailing on the Northern Sea Route, Part 1* (Oslo: Fridtjof Nansen Institute).

Dickey, T., 1991, Concurrent high resolution physical and bio-optical measurements in the upper ocean and their applications. *Reviews of Geophysics*, **29**, pp. 383-413.

Endresen, Ø., 1995, *CTD Report from KAREX 1995* (Oslo: Norwegian Polar Institute).

Hamre, T., 1995, *Development of Semantic Spatio-temporal Data Models for Integration of Remote Sensing and in situ Data in a Marine Information System* (Bergen: University of Bergen).

Hanzlick, D. and Aagard, K., 1980, Freshwater and Atlantic water in the Kara Sea. *Journal of Geophysical Research*, **85**, pp. 4937-4942.

Haugan, P.M. and Drange, H., 1995, Disposal options in view of ocean circulation. In: *Direct Ocean Disposal of Carbon Dioxide*, edited by Handa, N. and Ohsumi, T. (Terra Scientific Publishing), pp. 123-145.

Høkedaal, J., 1995, *CTD Data from Cruise with G/S: Pavel Bashmakov in the Kara Sea, 14 September to 16 October 1993* Report Series No. 89 (Oslo: Norwegian Polar Institute).

Johnson, D.R., McClimans, T.A., King, S., and Grenness, Ø., 1997, Freshwater masses in the Kara Sea during summer. *Journal of Marine Systems*, **12**, pp. 127-145.

Kemp, K.K. 1993, *Environmental Modelling with GIS: A Strategy for Dealing with Spatial Continuity*, NCGIA Technical Paper 93-3 (Santa Barbara: University of California).

Kemp, K.K., 1997, Fields as a framework for integrating GIS and environmental process models. Part 1: Representing spatial continuity. *Transactions in GIS*, **1**, pp. 219-234.

Kogler, J., Anselme, B. and Falk-Petersen, S., 1995, Some applications of AVHRR and CZCS satellite data in studies of the Barents and Kara Seas. In *Ecology of Fjords and Coastal Waters*, edited by Skjoldal, H.R., Hopkins, C., Erikstad, K.E. and Leinnas, H.P. (Amsterdam: Elsevier Science), pp. 219-228.

Lee, C.T., Johnson, P., and Hay, R., 1998, Mapping vegetation dynamics in the lower Colorado River delta using archival LANDSAT MSS satellite imagery. In *Proceedings 5th International Conference on Remote Sensing for Marine and Coastal Environments, Volume 1* (Ann Arbor: ERIM), pp.207-219.

Levitus, S., Burgett, R., and Boyer, T.P., 1994, World ocean atlas 1994, Volume 3: salinity, NESDIS 3 (Washington: U.S. Department of Commerce).

Li, R. and Saxena, N., 1993, Development of an integrated marine geographic information system. *Marine Geodesy*, **16**, pp. 293-307.

Lucas, A.E., 1996, Data for coastal GIS: issues and implications for management. *GeoJournal*, **39**, pp. 133-142.

Nygaard, E., 1995, *CTD Report from KAREX 94*, Report Series No. 90 (Oslo: Norwegian Polar Institute).

Pavlov, V.K. and Pfirman, S.L., 1995, Hydrographic structure and variability of the Kara Sea: Implications for pollutant distribution. *Deep-Sea Research 2*, **42**, pp. 1369-1390.

Pavlov, V., Timohov, L.A., Baskakov, G.A., Kulakov, M., Kurazhov, V.K., Pavlov, P.V., Pivovarov, S.V. and Stanovy, V.V., 1994, *Hydrometeorological regime of the Kara, Laptev and East Siberian Seas* (St. Petersberg: The Arctic and Antarctic Research Institute).

Pfirman, S.L., Kogler, J., and Anselme, B., 1995, Coastal environments of the western Kara and eastern Barents Seas. *Deep-Sea Research 2*, **42**, pp. 1391-1412.

Pfirman, S.I., Kogler, J., and Rigor, I., 1997, Potential for rapid transport of contaminants from the Kara Sea. *Science of the Total Environment*, **202**, pp. 111-122.

Robin, M.,1995, *La teledetection* (Paris: Nathan Universite).

Robinson, I.S. 1995. *Satellite oceanography: an introduction for oceanographers and remote sensing scientists* (Chichester: Wiley).

Salbu, B., Nitkin, A., Strand, P., Christensen, G., Chumichev, V., Lind, B., Fjelldal, H., Bergen, T., Rudjord, A., Sickel, M., Valetova, N., and Foyn, L., 1997, Radioactive comtamination from dumped nuclear waste in the Kara Sea: Results from the joint Russian-Norwegian expeditions in 1992-1994. *Science of the Total Environment*, **202**, pp. 185-198.

Schlosser, P., Swift, J., Lewis, D., and Pfirman, S.L., 1995, The role of the large-scale Arctic Ocean circulation in the transport of contaminants. *Deep-Sea Research 2*, **42**, pp. 1341-1367.

Siegel, H., Gerth, M. and Schmidt, T., 1996, Water exchange in the Pomerian Bight investigated by satellite data and shipboard measurements. *Continental Shelf Research*, **16**, pp. 1793-1817.

Stewart, R.H., 1985, *Methods of satellite oceanography* (Berkeley: University of California Press).

Victorov, S., 1996, *Regional satellite oceanography* (London: Taylor & Francis).

Walker, A. and Wilkin, J., 1998, Optimal averaging of NOAA/NASA Pathfinder satellite sea surface temperature data. *Journal of Geophysical Research*, **103**, pp. 12,869-12,883.

Warden, J.M., Lynn, N., Mount, M., Svinitsev, Y., Timms, S., Yefimov, E., Gessgerd, K., Dyer, R. and Sjoeblom, K-L., 1997, Potential radionuclide release from marine reactors dumped in the Kara Sea. *Science of the Total Environment*, **202**, pp. 225-236.

Wright, D.J. and Goodchild, M.F., 1997, Data from the deep: Implications for the GIS community. *International Journal of Geographical Information Science*, **11**, pp. 523-528.

Yang, L., Zhu, Z., Izaurralde, J. and Merchant, J., 1996, Evaluation of North and South America AVHRR 1km data for global environmental modelling. In *Proceedings Environmental Modelling and GIS* (Sante Fe, New Mexico: NCGIA).

Zujar, J., Rodriguez, E., Fernandez-Palacios, A. and Loder, J., 1997, Characterization of coastal waters for the monitoring of pollution by means of remote sensing; the use of satellite imagery to establish the appropriate pattern for timing and location of sampling in coastal waters. In *Proceedings of CoastGIS'97*, Aberdeen.

CHAPTER SIX

Applying Spatio-Temporal Concepts to Correlative Data Analysis

Herman Varma

6.1 ABSTRACT

Geographic information systems (GISs) and spatial analysis techniques have so far been developed for reasoning with respect to earth geometry (map algebra). There is, however, no reason to refrain from applying fusion methods and techniques to analyse and explore multidimensional aspects of spatial data related to the earth's surface over periods of time. Important concepts such as *neighbourhood* or *regionalisation* play important roles in understanding the spatial fusion of the related earth sciences. The key to such liberation of *"spatial reasoning"* is the treatment of each attribute as a dimension in itself.

All spatial data structures encode basic metric information about the shape and position of the spatial features being represented. This is usually implemented with complex indexing schemes such as B-trees, R-trees, quadtrees, K-D trees etc. (Li, 1999). Space is considered to be continuous, and requires mechanisms to create and identify irregular bounding surfaces in order to perform spatial interactions between diverse spatial data sets such as geology, tides, salinity, temperature, bathymetry, biomass etc. This is especially true in the formulation and use of complex topological relationships between geo-temporal features, which have been collected over different periods of time. The approach of the helical hyperspatial or HHCode (Varma *et al.*, 1990) explicitly stores topological constructs at a spatial level. This is done by associating feature attribution to the spatial *area* between specified time periods rather than to a point, establishing the concept of the area/timeline paradigm.

6.2 INTRODUCTION

Current use of GIS is limited to the analysis of points on a plane, referenced to the earth's surface. With the exception of a few geophysical research projects and Intergraph's Voxel Analyst, there are hardly any tools to examine a third or even fourth dimension (temporal). Most GISs deal only with two-dimensional manifolds, such as TINS or regular grids. Height or time are usually treated as attributes, and only Langran (Langran, 1992) mentions the reciprocal approach to view these parameters as additional dimensions.

The techniques used in two-dimensional (2-D) spatial reasoning have become commonplace and are, if not standardised, well agreed upon. While implementations differ, most systems now offer a similar set of operations for 2-D. This is a highly specialised domain; however, current GISs are still too cumbersome to be used by other domain specialists. Therefore, these techniques have not yet been applied in the analysis of multidimensional data. The minimal acceptance of GIS methods for correlative data analysis is reflected by the lack of more but the most primitive modules in standard statistical packages such as SAS or SPSS.

Yet GISs and related methods of spatial analysis have a lot to offer. Space is one of the fundamental human experiences. Cognitive studies prove that people tend to "spatulas" many aspects of their everyday life (Mark and Frank, 1988). As such, spatial metaphors are powerful means of categorisation; they help us to structure the complexity of reality. Research in multidimensional domains, such as in environmental applications, face similar problems of complexity; however, they do not yet employ such cogent concepts like neighbourhood, proximity and shape.

To overcome these problems, a formal, dimension-independent approach for building multiple, hierarchically-related representations of spatial objects is proposed. In 1990, the Canadian Hydrographic Service (CHS) forged ahead to create and implement a technology, termed HHCodes, for the management and manipulation of spatial data based on the Riemannian hypercube. HHCodes are used to encode multidimensions into binary streams.

The current CHS data model encodes time and space into this structure, using temporal hypercubes as spatial keys with attributes associated with them. This establishes a 2-dimensional manifold of the earth with time as the third dimension, creating a 3-dimensional spatio-temporal tessellation.

Multidimensional codes facilitate the fusion and interrelation of spatial data by the use of variable sized cells. These variable sized cells are locked to temporal periods. They are generated by aggregation techniques that form statistical surfaces or volumes based on selected attribution. Such techniques are very powerful, making a significant step towards interrelating diverse spatial data sets such as geology, temperature, salinity, bathymetry, biomass, tides, currents, etc.

6.3 A REVIEW OF HHCODES

In 1856 Georg Bernard Riemann defined multidimensional space in terms of "hypercubes" for his mentor Gauss (Kaku, 1994). In his brilliant paper on geometry, Riemann eclipsed Euclid's 2000-year legacy, resulting in the Riemannian tensor matrix (Spiegal, 1959). According to Riemann, the relationship of the hypercube diagonal and the related dimensions was quite simply an extension of the Pythagorean Theorem. For example, in two dimensions, $diag^2 = x^2 + y^2$ is the diagonal of a square. In 3 dimensions, $diag^2 = x^2+y^2+z^2$ is the diagonal of a cube, in n dimensions, $diag^2 = x^2 + y^2.........+n^2$ is the diagonal of the hypercube. This establishes a geometric relationship between the dimensional vectors in multidimensional space. HHCodes are bit representations of the Riemannian diagonal, ranging from 1 to n dimensions. When sorted, HHCodes cluster. In one dimension, they cluster like binary trees. In two dimensions, they cluster like quadtrees. In three dimensions, they cluster like octrees. In n dimensions, they cluster in a n dimensional construct. The masking of dimensions from the bit streams effectively allow HHCodes to exist in different dimensional states, allowing spatial comparisons of objects within these states.

Value (0-360)

40.625

HHCode

1CE38E38

Original point

x = 40.625

Figure 6.1 Binary subdivision of a vector.

A one-dimensional HHCode is the binary subdivision of a dimensional vector until a user specified level is reached (Figure 6.1).

The specified vector element resides within the boundary limits of the final level, establishing an uncertainty neighbourhood containing the point. On examining the subdivisions of the dimensional vector, one can see a self-similar fractal structure (Schroeder, 1990) as the line is continually broken down into sub lines until Cantorian dust is achieved. The vector element is contained in one of the Cantorian dust.

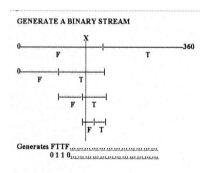

Figure 6.2 A True False tree being generated by binary subdivisions.

The binary subdivision of the dimensional vector generates a Truth table (Figure 6.2), creating a normalised logical representation of where the X element lies within the dimensional vector. This logical position is also constrained to an uncertainty neighbourhood that is determined by the resolution of the subdivision level. The binary formation of using Booleans of "false" and "true" is the generation of a linear binary tree (Pavlidis, 1982) in bit patterns, shown as a hexadecimal number representing the HHCode fractal.

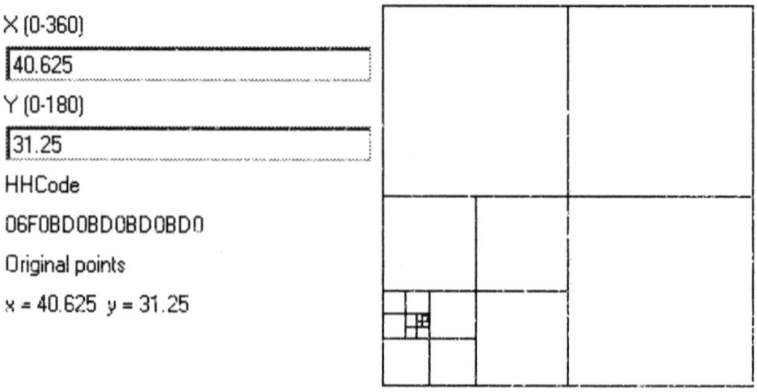

X (0-360)

40.625

Y (0-180)

31.25

HHCode

06F0BD0BD0BD0BD0

Original points

x = 40.625 y = 31.25

Figure 6.3 Two-dimensional fractals generated by interlacing two one-dimensional fractals.

A two-dimensional HHCode is the bit interlacing of two one-dimensional HHCodes. The interlacing of the two self-similar 1-D fractal structures creates a two-dimensional self-similar square structure which resembles a quadtree (Figure 6.3).

A quadtree can be thought of as a two-dimensional fractal that repeats itself after every subdivision. The intersection of the two vector elements resides in a square whose size is determined by the number of levels of subdivision of the individual dimensional vectors This also establishes an uncertainty neighbourhood area in 2-D space that contains the 2-D point.

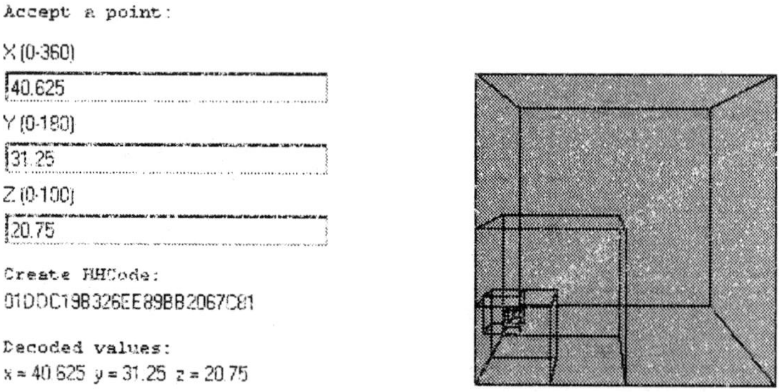

Accept a point:

X (0-360)

40.625

Y (0-180)

31.25

Z (0-100)

20.75

Create HHCode:

01D0C19B326EE89BB2067C81

Decoded values:

x = 40.625 y = 31.25 z = 20.75

Figure 6.4 Three-dimensional fractals generated by interlacing three one-dimensional fractals.

A 3-D HHCode (Figure 6.4) is the bit interlacing of three one-dimensional HHCodes. The interlacing of three 1-D self-similar fractal structures creates a 3-D self-similar cubic structure resembling an octree An octree can be thought of as a three-dimensional fractal that repeats itself after every subdivision. The intersection of the three vector elements resides in a cube whose size is determined by the number of levels of subdivision of the individual dimensional vectors. This also establishes an uncertainty neighbourhood volume in 3-D space that contains the 3-D point.

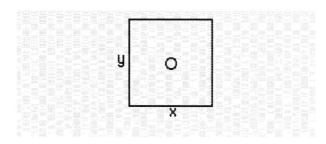

Figure 6.5 Point in a two-dimensional bucket.

HHCodes can be visualised as 2-D quadrants subdivided into smaller quadrants, resembling a quadtree-type construct until the point resides in a 2-D bucket of a given resolution determined by the subdivision level (Figure 6.5).

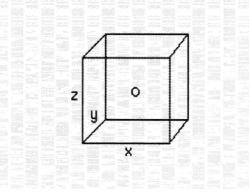

Figure 6.6 Point in a three-dimensional bucket.

HHCodes can be further seen as the orderly breakdown of 3-D space until the point resides in a 3-D bucket (Figure 6.6), whose size is determined by a number of subdivision levels, establishing an uncertainty neighbourhood in the form of a cube. The longer the code, the finer the cube neighbourhood resolution. This concept can be extrapolated to higher dimensions, where the points reside in multidimensional buckets whose neighbourhood size is determined by the number of subdivision levels.

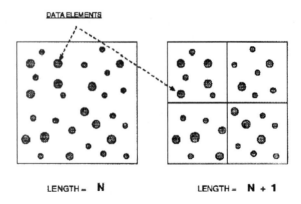

Figure 6.7 Neighbourhoods defined by substring length.

HHCodes can also be thought of two-dimensionally as interlocking variable sized cells, individually encoded into bit streams. The length of these codes defines the neighbourhood size of the spatial cells, which contain the points. Taking substrings of these bit streams allows the dynamic change of the neighbourhood size of the cells (Figure 6.7). Statistical aggregations of points can also in turn be used to drive the neighbourhood sizes of the variable sized cells, allowing homogeneous point aggregations to be grouped together in variable sized cells.

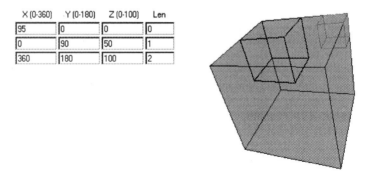

Figure 6.8 Dynamic cell topological interactions.

The substring capability establishes a mechanism for performing topological interactions (Kuratowski, 1977) between data sets by bit matching to given lengths. HHCodes can be used as a form of tesseral arithmetic, which allow the performance of set theoretic topological interactions in multidimensional space, such as disjoint, containment or adjacency (Figure 6.8). These substrings can also be visualised as containment fields. Sorting these codes effectively clusters the data spatially. In multidimensions, the constructs take on added complexity, yet they behave according to the logical progression of squares, cubes and hypercubes, and the data clusters accordingly when sorted.

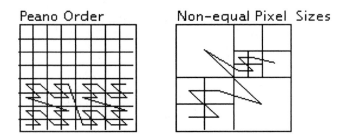

Figure 6.9 Peano order or Z order.

One must emphasise that HHCodes are bit representations of the Riemannian diagonal. In two dimensions, the data sorts in a Z or N pattern called the Peano order (Laurini and Thompson, 1992; Figure 6.9). In three dimensions HHCodes appear to sort in a helical pattern. In higher dimensions, the data is visualised as sorting in a helical pattern leading to the term *helical hyperspatial* as a name for the structure. This essentially clusters spatial data into homogeneous packets in the file and organises the data for fast access. The fast access is achieved by performing binary searches to the appropriate cluster, then doing sequential reads until the cluster is exhausted.

A subtle difference between Peano order (Samet, 1990) and HHCodes is that the binary subdivision of the bounded dimensional vectors normalises the data to a logical representation of the data. Once data is encoded in this form it can be logically decoded using different limits. For example, latitudes and longitudes that are encoded using range limits of -90 to 90 and -180 to 180, can be logically decoded using 0 to 1024 and 0 to 810 raster via the truth table. There exists a one-to-one mapping between the two data sets through normalisation.

On examining these structures closely, one recognises that quadtrees and octrees are simply interleaved binary trees. Doing binary subdivisions of each dimension until they converge to a given resolution generates the binary trees. This gives the added functionality of providing positional uncertainty to the vector element. The combinations of the interleaved binary trees in bit form generate structures such as quadtrees, cubes and hypercubes. Such logical sequences have been extrapolated to put multidimensions on to the HHCode string structure.

SELECT HHCODE, PRO_DEPTH FROM D33454015232

HHCODE	PRO_DEPTH
D330100112321311101312120531292120300010101235020302	8400
D330100112321311101312123021121110200333333021230300	8400
D330100112321311101312123120002333223320102032302220	8400
D330100112321311101312123123002201013123313111201	8400
D330100112321311101312310000333210012203033100312	8500
D330100112321311101323100112320332101322035582101120	8500
D330100112321311101323105011211025130221212101213	8500
D330100112321311101323112220033223103122302002330	8500
D330100112321311101323112322115330131131133222135	8500
D330100112321311101323131312322102003390021133230	8500
D330100112321311101332022112103130322130232312331	8600
D330100112321311101332023200201201013120101323213	8600
D330100112321311101332212003323511302110122311033	8500
D330100112321311101332212123220231302020012011231	8500
D330100112321311101332231011210322310030100239311001	8500
D330100112321311101332322020213335110320203231320	8600
D330100112321311101332303032121030022233323320211	8600
D330100112321311101332322312893130920322093122213	8600
D330100112321311031101102112120122121312122111012	8600
D330100112321311021101112002021001202221233001100	8600
D330100112321311021101113122130311300021301122312	8700
D330100112321311031110200211010103212123031230011	8700
D330100112321311031110201122333223101200300001033	8700
D330100112321311031110230112031338310050302213233	8600
D330100112321311031110322202201011111005221302	8600
D330100112321311031110322321303212023131101132200	8600
D330100112321311031112113300110122121102131230230	8600
D330100112321311031113022112035092312332311113300	8500
D330100112321311031113023113133113003010201222221	8600
D330100112321311031113211221320103210320221112128	8600
D330100112321311031113320211003031011310232121133	8900
D330100112321311031113321300101333132203233333010	8900

32 RECORDS SELECTED.

Figure 6.10 ASCII representations of HHCode bit structure.

Level	Nbr of Quadrants	Resolution
1	1	20,001.6 x 10,000.8 Km
2	4	10,000.8 x 5,000.4 Km
3	16	5,000.4 x 2,500.2 Km
4	64	2,500.2 x 1,250.1 Km
5	256	1,250.1 x 625.1 Km
6	1,024	625.1 x 312.5 Km
7	4,096	312.5 x 156.3 Km
8	16,384	156.3 x 78.1 Km
9	65,836	78.1 x 39.1 Km
10	262,144	39.1 x 19.5 Km
11	1,048,576	19.5 x 9.8 Km
12	4,194,304	9.8 x 4.9 Km
13	16,777,216	4.9 x 2.4 Km
14	67,108,864	2.4 x 1.2 Km
15	268,435,456	1,220.8 x 610.4 m
16	1,073,741,824	610.4 x 305.2 m
17	4,294,967,296	305.2 x 152.6 m
18	17,179,869,184	152.6 x 76.3 m
19	68,719,496,736	76.3 x 38.2 m
20	274,877,906,944	38.2 x 19.1 m
21	1,099,511,627,776	19.1 x 9.5 m
22	4,398,046,511,104	9.5 x 4.8 m
23	17,592,186,044,416	4.8 x 2.4 m
24	70,368,744,177,664	2.4 x 1.2 m
25	281,474,976,710,656	1,192.0 x 596.0 mm
26	1,125,899,906,842,624	596.1 x 298.1 mm
27	4,503,599,627,370,496	298.1 x 149.0 mm
28	18,014,398,509,481,964	149.0 x 74.5 mm
29	72,057,594,037,927,936	74.5 x 37.3 mm
30	288,230,376,151,711,744	37.3 x 18.6 mm
31	1,152,921,504,606,846,976	18.6 x 9.3 mm
32	4,611,686,018,427,387,904	9.3 x 4.7 mm

Figure 6.11 HHCode resolution.

_The true power of these codes is realised by matching bit patterns to given lengths. Pattern matches determine if the object compares spatially with another object within the given framework of space and time. This allows for spatial operations, such as spatial intersections and unions between objects.

In Figures 6.10 and 6.11, the depths at 8400 mm are grouped together at 23 levels of the HHCode string, establishing a cell neighbourhood size of 4.8 x 2.4

meters. Matching bit streams to common string lengths, establishing topological relationships by correlating the associated attributes does spatial interactions. In essence, HHCodes become hyperspatial keys with attributes within the data model.

6.4 TEMPORAL COMPONENTS OF HHCODE

Latitudes and longitudes are elements that designate space, but are not space themselves. On the other hand, HHCodes are a representation of space in multi-dimensions. The HHCode itself can be termed as a space/time key which is unique, because no one object can be in two spatial locations at the same time, or no two objects can occupy the same space at the same time, unless they are part of the same object. This forms the basis of doing interaction between spatial data sets on a temporal basis.

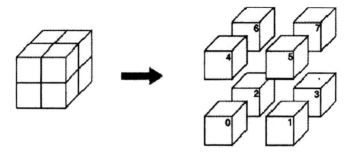

Figure 6.12 HHCodes subdivide into cubes.

_The approach for making *time* the third dimension a function of latitude or longitude, is to create a three-dimensional temporal cube (Kucera, 1996). It represents the translocation of a 3-D object, or voxel, through time. This object can be termed a *toxel,* an extension of the term temporal voxel.

The application of time as the third dimension to the structure does not change the logic. Instead, the patterns behave like octrees and the cells become temporal cubes (Figure 6.12). The temporal cubes are subdivided according to the same logic until a single temporal point (x,y,t) resides in a temporal cube. The other consideration is that if one sorts on this key, it effectively clusters the data in space and time. Retrievals are done by binary searches on the HHCode, rather than exhaustive sequential searches. String comparisons are subsequently done in parallel on a bit level, using bitstream masks.

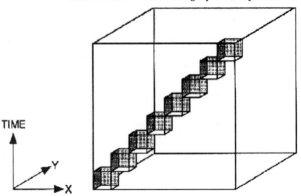

Figure 6.13 An object moving with time as a dimension.

The inclusion of the temporal element in HHCodes, conceptually speaking, creates space-time cubes where depth is the time dimension. Temporal topologies can then be constructed over spatial areas by using time lines (Chaudhuri, 1988), implying that objects have lifespans, during which their attributes can change spatially, yet they still remain the same object (Figure 6.13). Time lines defined over a specified spatial area provide the temporal boundary constraints. This creates continuity by constructing space-time volumes, which intersecting temporal topologies require for interrelationships.

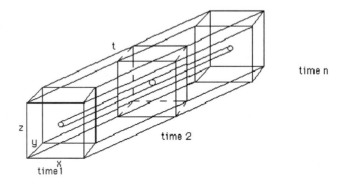

Figure 6.14 Temporal topology consists of the convex hull of the vertices of the hypercubes.

The idea of a "temporal topology" is a four-dimensional topological space including time. It can be visualised as an object moving in space between time 1 and time 2 (Figure 6.14). In effect, it forms a topological volume existing spatially in time. It is mathematically the convex hull of all the vertices of the temporal cubes from time 1 to time 2

An example of a temporal topology is the movement of a biomass through a particular region over a time-interval. The intersection of another biomass through the same region at an earlier time can lead to some interesting results. For example, the second biomass could release toxins or destroy organisms required by the first biomass, leading to the first biomass' extinction.

The two biomasses never actually touch; they meet spatially only in time. The intersection of two temporal topologies and the aftermath is termed as *causality* (Ota, 1996). One biomass causes the termination of the other biomass without ever having touched it. These types of topological operations are visualised as the intersection of two spatio-temporal convex hulls and can be extended to economics, ergonomics, business, population growth, aquaculture, disease control, etc.

6.5 APPLICATIONS

The potential number of applications that can use this structure is quite impressive; for instance statistics, trend analysis and topological surfaces can be generated from point data sets. The length of HHCode string designates the neighbourhood size of the cell, the shorter the string length the larger the cell. From this, one can create on the fly *dynamic cell* sizes. Surfaces and volumes can be generated using aggregation utilities based on user defined difference tolerances for specified attributes. Statistics such as standard deviations, medians, variance and means become easy to compute, as they are generated from the aggregated points in the variable sized cells (HHCodes). The more complex the area the smaller the cells.

In essence, a window in time has been created, which enables functionalities such as trend analysis. Each cell has time tags associated with it. For example, by stacking the cells temporally, one can monitor the rate of siltation in a river. This allows prediction analysis that can determine when dredging will be required in a river channel (temporal data has momentum) or for monitoring rate of increase of toxic wastes in our oceans. One can correlate spatial information in various manners, such as the rate of decrease of aquaculture with the variance of water temperatures over time or the rate of erosion of the coastline due to storm surges. The ability to do retroactive analysis to tune data models for doing predictive analysis becomes a reality.

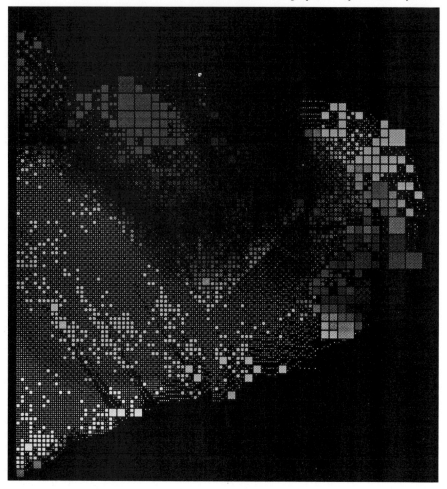

Figure 6.15 Aggregated depth data as statistical cells.

In Figures 6.15 and 6.16, the highlighted cell is selected and the associated attribute statistics are portrayed. The HHCode cell delineates the spatial area, while the min time and max time establish the time-line of the space time cube. Matching objects in this area between the temporal ranges allows the capability for topological interaction between diverse data sets over periods of time.

The attribution associated with each cell is the number of points, the range of min/max values of the depth attribute, the average, median, standard deviation of the depth attribute, the size of the cell and the minimum and maximum time values (decimal Julian day to the millisecond) of the depth elements within the cell. This

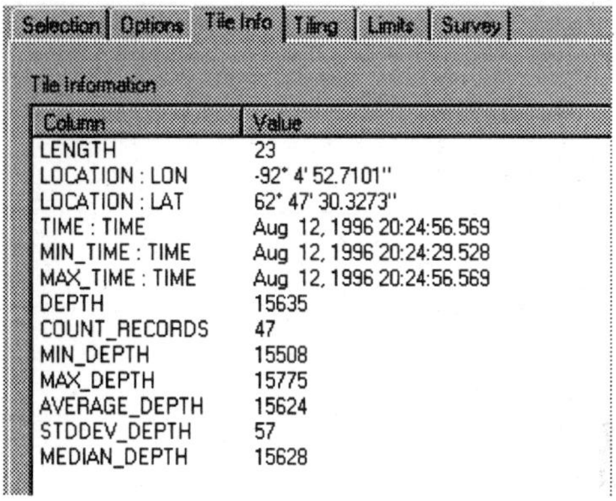
Selection
Options
Tile Info
Tiling
Limits
Survey
Tile Information
Column
Value
LENGTH
23
LOCATION : LON
-92° 4' 52.7101"
LOCATION : LAT
62° 47' 30.3273"
TIME : TIME
Aug 12, 1996 20:24:56.569
MIN_TIME : TIME
Aug 12, 1996 20:24:29.528
MAX_TIME : TIME
Aug 12, 1996 20:24:56.569
DEPTH
15635
COUNT_RECORDS
47
MIN_DEPTH
15508
MAX_DEPTH
15775
AVERAGE_DEPTH
15624
STDDEV_DEPTH
57
MEDIAN_DEPTH
15628

Figure 6.16 Statistical attributes associated with each tile.

establishes variable sized area cells with associated time-lines, which are required for data analysis.

This statistical surface with associated timelines is generated on a tile by tile basis using a tiling utility. The use of this type of tessellation gives a 10 to 1 data reduction by providing an aggregated statistical surface in lieu of a point surface.

6.6 DATA FUSION

Data fusion is the dynamic re-sectioning of cells, allowing multiple attribution to be associated with each cell. The capability of automatically fragmenting variable-sized cells to higher resolution allows the user to associate attributes on a neighbourhood resolution basis. The larger overlapping cells are fragmented to the same resolution as the associated smaller cells. This allows interaction of attribution based on a neighbourhood resolution basis. The fragmented cells from the large cell maintain the same attribute values as the large cell. Only the cells that are equal between the two sets of data are attributed with both sets of attributes or are allowed interoperations between the two attribute sets (Figure 6.17).

This particular operation has been designed to perform dynamic vertical and horizontal datum adjustments on a cell by cell basis, allowing temporally controlled datum adjustments. Generating two sets of cells does this. Using collected depths at a specific vertical datum, the lowest low water large tides

(LLWLT), locked to a specific epoch (1984-Jan-01), generated one set. A series of adjustment tiles were also generated based on transform models LLWLT to a geoidal datum, mean sea level (MSL), or higher high water large tides (HHWLT), also locked to particular epochs. The fusing of the bathymetric tiles with the adjustment tiles creates an "on the fly" datum adjusted surface without impacting the source collected data.

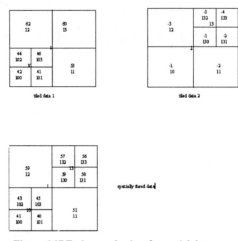

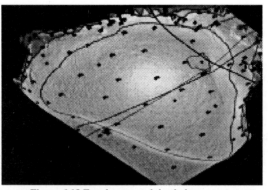

Figure 6.17 Fusing mechanism for spatial data.

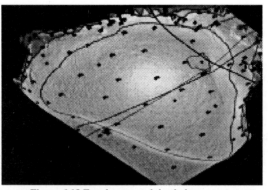

Figure 6.18 Fused raster and depth data.

The fusion of the raster colour attribute with the depth attribute on a common cell basis shows the power of data fusion with complementary data types (Figure 6.18). Detail is shown by the depth attribute and colour is shown by the raster chart colour.

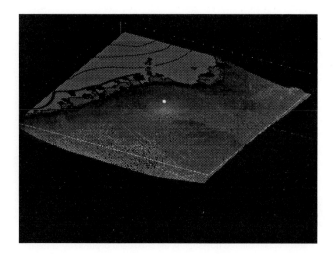

Figure 6.19 Merged raster and depth data sets.

Figure 6.19 shows a raster chart merged with multibeam data, giving another perspective view of data fusion. This allows two different types of data complementing each other, to coexist and be viewed simultaneously. Such techniques provide new methods and capabilities of examining different data sets in conjunction with each other.

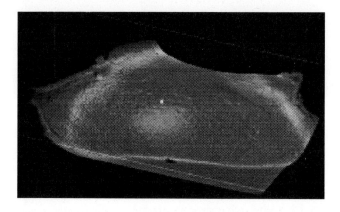

Figure 6.20 Z value is the depth and colour represents ramped depth value.

Figure 6.20 is a three-dimensional bathymetric view of the same fused data set with colour and the z-axis represented by the depth values.

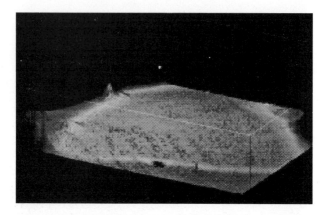

Figure 6.21 Standard deviation is the z-axis and depth value is the colour.

Figure 6.21 shows standard deviations of depths as the z-axis. The colour is represented by the ramped depth values. These techniques allow the capability to quality control data at a detail that previously could only be conceptualised.

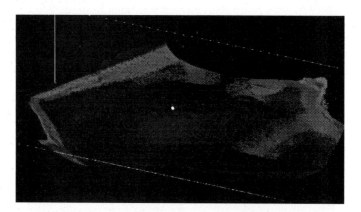

Figure 6.22 Depth value as the z-axis by and temporal value as colour.

Figure 6.22 shows depth values as the z-axis, the colour represents the ramped temporal values showing the currency of the data set.

The importance of associating data spatially and temporally must be stressed. Without such considerations a user would be unable to take advantage of powerful analysis tools for seeing arbitrary data of interest. In shorter words, *to see the unseen.* This generalised approach to data analysis is very valuable for correlative analysis of distinct complex multidimensional data sets in the earth and biological sciences. In the real world one is interdependent of the other. Different combinations of data show different relational aspects of the data.

6.7 DIFFERENCE ANALYSIS

Difference analysis is the other part of the equation. This operation is done by determining and portraying only the differences between data sets using a tolerance criteria that takes in account the noise errors in the data sets. It proves to be useful in detecting differences in objects that change over given areas or times. Such queries can detect anomalies between large data sets, as they zero in on questionable data without creating information overload to the users. This is done by portraying *only* the information that doesn't correlate over time periods within the noise tolerances, rather than by portraying all the collected information.

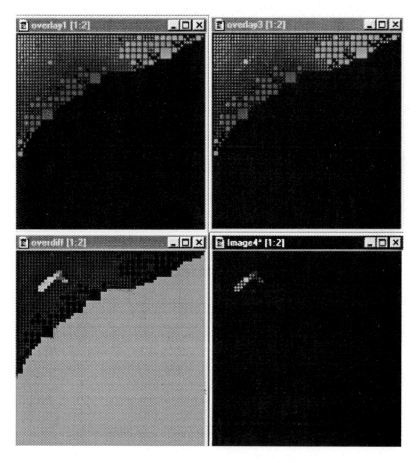

Figure 6.23 Difference analysis portrays depth differences between two data sets.

Figure 6.23 portrays differences between two bathymetric data sets collected at two different instances in time. The differences are formulated based on empirical tolerances that take in account errors due to noise in the data sets. The depiction in the lower right frame shows only the areas where the depth attribute differences exceed the noise error tolerances. The depicted differences are due to dredging in the respective areas between two time periods. Correlative data analysis using spatio-temporal constraints implies the ability to examine many different parameters from disparate data sets in the same fashion. This is the manner in which one generates information for generic modelling.

These operations show the possibilities of using the HHCode architecture to represent spatial data in terms of attributed tessellations. It opens up the possibilities of interrelating various types of data within the same paradigm. Raster images of the earth, based on systems such as Radarsat, Topex/Poseidon and laser infrared digital airborne radar (LIDAR) can be fused with collected point data sets. This allows multifaceted capabilities to analyse, generate or interrelate spatio-temporal information.

6.8 CONCLUSION

We are on the threshold of a new era in GIS. The use of multidimensional tools facilitates new approaches that can be used to generate perspective views for qualitative and dynamic (temporal) analysis. The use of correlative statistical analysis in evaluating the relationships between diverse spatio-temporal data, in turn generates its own class of information. The ability to fuse image data such as backscatter (sidescan) information, with collected point data creates new data sets, drawing on the strengths of the complementary data types, and allowing for the calibration of raster image information with point information. Applications such as draping raster images over digital terrain models or fusing raster images with point information provide further flexibility and functionality, and usher in the new marine and coastal GIS for the next millennium.

6.9 ACKNOWLEDGEMENTS

The author would like to thank Wayne Prime, Paul Holroyd, Ralph Renaud, Trevor Milne, Greg Kierstead, Andre Roy, Wai Choi and Mike McConnel for the use of the HHViewer.

6.10 REFERENCES

Chaudhuri, S., 1988, Temporal relationships in databases. In *Proceedings of the 14th VLDB Conference,* Los Angeles, California, pp. 161-162.
Kaku, M, 1994, *Hyperspace* (Oxford University Press).
Kucera, G.L., 1996, Temporal Extensions to Spatial Data Models, *U.S. Army Engineering Research Laboratory*, Contract DACA88-95-C-0013.

Kuratowski, K., 1977, *Introduction to Set Theory and Topology* (Oxford: Pergamon Press Ltd).

Langran, G., 1992, *Time in Geographic Information Systems* (London: Taylor & Francis).

Laurini, L. and Thompson, D., 1992, *Fundamentals of Spatial Information Systems* (London: Academic Press).

Li, R., Data models for marine and coastal geographic information systems, in this volume, Chapter 3.

Mark D. and Frank A., 1988, *Concepts of Space and Spatial Language* (Santa Barbara, California: National Center for Geographic Information and Analysis).

Ota, M., 1996, The linkage of spatial and temporal concepts. In *Document for ISOTC211 WG2 on Geospatial Data Models and Operators, Temporal Subschemas*, pp. 4-6.

Pavlidis, T., 1982. *Algorithms in Graphics and Image Processing* (New York: Computer Science Press).

Samet, H., 1990, *Applications of Spatial Data Structures* (New York: Addison-Wesley Publishing Co.).

Schroeder, M., 1990, *Fractals, Chaos, Power Laws* (New York: W.H. Freeman and Company).

Spiegel, M., 1959, *Theory and Problems of Vector Analysis* (New York: Schaum Publishing Co.).

Varma, H., Boudreau, H., and Prime, W., 1990, A data structure for spatio-temporal databases. In *IHO Review*, Monaco, LXVII (1), pp. 1-10.

CHAPTER SEVEN

Linear Reference Data Models and Dynamic Segmentation: Application to Coastal and Marine Data

Andrew G. Sherin

7.1 INTRODUCTION

The coast has for centuries been depicted as a line on maps. Tufte (1983) chose an example of cartography, carved in stone during the eleventh century AD in China, as an example of graphical excellence. The map displayed a "relatively firm coastal outline and extraordinary precision of the network of river systems." The map is a demonstration of the long history of visualising as a line where the sea meets the land. For the coast, we use words like coastline or shoreline. The length of the coastline or the shoreline is a defining property for a nation, tallied by national and international agencies and a topic for boasting.

Navigation along a linear path is also a widespread and time-honoured activity. The native peoples of North America used linear navigation along well travelled routes through forest or along waterways often using units of time as the measurement. Lewis (1998) summarises the cartography of traditional North American societies as comprised of networks, most commonly single path networks. Land and water distances were in experiential measures such as a day's journey or number of overnight stops. These networks of rivers and trails were topologically structured and functioned as routes and served as a structural base to mentally situate features.

Modern vehicle navigation systems have developed operator interfaces with non-map based instructions using distance, time and land marks. Displays in vehicles are best limited to turn arrows, the shape of the next intersection and distances (Zhao, 1997). Pauzié (1994) found that along route guidance information had positive effects upon driver route following.

The discussion above suggests that the linear modelling of geographic features such as coasts, rivers and highways is intuitively satisfying. The navigation of these features using measures along their length and landmarks along the route is also a well-established practice and is easily understood. A technical implementation of these intuitive concepts is the linear reference system (LRS).

In modern information systems, a LRS uses measurements along a line for georeferencing. This chapter describes LRS and an associated data display technique called dynamic segmentation and the use of both for coastal and marine applications of geographical information systems.

7.2 THE SUITABILITY OF MODELING THE COAST AS A LINE

The coast is a complex and dynamic feature of the earth. In tidal waters, the concept of a single line representing the contact between land and sea is an obvious over-simplification. Even in non-tidal waters there is water level variability in time frames of years and decades. Change in the alignment of the coast is continuous both from natural and anthropogenic processes of progradation and recession. Byrnes *et al.* (1991) suggest that the interaction of five factors at the shoreline make shoreline mapping difficult. The five factors identified are coastal processes, relative sea level, sediment budget, climate and human activities. A coastal map is only one model of the coastal zone to which this caveat would apply.

In most coastal studies, the length of coast considered is significantly longer than the width of coast investigated. When visualising the data from these studies the extremes of scale mean that features, which are represented as areas or polygons, appear as lines because boundaries coalesce.

Conventional approaches to representing the coast as a line use an arc-node data model whereby each line segment or arc is defined in relation to its starting and ending node (e.g., Environmental Systems Research Institute, 1994). Attribution attached to the line segment or arc applies to the whole line segment and the intervening vertices, which define the shape of the line, are essentially invisible to the system. If the attribute's value changes along the line segment, the line segment must be divided with a new node where the attribute changes. If there are many attributes that are being acquired for a length of coast represented by a line segment, the line segment must be divided into smaller and smaller segments to accommodate changes in the attribution or increasing types of attribution. The association of two or more pieces of information with the same location on a linear feature is a one to many relationship. Conventional arc-node approaches become complex for one to many relationships requiring finer and finer division of lines and inefficient use of space in the attribute tables because of null values for certain attributes.

Bartlett *et al.* (1997) also identified a further constraint of the conventional arc-node structure. Analysis of linear data is limited. Operations such as point-on-line or line-on-line overlay operations result again in a multiplicity of line segments representing different permutations or combinations of attributes.

A similar impact of the number of data base entities occurs when modelling the coast with areas or polygons, a large number of attributes and spatial variation will require a growing number of smaller and smaller polygons to capture the variability. There are, however, cases where polygonal, cell or raster based representation is necessary. Bartlett *et al.* (1997) suggest there are sound operational and conceptual reasons for seeking a hybrid data model for representing the coastal zone.

Bartlett *et al.* (1997) suggested it was possible to consider the shoreline as a special and simplified instance of a network. A network is a compound object consisting of a structured combination of line segments or links and nodes. A shoreline can be modelled as a network with only one link and only two nodes. This network view of the shoreline captures the inherent connectivity of the whole coast. Bartlett *et al.* (1997) also suggests that the network model for the coast support an attribute driven view placing an appropriate emphasis on changes in shoreline properties. This view of the coast conceptually divides or segments the line segment or link in our simple network into subunits using distances measured from an origin or topological node. The distances and attributes are stored in data base tables. This is the basic model on which LRS and dynamic segmentation are based. The simplification of the geometry using this approach is matched by an increase in the number of data base tables to be managed.

Mandelbrot (1967) demonstrated the difficulties in measuring distances along the coast. As finer and finer features are taken account of, the measured total length increases. The measured length of the coast is therefore dependent on the scale being examined. The consequences of this property to our simple network model of the coast is that it is necessary to take into account the source and scale of the line information being measured.

Fleming and Townend (1989) introduced the concept of a coastal reference string (CRS). The CRS is a line used to structure coastal data. No special geographical significance is attributed to the string, although Fleming and Townend (1989) recognised the value of using a line from a widely used cartographic source. This technique overcomes some of the reservations about a simple modelling of the coast as a line and is consistent with handling coastal data as a simple network or linear reference system model using dynamic segmentation. Fleming and Townend (1989) also used the CRS to link all the data for the coast whether represented by point, line, polygon or raster. Bartlett *et al.* (1997) also proposes this use as part of a hybrid data model for coastal data.

7.3 A DESCRIPTION OF LINEAR REFERENCE SYSTEMS AND DYNAMIC SEGMENTATION

The LRS data model is an abstract representation of real-world linear features and their associated characteristics. It consists of a line or lines, measures along the line or combination of lines (usually distance, but can be time or other type of measure) and tables of attributes. Instead of georeferenced positions for information with x, y co-ordinates, the LRS method defines a position in geographic space by a distance or other measure along a known feature. For example, Highway 102, Kilometre 102.5678 is the location of a culvert.

Figure 7.1 shows the components of a LRS data model for three types of information. Information on highways, rivers and streams and coastlines has traditionally used the LRS data model. The figures also show three different types of attribution: (1) attributes at a discrete point that has one distance; (2) attributes that are continuous along the feature and have one distance describing where the attribute changes; and (3) attributes with a beginning and end point that have two distances.

For display and analysis purposes the software that supports the dynamic segmentation methodology, divides the line, at the time the query is made into discrete segments. This procedure from which the technique, dynamic segmentation, gets its name is illustrated in Figure 7.2 using the same examples as Figure 7.1. The dynamic segmentation is initialised by a query. Figure 7.2 gives three examples of simple queries which could be made and shows the result as discrete line segments, portions of the measured line which meet the criteria established by the query. The complexity of the query is limited only by the sophistication of the query language used in the data base management system that stores the attributes.

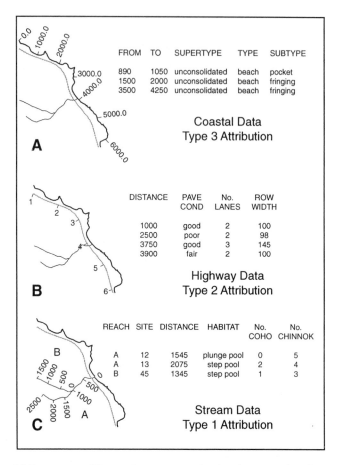

FROM	TO	SUPERTYPE	TYPE	SUBTYPE
890	1050	unconsolidated	beach	pocket
1500	2000	unconsolidated	beach	fringing
3500	4250	unconsolidated	beach	fringing

Coastal Data
Type 3 Attribution

DISTANCE	PAVE COND	No. LANES	ROW WIDTH
1000	good	2	100
2500	poor	2	98
3750	good	3	145
3900	fair	2	100

Highway Data
Type 2 Attribution

REACH SITE	DISTANCE	HABITAT	No. COHO	No. CHINNOK
A	12	1545	plunge pool	0	5
A	13	2075	step pool	2	4
B	45	1345	step pool	1	3

Stream Data
Type 1 Attribution

Figure 7.1 Components of linear reference system showing three examples of measured lines and three types of attribution: (A) a coastline with linear attribution; (B) a highway with continuous attribution; and (C) rivers and stream with point attribution.

The LRS methodology handles one to many relationships effectively. When the value of an attribute changes, it requires only a new entry in the attribute table, no change to the underlying line structure or topology is necessary. When an additional type of attribution is added, an additional attribute table is added to the structure, again with no change to the underlying line or arc structure. For example, in Figure 7.2, the addition of another beach to the shoreline description, a change in pavement condition or the addition of new salmon population measuring site would simply be the addition of a new record to the respective tables. Additional types of attribution could be added by building additional tables using the same along line measures, for instance, a table for visitor numbers could be added to the shoreline system, the status of ditching or drainage could be added to the highway system or water regulation information could be added to the stream system.

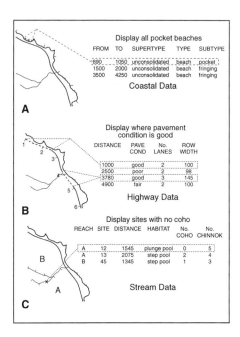

Figure 7.2 The results of the dynamic segmentation of the measured lines for three simple queries: (A) a pocket beach on the coast; (B) good pavement on the highway; and (C) sites on the river with no coho.

7.3.1 A Brief History of LRS and its Roots in GIS-T

The use of LRS and dynamic segmentation for the storage and analysis of georeferenced data was pioneered in the use of GIS for transportation (GIS-T). The technique was applied for three principle reasons:

1. Highways could easily be modelled as lines within a GIS.

2. The geometry of the lines was relatively stable.

3. Traditionally data about highways had been captured using "milepost" or distance references.

Beskpalko *et al.* (1997) described LRS and dynamic segmentation as the only alternative for representing more than one event along a linear feature before the wide use of GPS technologies. He described current GIS-T systems built on three layers:

1. A fundamental arc-node layer derived from traditional systems. This
    model is linked to tabular details through a relational database manager.
    The arc-node layer is measured and calibrated to the linear reference
    system (LRS).

2. A 1-D offset measurement technique known as a linear reference system
    (LRS), for example highway mileposts.

3. Dynamic segmentation as an enabling tool for assigning multiple
    attribute sets over a single linear event.

The use of LRS and dynamic segmentation has proliferated throughout
transportation related organisations and is now used extensively. Nyerges (1990)
suggests that different linear referencing schemes used by different transportation
organisations makes it difficult to develop integrated highway information systems.
Nyerges (1990) also suggests further research is needed into the applicability of
linear reference models to multi-level transportation networks and lineal operators
for network analysis. Beskpalko *et al.* (1997) are more conclusive about the
appropriateness of linear reference models for transportation data citing many
pathologies which appear in existing systems due to the lack of a three-dimensional
(3-D) model. They suggest that what is needed is a 3D GIS model based upon
accurate Global Positioning System (GPS) measurements. These issues are also
relevant to the application of this technology to coastal and marine data and will be
discussed later.

7.3.2 Previous Applications of LRS to Coastal and Marine Data

LRS and dynamic segmentation have more recently been applied to non-
transportation information. Cowie (1997), Vanzwol (1995), and California
Department of Fish and Game (1996) have made extensive use of the technique for
data related to streams and rivers. Vanzwol (1995) describes the Washington Surface
Water Identification System that provides query and plotting tools to support data
browsing, reporting, conversion, and cartographic output. It and other systems in
the United States of America (U.S.) are linked to a national standard for LRS, the
Environmental Protection Agency's River Reach File System. A new standard, the
National Hydrographic Data Set, is under development under the umbrella of the
National Spatial Data Infrastructure initiative in the U.S. (Dewald and Roth (1998).
    Wong *et al.* (1995) describe the use of the inherent time-based LRS in marine
survey data and dynamic segmentation for the management and analysis of marine
geoscience data. Data gathered during a marine survey are commonly geographically
located by referencing their time of collection with the time of a position from
GPS or other navigation system. Wong *et al.* (1995) demonstrated that using time
as the measure has the added benefit of maintaining the start-to-end continuity of
survey track lines that is ordinarily fragmented in a topological arc-node
environment. Figure 7.9 presents similar work but where time was used as a key to
linking along-track attributes with a distance based LRS.

Fricker and Forbes (1988) initially used the application of LRS concepts in their development of a coastal information system. Their system was developed solely within a data base management system. Bartlett (1993) suggested the use of dynamic segmentation for coastal data in his discussions on suitable data models for coastal data. Sherin and Edwardson (1994, 1996), McCall (1995), Bartlett *et al.* (1997) and Liu (1998) subsequently implemented systems using the technique. Bartlett *et al.* (1997) demonstrated their application in southern and western Ireland and included both environmental and socio-economic and cultural data and extended it to the development of sensitivity maps. Sherin and Edwardson (1994, 1996) and McCall (1995) limit the data handled to a physical coastal classification, McCall (1995) for Ireland and Sherin and Edwardson (1994, 1996) for selected areas of Canada's east coast. Liu (1998) also included coastal classification but extended the use of dynamic segmentation to coastal erosion rates and calculation and display of predicted coastal placement for parts of the U.S. shoreline of Lake Erie.

7.4 EVALUATION OF LRS AS A DATA MODEL FOR GIS

Van Oosterom (1993) sets out six requirements for a data model for GIS.

1. Geometric, topological, and attribute data must be stored in a single undivided storage system.

Although current implementations of LRS and dynamic segmentation still use separate systems for storing geometry and attribute data, the LRS and dynamic segmentation model stores most of its information in tables of attribute information usually managed by a relational data base management system (RDBMS). Vendors of GIS are also now offering systems which store all of the data in the RDBMS (e.g., ESRI Spatial Database Engine or SDE) and vendors of RDBMS are offering software tools with geospatial functionality (e.g., ORACLE SDO).

2. The data model must possess good spatial capabilities in three fundamental areas, display, selection and geometric calculations.

Dynamic segmentation requires an intermediate process to be run before the display of the data to divide the line into discreet segments meeting the criteria of the query. Attribute selection is limited only by the power of the RDBMS query engine. Spatial queries can be limited since the spatial referencing uses distance along the line instead of traditional geospatial co-ordinates.

3. The data model must be able to store the data in a form suitable for use at several levels of detail.

LRS and dynamic segmentation rely on the linear feature upon which the LRS is based upon being temporally stable. This would suggest that LRS and dynamic segmentation would not be suitable for presentation at several levels of detail geometrically, since generalisation would impact the 1-D distances.

Generalisation post-segmentation would of course still be possible. Nyerges (1990) describes how in highway systems distance factors are introduced to adjust for road realignments. Such a technique could permit the display of segments using a linear feature at a different level of detail. The transfer of attribute information from one scale to another is discussed later in this chapter. Display of attribute data at different levels of detail is dependent on the structure of the attribute data. For example, Sherin and Edwardson (1996) in their coastal geomorphologic classification store data with a three level hierarchy permitting the display of geomorphic types at the supertype, type and subtype levels.

4.   The data model must store data in layers of geographic entities.

LRS and dynamic segmentation are particularly useful in storing data in layers. Many tables of attribute information can be linked to the LRS and displayed in any combination. In Figure 7.3, an entity relationship diagram is presented after Sherin and Edwardson (1996). In addition to the geomorphic form and material, the data structure can be extended to include many other attribute layers. Bartlett *et al.* (1997) includes degree of exposure, density of visitors and the all-important etc. in their discussion of the attribute view of coastal information. Liu (1998) includes coastal erosion rates and accumulated shoreline offsets due to coastal recession or progradation.

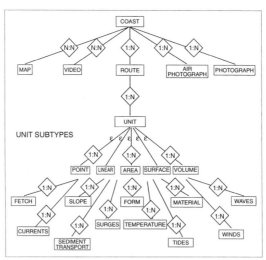

Figure 7.3 A sample entity-relationship diagram for a coastal information system, after Sherin and Edwardson (1996), and following the format of Kronke (1992).

5.   The data model must support persistent storage.

All commercial GIS that support LRS and dynamic segmentation support the long-term storage of the data.

6.   The data model must be dynamic.

As shown in item 4 above, the LRS dynamic segmentation model is fully extensible for attribute data. However, the geometric data is somewhat less dynamic due to the reliance of the technique on the distance measure along a particular linear feature. Methods of compensating for this have already been described in item 3.

Evaluation against van Oosterom's (1993) criteria would suggest some significant shortcomings with LRS and dynamic segmentation as a GIS data model. There are however some special advantages to LRS and dynamic segmentation for geographic entities that can be easily modelled as lines.

LRS and dynamic segmentation reduces the quantity and complexity of the storage and maintenance of geometric and topological information. The GIS only needs to store the co-ordinates and arc-node information for a limited number of lines, one line for each feature on which distances are measured. For example Sherin and Edwardson (1996) stores one line for the mainland shoreline and one line for island shorelines.

In a conventional GIS in our earlier discussion on arc-node structure where the geographic extent of a record needed modification, a graphic editing process needs to be invoked and topological relationships updated. For LRS, the extent can be changed by simply modifying the distance or distances in the relational data base table.

As shown in Figure 7.3, it is very easy to add more types of attribute data by linking additional relational tables to the LRS. If the data is in geographic co-ordinates, vendors supply tools that will measure the distances needed for the LRS and add the distances to the data tables.

A unique capability provided by LRS is along-feature analysis. Non-LRS analysis assumes two dimensions when performing overlay, proximity or adjacency operations. This assumption can be inappropriate for selection or analysis in convoluted shorelines, dense stream networks or closely spaced highways. Two examples of along-feature analysis are discussed later in this paper, the analysis of beach type and length for two different coasts and the analysis of coastal geomorphic form adjacency for three barrier systems on the same coast.

7.4.1 Coastal Mileposts

As discussed above, the attractiveness of LRS and dynamic segmentation to the transportation sector is obvious since much of their data was traditionally collected using distances or mileposts. This technique has also been extended to streams and rivers by the development of a standard river distance system for example in the U.S., Dewald and Roth (1998). For the coastal zone such standardisation does not exist, nor are there mileposts along the coast. The construction of linear event databases can be very time consuming McCall (1995). The time taken is directly related to the heterogeneity of the coastline. Sherin and Edwardson (1996) have developed a technique for the establishment of virtual mileposts using an interactive data acquisition tool that allows for the capture of distance measures for point and linear coastal features.

The technique involves displaying the digital coastline along with other map or image information to assist the operator's geographic orientation. The operator chooses points on the digitally displayed coastline that represent the location, beginning or end of a coastal feature. Placement is determined by matching of the position of the feature on a source document, a video, air photograph or manuscript map with the digital display. Upon confirmation of correct placement, the operator is presented with a form to capture attribute information.

Just as Beskpalko *et al.* (1997) suggested the use of GPS as a tool for capture of transportation data; GPS is an important tool for the capture of coastal data. GPS combined with inertial navigation systems has been used to georeference vertical aerial video (e.g., Quinn, R., Terra Surveys Ltd., Videomap system, pers. comm., 1996). GPS is also being used as control points for georeferencing air photos (e.g., Covill *et al.*, 1995; Forbes and Liverman, 1996). The video and air photos have then been used to capture information about the coast by heads up digitising techniques. GPS can also be used in direct survey methods to capture the position of coastal features, although this is practical only for data capture at large scales. This georeferenced data can then be transformed into a LRS using vendor-supplied tools.

7.4.2 Changing Lines

As discussed earlier, coastal alignment undergoes continuous change. Coastlines are also mapped at different scales and it is often necessary to present data at different levels of generalisation. Figure 7.4 shows the same part of the East Coast of Nova Scotia, Canada from cartographic products at 1:50,000 and 1:10,000 scales. The coastlines were generated using photogrammetric techniques at different times and using air photographs of different scales.

The displayed coastal units representing the location of beaches on both coastlines were originally mapped on to the 1:50,000 line. Using a Theissen

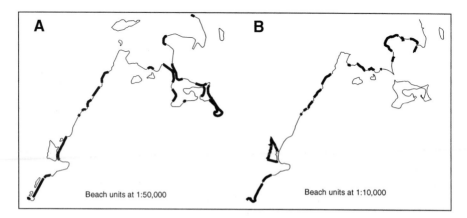

Figure 7.4 Beach units from 1:50,000- and 1:10,000-scale shorelines presented at the same scale.

polygon technique described by Hargrove (1995), the 1:10,000 line was calibrated at every 10 metres to the along line distances used for the 1:50,000 line. A similar method is used in transportation systems to accommodate changes in road alignment due to reconstruction. These techniques can also be used to overcome differences in length measures that Mandelbrot (1967) described and which were discussed earlier. Some measurement pathologies such as measurement reversals were introduced during the calibration and topological differences were not accommodated. Manual adjustments would be required to correct for these errors and inadequacies. These topological differences include a peninsula in the 1:50,000 data set being an island in the 1:10,000 data set and a lagoon being connected to the ocean at 1:10,000 and not connected at 1:50,000. The substantial success of the method, especially on the straight piece of coast between the headlands, does demonstrate that coastal data referenced to one LRS can be transferred to another LRS. This will allow for presentation of data at different levels of generalisation and accommodate changes in the alignment of the coast within the same LRS.

7.5 CASE STUDIES

The following case studies will demonstrate some specific applications of LRS and dynamic segmentation relevant to coastal and marine data. The case study on maps shows three examples of coastal geomorphology showing both linear and points features. The analytical case study shows two examples of the use of an LRS system for analysis of coastal geomorphology. Beyond coastal applications, the marine survey case study presents a use of LRS and dynamic segmentation in slope water depths for preliminary mapping of surficial sediment properties. The case study on geological profiling is another example of the use of the technique for marine geoscience applications. Finally an example of the use of LRS and dynamic segmentation for streams and rivers in New Brunswick, Canada is shown.

7.5.1 Coastal Mapping

LRS and dynamic segmentation can be used for the production of maps. Three examples of maps are shown in Figure 7.5 from one of the eight areas mapped using the system described by Sherin and Edwardson (1996). Difficulty in cartographically representing the complexity stored in the database requires the examples to be simple. Figure 7.5a displays information about beaches and slopes (unconsolidated sloping shorelines) from Cape Broyle Harbour in Newfoundland (see Figure 7.8 for location). Using line symbology, which is offset, but conforming to the shape of the shoreline both the type of beach and type of slope are displayed. The material type for the beaches is displayed using text. Figure 7.5b shows information about cliffs in the same area as Figure 7.5a. Again offset lines display whether the cliff is composed of consolidated or unconsolidated materials. The relative stability of the cliff and the height class of the cliff are presented with the use of text. Figure 7.5c is an example of coastal features represented by points in the database.

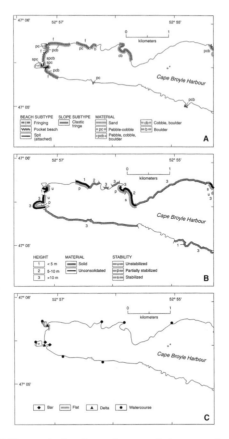

Figure 7.5 Three examples of coastal geomorphology maps for Cape Broyle Harbour, Newfoundland, Canada, generated using dynamic segmentation: (A) beach and slope subtypes with their materials; (B) cliff material supertype with cliff height class and stability of cliffs composed of unconsolidated materials; and (C) coastal geomorphic features represented as points.

7.5.2   Coastal   Analysis

LRS also supports along-feature analysis. Figures 7.6 and 7.7 show two examples. Figure 7.6 from Sherin and Edwardson (1996) results from a query of beach length for two sets of beaches, one set on the west coast of Newfoundland (see location B in Figure 7.8) and one set on the east coast of Cape Breton Island, Nova Scotia (see location A in Figure 7.8). The figure shows histograms of the number of sand and gravel beaches per length class for the two study areas and the total length of sand and gravel beaches per length class. This analysis suggests similar distributions of gravel beaches in the two areas but distinct differences in the distribution of sand beaches. A few large sand beaches dominate the distribution in the Cape Breton data set.

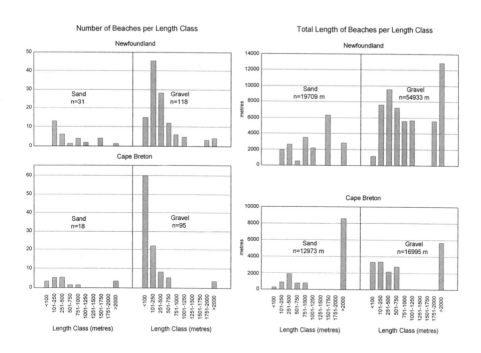

Figure 7.6 Length class comparisons of sand and gravel beaches for Cape Breton, Nova Scotia, Canada (location A in Figure 7.8), and southwest Newfoundland, Canada (location B in Figure 7.8).

Figure 7.7 from Edwardson *et al.* (1997) explores adjacency relationships between coastal features for three barrier systems in Newfoundland (see location C in Figure 7.8). Binary sequence relationships were analysed for coastal form type (i.e. beach, cliff, etc.) and general form material or material supertype (i.e. solid or unconsolidated). In Figure 7.7, the relationship *csos* (cliff solid beside outcrop solid) occurs most frequently in two bays, Holyrod and Biscay. In Mall Bay, however, the relationship ***bucs*** (beach unconsolidated beside cliff solid) is the most predominant reflecting the greater number of pocket beaches. Relationships containing unconsolidated materials principally unconsolidated cliffs (***cu***) suggest that these cliffs are potential sources for the barriers.

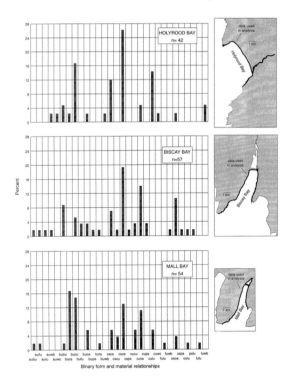

Figure 7.7 Binary adjacency form and material relationships for three barrier beach systems, Newfoundland, Canada (location C in Figure 7.8). Form types are: a = anthropogenic, b = beach, c = cliff, f = flat, o = outcrop, p = platform, l = slope, wb = waterbody, and wc = watercourse. Material supertypes are s = solid and u = unconsolidated. For example, "cuos" = unconsolidated cliff beside solid outcrop.

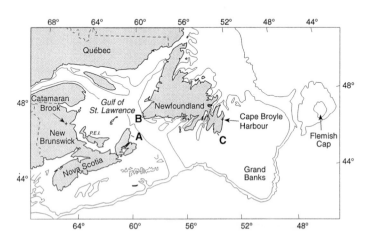

Figure 7.8 Location map for Figures 7.5, 7.6, 7.7, 7.9 and 7.12 showing Cape Broyle Harbour (Figure 7.5), Flemish Cap (Figure 7.9) and Catamaran Brook (Figure 7.12). A is Cape Breton (Figure 7.6), B is southwestern Newfoundland (Figure 7.6) and C is the location of the three barrier beach systems in Newfoundland used in Figure 7.7.

The statistical analysis for Figure 7.6 was conducted outside the GIS. The relationship analysis shown in Figure 7.7 was conducted using custom-built programs within the GIS. This experience suggests that more research is necessary for the development of lineal operators for along coast analysis just as Nyerges (1990) suggested for GIS-T.

LRS also shows great potential for the inclusion of fuzzy boundaries within a data structure for the coast. For example, Drosesen and Geelen (1993) applied fuzzy sets to hydrological expert modelling using a raster based system. The incorporation of class membership information in the relational data base component of a LRS application makes fuzzy queries possible. The use of LRS to construct the equivalent to a linear raster representation of coastal data will enable similar fuzzy functionality for linear coastal analysis as enjoyed by two-dimensional raster representations. For example, Table 7.1 shows a hypothetical length of beach for the parameter pebbliness where 1.0 would be a beach composed entirely of pebble-sized material and 0.0 would be a beach without any pebble-sized material.

Table 7.1 Pebbliness along a hypothetical length of beach.

From Distance	To Distance	Pebbliness
100	110	0.6
110	121	0.75
121	155	0.6
155	180	0

Table 7.2 Linear raster presentation of pebbliness for a hypothetical length of beach.

From Distance	To Distance	Pebbliness
100	105	0.6
105	110	0.68
110	115	0.68
115	120	0.65
120	125	0.6
125	130	0.6
130	135	0.6
135	140	0.6
140	145	0.6
145	150	0.6
150	155	0.6
155	160	0.4
160	165	0.2
165	170	0
170	175	0
175	180	0

The boundaries between pebbliness classes are abrupt. Table 7.2 shows the same length of beach in a linear raster form with a 5 m "cell" size where pebbliness is stored as a continuous class membership value. This kind of table connected to a LRS would enable query and analysis along the coast at the resolution of the segment length without explicitly defined boundaries.

7.5.3 Marine Geoscience Mapping

Wong *et al.* (1995) demonstrated the use of LRS for marine geological mapping exploiting the inherent linear reference system of elapsed time in a marine survey. Figure 7.9 shows along the ship's track where the geologist interpreted well-stratified sediments from low frequency echosounder data. The technique used for Figure 7.9 differs slightly from that of Wong *et al.*   (1995),wheretime was used directly as the LRS measure. This technique uses time only as a link to a distance measure along track. In this application on the continental slope of Canada east of Newfoundland, on the slopes of Flemish Cap (see Figure 7.8), the database also includes linear events for subsurface character and geomorphic features. LRS is useful for preliminary marine geological mapping in all water depths and particularly useful in areas of sparse data like slope and abyssal plain water depths where the interpolation of sediment types and features between lines is not appropriate.

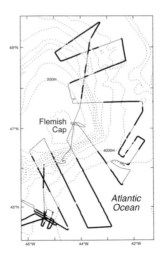

Figure 7.9 Well stratified sediments near Flemish Cap, offshore Newfoundland, Canada. The thin line is the ship's track. The thick line shows the location of well stratified sediments along the track.

7.5.4 Profiling

An extension of the technique for preliminary marine geological mapping, is the use of LRS for constructing subsurface profiles. The technique attaches vertical measurements to horizons interpreted from acoustic records to distances measured along the track of the geophysical survey. The vertical measures are combined with the distance measures to create co-ordinates for the horizons in a vertical plan. These co-ordinates can then be edited to create a geological profile, an example of which is shown in Figure 7.11. Figure 7.11 is a profile constructed from high-resolution seismic reflection data collected in Severn Sound, Georgian Bay, Canada (Sherin (1997).

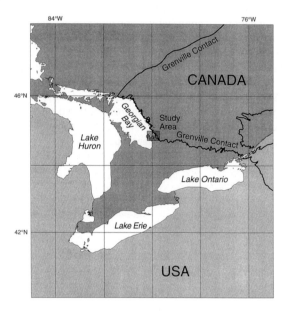

Figure 7.10 Location map for Figure 7.11.

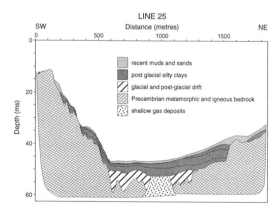

Figure 7.11 Geological cross-section from Severn Sound, Georgian Bay, Canada, developed from high-resolution seismic reflection intepretations. The x-axis is the distance in metres along the seismic reflection survey line. The y-axis is the depth measured as acoustic travel time in milliseconds.

7.5.5 Stream and River Mapping

Cowie (1997) uses LRS and dynamic segmentation to map salmon habitat and recreational fishing management for rivers and streams in the New Brunswick, Canada. Figure 7.12 is an example of habitat mapping for the Catamaran Brook, a tributary of the Mirimachi River in New Brunswick (see Figure 7.8). Figure 7.12 shows the location of sections of the stream where the stream type is riffle. A riffle is a ridged shallow that extends across the river and creates broken water. These sections of stream represent good habitat for juvenile Atlantic salmon. These data are stored as database tables with to and from distances. Figure 7.12 also shows point type records of water temperature measurements as open triangles. Figure 7.12 also demonstrates a commonly used feature of LRS and dynamic segmentation, the offset of lines to show an additional attribute. The offset line in Figure 7.12 is symbolised to show the type of riffle. Cowie (1997) also uses LRS and dynamic segmentation for the mapping of regulated waters. Closing stretches of river to all angling is common in most jurisdictions, but New Brunswick uses other unique ways of allocating or managing the fishery resources, specifically through leasing Crown-owned waters, and several types of Crown reserves.

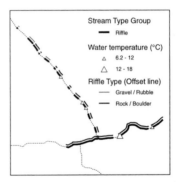

Figure 7.12 An example of an Atlantic salmon habitat map showing the location and type of riffles and water temperature measurements, Catamaran Brook, New Brunswick, Canada.

7.6 CONCLUSION

The use of LRS in the coastal and marine applications described above is a useful technique which parallels the intuitive view of the geography, enables special analysis along the coast, and reduces the complexity of data management. LRS and dynamic segmentation should be used in conjunction with other data structures, such as polygonal, surface, raster and 3-D object models, in an integrated GIS environment, as suggested by Sherin and Edwardson (1994, 1996), McCall (1995) and Bartlett *et al.* (1997). These integrated approaches demonstrate the benefits that LRS and dynamic segmentation bring to coastal and marine applications, in addtion to traditional GIS methods.

   McCall (1995) and Bartlett *et al.* (1997) refer to the potential of LRS and dynamic segmentation for introducing *temporal* aspects to a coastal GIS as well.

Langran (1993) reviews the temporal capability of existing data base management systems for the management of aspatial data and concludes that rudimentary tools and components exist to assemble a temporal GIS. Langran (1993) also describes a conception of time for use in the GIS, the space-time composite. This technique reduces three dimensions (in this case, two spatial dimensions and one temporal dimension) to two and permits the treatment of space atemporally and of time aspatially. This construct fits well with an LRS environment allowing temporal information to be maintained within the data base component of the system. Research into the use of LRS and dynamic segmentation to introduce fuzzy reasoning and time into coastal GIS should provide considerable rewards in fashioning an information system that better models the real coast.

7.7 ACKNOWLEDGEMENT

Geological Survey of Canada Contribution No. 1999040.

7.8 REFERENCES

Bartlett, D., 1993, Space, time, chaos and coastal GIS. In *Proceedings, 16th International Cartographic Conference and the 42nd Deutscher Kartographentag*, Koln, Germany, May 3-9, 1993, pp. 539-551.

Bartlett, D., Devoy, R., McCall, S. and O'Connor, I., 1997, A dynamically segmented linear data model of the coast. *Marine Geodesy*, **20**, pp. 137-151.

Beskpalko, S.J., Wyman, M.M. and Sutton, J.C., 1997, The need for a formal GIS transportation model, In *Proceedings of Interop '97*, http://bbq.ncgia.ucsb.edu/conf/interop97/program/papers/bespalko.html (Sept. 17, 1998).

Byrnes, M.R., McBride, R.A., and Hiland, M.W., 1991, Accuracy standards and development of a national shoreline change database. In *Coastal Sediments '91 Proceedings*, Seattle, Washington, 25-26 June 1991, pp. 1027-1042.

California Department of Fish and Game ,1996, Department of Fish and Game GIS Metadata, http://ice.ucdavis.edu/wits/meta/bioinv.txt (Sept. 17, 1998).

Covill, R., Forbes, D.L., Taylor, R.B. and Shaw, J. 1995, Photogrammetric analysis of coastal erosion and barrier migration near Chezzetcook Inlet, Eastern Shore, Nova Scotia. *Geological Survey of Canada, Open File*, **3027**, 1 sheet [poster].

Cowie, F., 1997, New Brunswick Aquatic Resources Data Warehouse. (Doaktown, New Brunswick: Atlantic Salmon Museum). 170 pp. + apps.

Dewald, T.G., and Roth, K.S., 1998, The National Hydrography Dataset: Integrating the U.S.EPA Reach File and U.S.GS DLG. In *Proceedings of the 1998 ESRI Users Conference*, San Diego, California, http://www.esri.com/library/ userconf/proc98/PROCEED/TO500/PAP477/ P477.HTM (Mar. 4, 1999).

Drosen, W. and Geelen, L. 1993, Application of fuzzy sets in ecohydrological expert modelling. In *HydroGIS 93: Application of Geographic Information*

Systems in Hydrology and Water Resources, Vienna, April 1993. IAHS Publication, **211**, pp. 3-11.

Edwardson, K.A., Sherin, A.G. and Horsman, T., 1997, The use of dynamic segmentation in the coastal information system: adjacency relationships from southeastern Newfoundland, Canada. In *CoastGIS '97 International Symposium on GIS and Computer Mapping for Coastal Zone Management, Proceedings*, Aberdeen, Scotland, edited by Green, D.R. and Massie, G., np.

Environmental Systems Research Institute, 1994, *ARC/INFO Data Management: Concepts, Data Models, Database Design, and Storage* (Redlands, California: Environmental Systems Research Institute).

Fleming, C. and Townend, I., 1989, A coastal management database for East Anglia. In *Coastal Zone '89, Proceedings of the Sixth Symposium on Coastal and Ocean Management*, Charleston, South Carolina, American Society of Civil Engineers, pp. 4092-4107.

Forbes, D.L., and Liverman, D.G.E., 1996, Geological indicators in the coastal zone. In *Geoindicators: Assessing Rapid Environmental Changes in Earth Systems* edited by Berger, A.R. and Iams, W.J. (Rotterdam: A.A. Balkema), pp. 173-192.

Fricker, A. and Forbes, D.L., 1988, A system for coastal description and classification. *Coastal Management* **16**: 111-137.

Hargrove, W., Winterfield, R. and Levine, D., 1995, Dynamic segmentation and Thiessen polygons: a solution to the river mile problem. In *Proceedings 1995 ESRI User Conference*, Palm Springs, California, http://www.esri.com/library/userconf/proc95/to150/p114.html (Mar. 5, 1999).

Kroenke, D.M., 1992, *Database Processing: Fundamentals, Design and Implementation, 4th Edition* (New York: Macmillan Publishing Co.).

Langran, G., 1993, *Time in Geographic Information Systems* (London: Taylor & Francis)

Lewis, G.M., 1998, Maps, mapmaking and map use by native North Americans. In *The History of Cartography, Volume 2, Book 3, Cartography in the Traditional African, American, Arctic, Australian and Pacific Societies* (Chicago: University of Chicago Press), pp. 51-182.

Liu, J-K, 1998, Developing geographic information system applications in the analysis of responses to Lake Erie shoreline changes. M.Sc. Thesis, The Ohio State University.

Mandelbrot, B., 1967, How long is the coast of Britain? Statistical self-similarity and fractional dimension. *Science*, **156**, pp. 636-638.

McCall, S., 1995, The application of dynamic segmentation in the development of a coastal geographic information system. In *CoastGIS '95 International Symposium on GIS and Computer Mapping for Coastal Zone Management, Proceedings*, Cork Ireland, 5-8 February 1995, edited by Furness, R., pp. 305-312.

Nyerges, T.L., 1990, Locational referencing and highway segmentation in a geographic information system. *ITE Journal*, March 1990, pp. 27-31.

Pauzié, A., 1994, Human interface of in-vehicle information systems. In *1994 IEEE Vehicle Navigation and Information Systems Conference Proceedings*, plenary, pp. 35-40.

Sherin, A.G., 1997, Integrating geoscience data for environmental planning in Severn Sound. In *Proceedings 1997 ESRI User Conference*, San Diego, CA, http://www.esri.com/library/userconf/proc97/TO650/PAP609/P609.htm (Mar. 4, 1999).

Sherin, A.G. and Edwardson, K.A., 1996, A coastal information system for the Atlantic Provinces of Canada. *Marine Technology Society Journal* **30**, pp. 20-27.

Sherin, A.G. and Edwardson, K.A., 1994, Using GIS and dynamic segmentation to build a digital coastal information database. In *Proceedings (abstract), Coastal Zone Canada '94,* Halifax, Nova Scotia, September 20-23, edited by Wells, P.G. and Ricketts, P.J., p. 2378.

Tufte, E., 1983, *The Visual Display of QuantitativeInformation* (Cheshire, Conn.: Graphics Press).

van Oosterom, P.J.M., 1993, *Reactive Data Structures for Geographic Information Systems* (New York: Oxford University Press).

Vanzwol, C. 1995, Washington surface water identification system. In *Proceedings 1995 ESRI User Conference*, Palm Springs, California, http://www.esri.com/library/userconf/proc95/to350/p346.html (Mar. 8, 1999).

Wong, F.L., Hamer, M.R., Hampton, M.A. and Torresan, M.E., 1996, Bottom characteristics of an ocean disposal site off Honolulu, Hawaii: Time-based navigational trackline data managed by routes and events. In *Proceedings of the 1996 ESRI Users Conference*, Palm Springs, California, http://www.esri.com/library/userconf/proc96/TO150/PAP114/PAP114.htm (Mar. 4, 1999).

Zhao, Y. 1997, *Vehicle Location and Navigation Systems* (Boston: Artech House Inc.).

CHAPTER EIGHT

Spatial Reasoning for Marine Geology and Geophysics

Dawn J. Wright

8.1 INTRODUCTION

Despite the general success of GIS in marine science as evidenced by the chapters in this volume, there are still remain constructive critics who are asking the "hard" questions, among them:

- *What good is GIS beyond colourful display of multiple types of data?*
- *What is "spatial analysis" in GIS and how will it inform the oceanographic science that I'm already established in?*
- *Why should I use GIS when there is a "home-grown" or public-domain mapping software already available, such as Generic Mapping Tools (GMT; Wessel and Smith, 1991, 1995) that accomplishes many, if not most, of my tasks?*

These questions are being raised particularly in the marine geology and geophysics (MG&G) community, where GIS appears to be at a crossroads. There is now a basic understanding of the utility of GIS for display, management, and mapping, particularly at sea where effective decisions may save thousands of dollars of ship time (e.g., Wright, 1996; Fox and Bobbitt, 1999; Hatcher, 1999; Su, 1999). However, many still wonder what all the excitement is about beyond this. Some answers to the above questions and dilemmas may be found in more cross-disciplinary communication. In other words, the challenge may be how best to communicate the concepts of geographic information *science* to practitioners in the newest application domains of GIS, deep-water marine science included. GISci is the "science behind the systems," including questions of spatial analysis (special statistical techniques variant under changes of location), spatial data structures, accuracy, error, meaning, cognition, visualisation, and more (for the most comprehensive treatment of GISci see Longley *et al.*, 1999). Pursuant to GISci is the notion of "spatial reasoning," defined by Berry (1995) as a situation where the process and procedures of manipulating maps transcend the mere mechanics of GIS interaction (input, display and management), leading the user to think spatially using the "language" of spatial statistics, spatial process models, and spatial analysis functions in GIS. This is how to move beyond mere display to see an additional or greater value of GIS

It is the purpose of this chapter to *briefly* introduce, primarily to those MG&G practitioners not intimate with GIS software documentation or the GISci literature, the notion of spatial reasoning by way of outlining the analytical functionality common in most commercial packages. As MG&G is the speciality of the author, the examples presented in the paper are of that nature. Recommended studies for other subdisciplines include Manley and Tallet (1990), Mason *et al.*

(1994) and Lucas (1999 and references therein) for physical oceanographers; Hansen *et al.* (1991) and Bobbitt *et al.* (1997) for chemical oceanographers; and May *et al.* (1996), Meaden and Chi (1996), Bobbitt *et al.* (1997), and Moses and Finn (1997) for marine biologists.

8.2 A CAVEAT ON APPLICATION DOMAINS IN GIS

Recent trends in GIS software architecture design (i.e., the framework for how the input/capture, management, manipulation/analysis, output/display subsystems of the GIS are constructed) have focused more on application domains than on a generic GIS. The architecture of each subsystem of the GIS is based on a particular data model, which provides a formal means of representing information. Peuquet (1988) defines a data model more specifically as a general description of the specific sets of entities (the phenomena of interest in reality) or objects (the phenomena of interest as digitally represented in the database) and the relationships between these sets. Geographical data models are therefore the sets of entities and relationships between them used to represent geographic variation in the discrete, digital world of a computer database (Goodchild, 1992). The geographical data model determines the constructs for storage, the operations for manipulation, and the integration constraints for determining the validity of data to be stored within the GIS. Differing ways of viewing the occupation of geographic space have resulted in the layer (field), object, or network data models for a GIS (Goodchild, 1992; Nyerges, 1993). Li (1999), Gold (1999) and Varma (1999) address the nature of data models and the manifestations of them most suitable for marine applications.Unfortunately, these alternative data models are still not readily available in most commercial GIS packages.

The tension between domain-dependent and domain-independent GIS lies in the fundamental differences between chosen data models (e.g., Knapp, 1991). On one hand, the commercial sector, which must of course consider commercial profit, caters to domain-specific niches in the GIS market (a transportation GIS, a hydrology GIS, an epidemiology GIS, etc.) and stays with one class of data model to fit the chosen applications of its best customers. For example, the layer (field) data model of Arc/INFO™ has remained basically unchanged, with features such as TIN, GRID and dynamic segmentation added to it over the years. Analysis modules for hydrology, location-allocation, and land-based spatial statistics, among others, have been included in recent revisions of Arc/INFO, within the layer data model context, to meet the needs of the largest, most profitable application domains. At the other end of the spectrum is the creation of a superstructure, standardised GIS data model that would provide linkages across narrow application domains. One of the challenges here is the development of a data model that is representative of a myriad of data sets, while at the same time simple to understand, flexible and capable of answering new types of queries without changing data structures. This is currently being addressed by the Open GIS Consortium, Inc., the objectives of which are to provide a shared data environment through the implementation of a generic GIS data model and a user workbench with needed tools and data to support a variety of applications (Buehler, 1994; http://www.opengis.org).

Linking data models to the data in question (terrestrial or marine) depends on the nature of GIS architecture. For example, with MG&G data sets come from a variety of sources, including a number of different instruments (Figure 8.1), all with different attributes, making it difficult to define a uniform data structure applicable to all. Developing or changing data models within a GIS to better fit the data can be accomplished if the tools or commitment is there. However, it appears that the tools are developed only if the marketplace acts to spur on that development. There is a dilemma with such development for advanced marine science oriented systems, as such systems use suffers from a lack of marketplace and benefactor. Therefore, the best alternative currently is to try to make one's data fit within the confines of the existing data model and functionality of the GIS. However despite the challenges involved in linking data to data model and system architecture (Burrough, 1992), many researchers have successfully integrated remotely-sensed and *in-situ* MG&G data into a GIS and found the application to be not only feasible but essential for the success of the research or project in question. Those who are most successful have discovered functionality that is somewhat buried within the GIS. Having found it though, has reaped benefits, particularly when it can be linked to data analysis packages such as GMT or Matlab, or to scientific visualisation packages (e.g., Goldfinger, 1999).

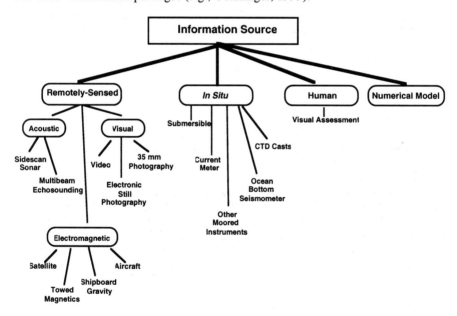

Figure 8.1 Hierarchy of possible MG&G information sources for GIS.

8.3 A SAMPLING OF MG&G PROBLEMS FOR SPATIAL REASONING

Below are three major examples of MG&G spatial problems that are useful to approach with spatial reasoning in GIS. They come mainly from insights gained from several decades of investigation along fast-spreading portions of the global mid-ocean ridge (particularly two of the best-studied sections of the East Pacific Rise or EPR at 9°-10°N and 17°-18°S; Figure 8.2). These investigations have provided a wealth of geographically referenced data, results and data-driven theoretical (often numerical) studies.

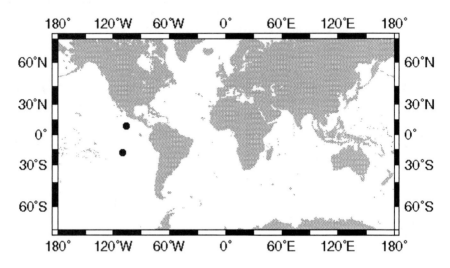

Figure 8.2 Location of the main northern (9°-10°N) and southern (17°-18°S) study sites on the East Pacific Rise.

1. The exploration of large volumes of spatially and temporally variable data to facilitate previously undiscovered relations between already-existing data. For example, during a recent results-based workshop for the EPR 9°N region (RIDGE Symposia on the Results of Field Studies Along the East Pacific Rise, 9°-10°N, 1998), it was apparent to all involved that the amount of spatially-reference data and accompanying results already in existence for this region of the ocean floor are: (1) staggering in its amount; and (2) diffusely distributed in the minds and journal articles of individual scientists (D. Toomey, pers. comm., 1999).

2. Detecting changes in observables that span several orders of spatial and temporal magnitude at the crests of mid-ocean ridges, particularly discrete magmatic events and the resulting distribution of hydrothermal vent sites.

3. Determining the relations between the subseafloor expression of magmatic systems and the nature of diking and eruptive events. Two topics of particular interest in the MG&G community are the physical structure of axial magma chambers (e.g., Sinton and Detrick, 1992) and the nature of hydrothermal convection (e.g., Davis *et al.*, 1996). However, studies of these entities are

usually conducted independently of each other. For instance, seismologists have focused on imaging seismic velocity and attenuation of the ridge crest in order to map the spatial dimensions and physical properties (temperature and melt fraction) of axial magma chambers (e.g., Toomey *et al.*, 1994; Dunn and Toomey, 1997). Geologists and geochemists have been equally focused on mapping the surface expression and chemistry of active vent fields (e.g. Haymon *et al.*, 1991, 1997), from which attempts are made to infer how deep and in what geometry fluids are penetrating the crust (e.g., Wilcock, 1998). Such independently conducted studies are linked by a common process (the interaction of hydrothermal fluids with an evolving magmatic system) and should be linked in a common spatial analysis environment.

8.4 GIS ANALYTICAL FUNCTIONALITY FOR MG&G

In addressing the above spatial problems mapping is an essential, sometimes formidable, first step. However, if a user is to move beyond mapping ("What is where and how do I change the colours in my legend, zoom in, zoom out, etc?") to spatial reasoning ("Why is it there?"), a road map to GIS usability is extremely helpful. In other words, what are the functions or operations that enable a specific application to solve problems and/or aid in interpreting data? Five major classes of GIS functionality (in the order that are often employed) are data input/capture, data storage/management, data manipulation, data analysis, and data output/display. Nyerges (1993), for example, provides an exhaustive compilation of GIS functions suitable for environmental applications.

The all-important data analysis functions focus on developing and synthesising spatial relationships in data in order to solve spatial problems, to answer scientific questions. They range from simple models integrated within a GIS context (i.e., fully embedded within GIS architecture) to elaborate models that are coupled or "hot-linked" to a GIS environment. Many of the current concerns with the analytical functions in GIS are with the design tradeoff between integration and/or coupling by way of a special language interface (Densham, 1991; Burrough, 1988; Fedra, 1993; Nyerges, 1993).

Four commonly used spatial analysis procedures in GIS are data interpolation, contour generation, buffer zone generation and theme merging. MG&G practitioners have found the inverse distance weighted (IDW) and triangulated irregular network (TIN) functions in GIS to be very useful for interpolating irregularly spaced cruise and ROV dive data and matching them to evenly distributed grid data (e.g., Su, 1999). Quick contour generation from grids of bathymetry, temperature, salinity, etc. make for efficient comparison to point and line observations already loaded into the GIS. Buffering (the creation of polygons around existing points, lines, or areas) of sampling sites or ship track lines are integral to estimating the accurate spatial extent of data collection (e.g., Wright *et al.*, 1995). Merge functions allow two or more individual or cruise data sets to be combined for further analysis (e.g., Hatcher and Maher, 1999; Su, 1999). GIS packages now come with an additional host of spatial analysis functions (e.g., point pattern analysis, spatial interaction models, calculations of Geary's and Moran's indices of spatial autocorrelation, network analyses) that have been traditionally used in terrestrial applications but hold great promise for MG&G studies, particularly at hydrothermal vent sites. Examples of primary data GIS analysis functions for MG&G applications are provided in Table 8.1.

Table 8.1 GIS analytical functionality for spatial problems in MG&G (listed alphabetically by likely names of commands, modules, or index terms in documentation of various GIS packages; actual Arc/INFO commands or functions listed in capital letters). Categories guided by Nyerges (1993).

Azimuth: compute azimuth, bearings, and geographic point locations.
Buffers: compute distances from point, line, and polygon.
Descriptive nonspatial statistics: e.g., frequency analysis, measures of dispersion (variance, standard deviation, confidence intervals), measures of central tendency (mean, median, mode), range, percentile.
Distance analysis: calculate distances between features, create grids of distances from source, create polygones of distance zones (e.g., BUFFER, REGIONBUFFER, NEAR, POINTDISTANCE, NODEDISTANCE, EUCLIDEANDISTANCE, PATHDISTANCE)
Group mathematical operations: perform addition, subtraction, multiplication, division, minimum, maximum on the grid cell for two or more data categories.
Inferential nonspatial statistics: e.g., correlation, regression, analysis of variance, and discriminant analysis (e.g., CORRELATION, REGRESSION, SCATTERGRAM, HISTOGRAM, STATISTICS).
Inferential spatial statistics: e.g., trend analysis, autocorrelation, Geary and Moran indices, Pearson's correlation coefficient (e.g., GEARY, MORAN).
Interpolation: Inverse distance weighted, spline, trend interpolation (e.g., IDW, SPLINE, TOPOGRID, TREND), conversions of TINs instead of points to grids (e.g., TINLATTICE).
Kriging computations: ordinary, stationary, and nonstationary kriging, kriging in 1-D, 2-D, and 3-D for scattered points, grids of points, and irregular grids. (e.g., KRIGING).
Merging: merge attribute information automatically or manually as a result of composite process; two attributes associated with same area.
Model structuring: model-structuring environment that links parts of numerical/process model to the GIS environment through a special language interface.
Multivariate statistics: output regression coefficients for a regression model, maximum likelihood classification(e.g., REGRESSION, SAMPLE, STACKSTATS, DENDEROGRAM, MLCLASSIFY).
Nearest neighbour: compute closest geographic phenomenon to each phenomenon of a particular kind.
Overlay operators: point, line, area object on/in point, line, and area object, data operators for Boolean AND, OR, NOT; for point in polygon point on line, line in polygon, and polygon in polygon.
Replacement: replace cell values with new value reflecting mathematical combination of neighbourhood cell values such as average, maximum, minimum, total, most frequent, least frequent, standard deviation.

Significance tests: t-test, chi-square.

Simulation: conditional simulation in 2-D and 3-D.
Single mathematical operations: perform exponential, logarithm, natural logarithm, absolute value, sine, cosine, tangent, arcsine, arc cosine, arctangent on the grid cell.
Slope/aspect: computer slope and aspect based on a fitting a plane (e.g., SLOPE, ASPECT), compute slope and aspect based on fitting a curve (e.g., CURVATURE).

Spatial interobject measurement: interobject calculations for distance and direction point to point, point to line, polygon perimeter, percent of total area, percentiles, range, midrange.
Spatial intraobject measurement: individual object calculation for line length, polygon area, surface volume, polygon perimeter, percent of total area, etc.

Superimpose: superimpose one feature on another with replacement.

Territory designation: creation of polygons around points, calculation of distance territories, (e.g., THEISSEN, ALLOCATE, EUCLIDEANALLOCATION).
Triangulated Irregular Network (TIN) production: TIN modelling with geologic fault specification.

Traverses: define open and closed traverses, trace flow through networks of connected nodes and arcs (e.g., TRACE).
Variogram: variogram calculations (e.g., SEMIVARIOGRAM).

Visibility: identify visual exposure and perform viewshed analysis on a surface (VISIBILITY, VIEWSHD, VISDECODE, SURFACEVIEWSHED, SURFACESIGHTING).
Weighting: weight features by category in the overlay process.

Zonal Functions: calculate sizes and shapes of regions (ZONALAREA, ZONALPERIMETER, ZONALTHICKNESS, ZONALSTATS, ZONALGEOMETRY)

It has been established quite readily that the data entry/capture, data management, and data display/output functions in a GIS can support MG&G. These are probably the most obvious of the support functions, and oftentimes are the only capabilities of a GIS that are recognised. Although not the focus of this chapter, examples of primary GIS data manipulation functions, often the direct precursors to analysis, are listed Table 8.2 for reference.

Table 8.2 GIS data manipulation functions for spatial problems in MG&G (listed alphabetically by likely names of commands, modules, or index terms in the documentation of various GIS packages; categories guided by Nyerges, 1993).

Spatial   Co-ordinate   Manipulation

Adjustments: mathematical adjustment of co-ordinate data using rotation, translation, or scaling.
Compression or decompression: of raster-formatted data with techniques such as run length encoding or quadtree.
Co-ordinate geometry: generate boundaries from survey data; intersect liens, bisect lines.
Co-ordinate recovery: recovery of geographic seafloor co-ordinates from photographic data using single photo resection/intersection together with bathymetric data.
Locational classification: grouping of data values to summarise the location of an object such as a deepsea vehicle trackline of a particular line number.
Locational transformation: transformation of seafloor survey bearing and distance data to geographic co-ordinates using least-squares adjustment of traverse to ground control; e.g., from latitude/longitude to transponder net co-ordinates and vice versa.

Merging: interactive or automatic merging of geometrically adjacent spatial objects to resolve gaps or overlaps with user-specified tolerance.
Resampling: modify grid cell size through resampling.

Smoothing: smooth line data to extract general sinuosity.
Spatial selective retrieval: retrieval of data based on spatial criteria such as rectangular, circular, or polygonal window, point proximity, or feature name.
Structure conversion: conversion of vector to raster, raster to vector, or quadtrees to vector with locational referencing.

Attribute   Manipulation

Arithmetic calculation: calculate an arithmetic value based on any other values (i.e., add, subtract, multiply, divide).

Attribute classification: grouping of attribute data values into classes.

Attribute selective retrieval: retrieval of data based on thematic criteria such as attribute of vehicle trackline or Boolean combination of attributes.
Class generalisation: grouping data categories into the same class based on characteristics of those categories.

Spatial   Co-ordinate   and   Attribute Manipulation

Contouring: generate contours from irregularly or regularly spaced data; constrain contour generation by break lines; e.g., shoreline, faults, ridges.
Gridding: generate grid data from TIN; generate a grid from contours.

Object conversion: point, line, area, grid cell or attribute conversion to point, line, area, grid cell or attribute.

Rescaling: rescaling of raster data values.

8.5 CONCLUSION

In working with the spatial analytical functionality outlined above the GIS may raise as many questions as it answers. However, this should lead not only to the increased spatial awareness of the user but in improvements and extensions of the GIS. Indeed this is a natural component of spatial reasoning: the realisation that improvements in the mechanics of GIS interaction are challenging the user to think more intently about the conceptual process and actual procedures of manipulating and analysing maps. Spatial reasoning is moving beyond mapping and spatial database management to modelling of relationships between and among mapped variables. It is looking at maps as numbers and relationships first, and as pictures second. Spatial reasoning is also facilitated by spatial dialogue which is important for MG&G practitioners who often collaborate with many other colleagues while at sea or on shore. A GIS that is "intelligent" can help achieve this dialogue by producing a variety of outcomes that make users examine the differences and reasons for the differences in various outputs. The intelligent GIS (characterized by a fairly "intelligent" user interface as in Su, 1999) will further facilitate spatial reasoning by guiding users through the spatial operations and models they use, and by encouraging them to try various routes to a similar end.

Goodchild (1999) discusses a range of advantages and disadvantages of GIS as a software tool for global change research that were identified at the Specialist Meeting of NCGIA Research Initiative 15 (Goodchild *et al.*, 1995). There are many, many parallels here to the MG&G community and it may be useful to consider some of the points of consensus that were reached among global change scientists (many of whom are oceanographers). One point was that a larger academic perspective of GIS might be important in order to understand all of the ramifications of its current usage, and to help users to understand and implement the notion of spatial reasoning. There have now emerged now at least three distinct perspective that are identifiable in the current use of the term "GIS" (Wright *et al.*, 1997; Goodchild, 1999):

- GIS as a class of software with a high level of functionality to handle specific types of information for specific tasks (i.e., GIS as "tool").
- GIS as an umbrella term for all aspects of computer handling of geographically-referenced data, including, the advancement of a software package's capabilities and ease of use, data, and the community of tool-makers and researchers (i.e., GIS as "tool-making").
- GIS as GISci, a set of research on a basic set of problems raised by GIS activities (i.e., GIS as "science").

Of the three the second perspective of GIS may be the most constructive with regard to MG&G as GIS is joined by other geographic information technologies such as mapping packages, remote sensing processing packages, in a broader category. As Goodchild (1999) puts it for the global change community: "the use of GIS in no longer an issue... research has no choice but to use computers and digital data; and the vast majority of the types of data needed... are geographically referenced." This raises a final important point with regard to spatial reasoning: the understanding not only of the basic questions that a GIS (the tool) can answer (Berry, 1995), but also a knowledge of the best combinations of GIS with other data analysis tools. This is not to suggest that GIS will always be the best or the only tool to use. For many scientists there are various statistical and mathematical packages, or home-grown analytical programs, that are much more supportive of complex process modelling than GIS. However, they may not have robust mapping and data integration capabilities. Again, it is the "science" that should supersede, but be supported by, the "system."

8.6 ACKNOWLEDGEMENTS

This paper benefited greatly from numerous discussions with and input from Douglas Toomey, an MG&G practitioner at the University of Oregon, and computer scientists Janice Cuny (University of Oregon), and Judy Cushing (Evergreen State College). The support of the National Science Foundation (grant OCE-9521039) is greatly acknowledged.

8.7 REFERENCES

Berry, J.K., 1995, *Spatial Reasoning for Effective GIS*, (Fort Collins, Colorado: GIS World, Inc.).

Bobbitt, A.M., Dziak, R.P., Stafford, K.M. and Fox, C.G., 1997, GIS Analysis of oceanographic remotely-sensed and field observation data. *Marine Geodesy*, **20**, pp. 153-161.

Buehler, K., 1994, OGIS paves the way to GIS interoperability. *GIS World*, **7**, pp. 40.

Burrough, P.A., 1992, Are GIS data structures too simple minded? *Computers and Geosciences*, **18**, pp. 395-400.

Burrough, P.A., van Deursen, W. and Heuvelink, G., 1988, Linking spatial process models and GIS: A marriage of convenience or a blossoming partnership? In *Proceedings, GIS/LIS '88*, Falls Church, Virginia (Falls Church, Virginia: ASPRS/ACSM), pp. 598-607.

Davis, E.E., Chapman, D.S. and Forster, C.B., 1996, Observations concerning the vigor of hydrothermal circulation in young oceanic crust. *J. Geophys. Res.*, **101**, pp. 2927-2942.

Densham, P.J., 1991, Spatial decision support systems. In *Geographical Information Systems: Principles and Applications*, Vol. 1, edited by Maguire, D.J., Goodchild, M.F. and Rhind, D.W. (New York: John Wiley & Sons), pp. 403-412.

Dunn, R.A. and Toomey, D.R., 1997, Seismological evidence for three-dimensional melt migration beneath the East Pacific Rise. *Nature*, **388**, pp. 259-262.

Fedra, K., 1993, GIS and environmental modeling. In *Environmental Modeling with GIS*, edited by Goodchild, M.F., Parks, B.O. and Steyaert, L.T. (New York: Oxford University Press), pp. 35-50.

Fox, C.G. and Bobbitt, A.M., 1999, NOAA Vents Program marine GIS: Integration, analysis and distribution of multidisciplinary oceanographic data, in this volume, Chapter 12.

Gold, C.M., 1999, An algorithmic approach to marine GIS, in this volume, Chapter 4.

Goldfinger, C., 1999, Active tectonics: Data acquisition and analysis with marine GIS, in this volume, Chapter 18.

Goodchild, M.F., 1992, Geographical data modeling. *Computers and Geosciences*, **18**, pp. 401-408.

Goodchild, M.F., 1999, Multiple roles for GIS in global change research. In *Geographic Information Research: Trans-Atlantic Perspectives*, edited by Craglia, M. and Onsrud, H. (London: Taylor & Francis), pp. 277-295.

Goodchild, M.F., Estes, J.E., Beard, K.M., Foresman, T. and Robinson, J., 1995, *Research Initiative 15: Multiple Roles for GIS in U.S. Global Change Research: Report of the first Specialist Meeting, Santa Barbara, California*, Technical Report 95-10 (Santa Barbara, California: National Center for Geographic Information and Analysis).

Hansen, W., Goldsmith, V., Clarke, K. and Bokuniewicz, H., 1991, Development of a hierarchical, variable scale marine geographic information system to monitor water quality in the New York Bight. In *GIS/LIS '91 Proceedings, Atlanta, Georgia* (Atlanta, Georgia: ACSM-ASPRS-URISA-AM/FM), pp. 730-739.

Hatcher, G. and Maher, N., 1999, Real-time GIS for marine applications, in this volume, Chapter 10.

Haymon, R.M., Fornari, D.J., Edwards, M.H., Carbotte, S., Wright, D. and Macdonald, K.C., 1991, Hydrothermal vent distribution along the East Pacific Rise Crest (9°09'-54'N) and its relationship to magmatic and tectonic processes on fast-spreading mid-ocean ridges. *Earth and Planetary Science Letters*, **104**, pp. 513-534.

Haymon, R.M., Macdonald, K.C., Baron, S., Crowder, L., Hobson, J., Sharfstein, P., White, S., Bezy, B., Birk, E., Terra, F., Scheirer, D., Wright, D.J., Magde, L., Van Dover, C., Sudarikov, S. and Levai, G., 1997, Distribution of fine-scale hydrothermal, volcanic, and tectonic features along the EPR crest, 17°15'-18°30'S: Results of near-bottom acoustic and optical surveys. *Eos, Transactions of the American Geophysical Union*, **78**, pp. F705.

Knapp, L., 1991, Volumetric GIS: generic or domain specific. In *GIS/LIS '91 Proceedings, Atlanta, Georgia* (Atlanta, Georgia: ACSM-ASPRS-URISA-AM/FM), pp. 776-785.

Li, R., 1999, Data models for marine and coastal geographic information systems, in this volume, Chapter 3.

Longley, P.A., Goodchild, M.F., Maguire, D.J. and Rhind, D.W. (editors), 1999, *Geographical Information Systems. Volume 1: Principles and Technical Issues*, Vols. 1 and 2 (New York: John Wiley & Sons).

May, L.N., Leming, T.D. and Baumgartner, M.F., 1996, Remote sensing and geographic information system support for the Gulf Cetacean (GULFCET) project: A description of a potentially useful GIS system for ichthyoplankton studies in the Gulf of Mexico. *Collective Volume of Scientific Papers, International Commission for the Conservation of Atlantic Tunas*, **XLVI**, pp. SCRS/96/55.

Meaden, G.J. and Do Chi, T., 1996, *Geographical Information Systems: Applications to Marine Fisheries*, Fisheries Technical Paper No.356 (Rome: Food and Agriculture Organisation of the United Nations).

Moses, E. and Finn, J.T., 1997, Using geographic information systems to predict North Atlantic right whale (*Eubalaena glacialis*) habitat. *Journal of Northwest Atlantic Fisheries Science*, **22**, pp. 37-46.

Nyerges, T.L., 1993, Understanding the scope of GIS: Its relationship to environmental modeling. In *Environmental Modeling with GIS*, edited by

Goodchild, M.F., Parks, B.O. and Steyaert, T. (New York: Oxford University Press), pp. 75-93.

Peuquet, D.J., 1988, Representations of geographic space: Towards a conceptual synthesis. *Annals of the Association of American Geographers*, **78**, pp. 375-394.

Sinton, J.M. and Detrick, R.S., 1992, Mid-ocean ridge magma chambers. *Journal of Geophysical Research*, **97**, pp. 197-216.

Su, Y., 1999, A user-friendly marine GIS for multi-dimensional visualisation, in this volume, Chapter 16.

Toomey, D.R., Solomon, S.C. and Purdy, G.M., 1994, Tomographic imaging of the shallow crustal structure of the East Pacific Rise at 9°30'N. *Journal of Geophysical Research*, **99**, pp. 24,135-24,158.

Varma, H., 1999, Applying spatio/temporal concepts to correlative data analysis, in this volume, Chapter 6.

Wessel, P. and Smith, W.H.F., 1991, Free software helps map and display data. *Eos, Transactions, American Geophysical Union*, **72**, pp. 441.

Wessel, P. and Smith, W.H.F., 1995, New version of the Generic Mapping Tools released. *Eos, Transactions, American Geophysical Union*, **76**, pp. 329.

Wilcock, W.S.D., 1998, Cellular convection models of mid-ocean ridge hydrothermal circulation and the temperatures of black smoker fluids. *Journal of Geophysical Research*, **103**, pp. 2585-2596.

Wright, D.J., 1996, Rumblings on the ocean floor: GIS supports deep-sea research. *Geo Info Systems*, **6**, pp. 22-29.

Wright, D.J., Haymon, R.M. and Fornari, D.J., 1995, Crustal fissuring and its relationship to magmatic and hydrothermal processes on the East Pacific Rise crest (9°12'-54'N). *Journal of Geophysical Research*, **100**, pp. 6097-6120.

Wright, D.J., Goodchild, M.F. and Proctor, J.D., 1997, Demystifying the persistent ambiguity of GIS as "tool" versus "science". *Annals of the Association of American Geographers*, **87**, pp. 346-362.

CHAPTER NINE

2.5- and 3-D GIS for Coastal Geomorphology

Jonathan Raper

9.1 INTRODUCTION

Since geomorphological processes in the coastal zone are highly sensitive to absolute elevation and surface form, most geomorphological mapping and modelling requires a 2.5-D (surface) or 3-D (solid) GIS approach. The use of a 2.5-D approach allows the surface representation of seabed, estuarine, nearshore and onshore topographies while a 3-D approach allows the solid reconstruction in solid form of subsurface and subsea phenomena. The use of these techniques is now becoming increasingly common in coastal monitoring, management and modelling. However specific coastal representation issues need to be solved in each case before practical applications can be built.

This chapter will look at these representational issues and review the previous work in this area before focusing on the techniques and applications of 2.5-D and 3-D GIS, which have been used in coastal geomorphological studies.

9.2 REPRESENTATION IN COASTAL GEOMORPHOLOGIC GIS

The key representational issues to be addressed when implementing GIS in coastal geomorphology are: the definition of the many poorly bounded phenomena found in the coastal zone; the handling of highly dynamic change to those phenomena; and the problems of establishing a suitable datum for any phenomena crossing the intertidal domain. These issues make the data-modelling phase of coastal GIS an especially important one for coastal geomorphological applications.

To tackle the problem of defining the poorly bounded phenomena found in the coastal zone, all coastal GIS need to specify a formal or informal ontology of the phenomena being studied. Many coastal GIS use an informal ontology: vector boundaries or raster cell values are defined empirically among small user groups who all understand precisely what phenomena (salt marshes, channels, beaches etc.) are being represented. These informal schemes are sometimes published as a means of providing widespread circulation (Welch *et al.*, 1992). Such ontologies are no less powerful for being informal, but cannot easily (safely) be transferred outside the domain of the creators. Formal ontologies are those which carry either an explicit metadata description or which identify a source for the phenomena definitions which have widely understood definitions. The most widely known form of a formal ontology in the coastal zone would be the phenomena marked on nautical charts. As yet there is no internationally standardised "meta-ontology" to which investigators can turn for a framework for coastal phenomena definition.

When using GIS to carry out surface and solid modelling key elements of such an ontology relate to the tideline definitions of the intertidal zone and these are often defined nationally or locally.

The reality of dynamic change to coastal phenomena also poses challenges to the creation of a coastal GIS, especially where an informal ontology of phenomena is in use. If phenomena are identified and defined in the GIS in either raster or vector form then the process needs to be time-stamped and documented so that later users can "map" their own ontology onto the earlier one. Investigators currently establishing monitoring systems for coastal change in GIS have a responsibility to future users to provide this metadata so that current work can have an enduring value. Where a GIS of coastal change is being created and based on a formal ontology it is essential to check the horizontal and vertical datums being used on charts or plans through time. Periodically, new definitions are adopted for phenomena that should always be checked carefully.

In surface and solid modelling the definitions of tidelines and elevation datums have a crucial importance for the creation of models which are to be investigated for geomorphological change. Terrestrial datums are usually based on mean high tide while marine charting is usually based on the conservative water depths associated with the lowest astronomic tide (LAT). In order to build any intertidal surface and solid models it is necessary to convert all elevation data to either the terrestrial or marine datum by using a correction factor equal to maximum tidal range. In regional studies where tidal range varies spatially then the correction factor must also vary spatially and be interpolated between known points. It is also possible to map the position of the LAT line over an intertidal surface model using a terrestrial datum (Pater, 1997). In this case the LAT line will not be at a constant elevation regionally.

9.3 SURFACE AND SOLID MODELLING IN COASTAL GIS

There is a growing literature on the use of surface and solid modelling in coastal geomorphology that demonstrates the usefulness of GIS techniques. This literature is currently drawn from a very diverse set of research communities using specific combinations of software tools and data and few attempts have been made to draw them together. The following sections review some of the key developments relevant to coastal surface and solid modelling with GIS.

9.3.1 Incorporation of Coastal Data into GIS

There are special problems associated with the capture of data referenced to x, y and z dimensions in coastal geomorphology which are not encountered in terrestrial applications surveyed by Cornelius *et al.* (1994). Welch *et al.* (1992) provided an early case study of the difficulties of data capture and integration for Sapelo Island, GA, USA. Recently some integrated coastal GIS systems oriented towards coastal data capture have emerged such as MIKE INFO Coast described by Andersen (1997).

The first difficulty for data capture is usually the extremely poor availability of control for georeferencing. Benchmarks are often either absent or have been destroyed by coastal change. When looking for control to georeference imagery there is often a one-sided constraint in that the rich control of an inhabited coastal strip contrasts with a zone of no control along dunes or beaches to the seaward of the inhabited zone. This constraint means that there is often a very poor distribution of control for ortho-rectification of imagery or the aerotriangulation of aerial photography.

A second key difficulty for surface and solid modelling is the source of the vertical control. Elevation benchmarks on the land are usually expressed relative to a national datum: trans-boundary coastal projects (e.g. those in the Dutch and German Wadden Seas) can have problems with reconciling different models of mean sea level (Pugh, 1987). Where the Global Positioning System (GPS) is used to carry out surveying it should be noted that mean sea level and the height above the geoid as determined by GPS receivers can be quite significantly different and corrections are required (Leick, 1995). In some locations GPS height appears to show below sea level elevations as a consequence of the negative differences between mean sea level and the geoid.

A final data capture problem for surface and solid modelling data capture is the time sensitivity of imaging in the intertidal zone. Available images from satellite remote sensing are rarely if ever coincident with high or low tides requiring the user to model the precise height of the tide at the time of the image with correction from tide gauges in order to be able to integrate such imagery with other data. Mason *et al.* (1998) used this technique to create a surface model for The Wash embayment in east England by using 13 ERS-1 SAR images taken at different stages of the tide to delineate "waterline contours".

9.3.2 Photogrammetry and Videography in Coastal GIS

In order to obtain imagery of the coastal zone at high resolution and a specific state of the tide, many coastal researchers have commissioned airborne remote sensing surveys using photography, videography and spectral imaging techniques. These images have been used to create surface and solid models of the coastal zone.

Livingstone *et al.* (in press) shows how digital photography and videography can be draped over ground surveyed elevation data or stereo-correlated to produce surface models. The advantage of these new aerial imaging formats is that they produce data in machine-readable format. Gould and Arnone (1998) showed how Compact Airborne Spectral Imager (CASI) data could be post-processed with a single cross-shore ground-truth profile to produce sea depth estimates in the nearshore zone. This method provided information on the water column temperature structure over surf zone landforms.

Newsham *et al.* (1997) used high-resolution aerial photography to create photogrammetric solid models of eroding cliffs on the Holderness coast of England which were integrated with nearshore bathymetric data. By repeating the exercise it was possible to difference the two solid models to determine the loss of sediment from the cliffs over the time interval between the photography.

9.3.3 Analysis of Coastal Change Records with GIS

One of the most common uses of surface and solid modelling in coastal geomorphology is for the assessment of temporal change in coastal landform morphologies. Approaches using surface and solid modelling in GIS have mostly been based on beach profiles for extensive monitoring and elevation data for intensive studies. Many coastlines experiencing rapid geomorphological change are now being monitored constantly by national agencies. In The Netherlands such programmes have been carried out for over 100 years generating a rich data set of coastal change for exploration by geomorphologists using GIS (Wijnberg and Terwindt, 1993).

Extensive monitoring programmes based on coastal cross section surveying at regular intervals have been analysed by many researchers. For example, Allen *et al.* (1995) reconstructed sediment volume change in Sandy Hook spit in NJ, USA while Humphries and Ligdas (1997) reconstructed beach morphodynamics through time in NE England. Most researchers have used triangulated irregular network (TIN) techniques to connect points along the cross sections to each other in to make a surface model (e.g., Brown *et al.*, 1997).

Intensive modelling of landform change usually involves the creation of surface models from bathymetric or photogrammetric sources. Some historical studies have created surface models from old charts in order to estimate the volumes of sediment flux and the changes in coastal configuration (e.g. Sims *et al.* 1995). Contemporary intensive monitoring programmes create surface models from intertidal or bathymetric survey in order to assess change in landforms like spits (e.g., Riddell and Fuller, 1995; Zujar *et al.*, 1997; Raper *et al.*, 1999) or to study changes in seafloor configurations in estuaries and deltas (Wang *et al.*, 1995).

9.3.4 Examining Large Scale Coastal Behaviour with GIS

Surface and solid modelling techniques in GIS have also been used in studying large-scale coastal behaviour (LCSB) in the last few years. Early attempts employed surface modelling in a descriptive fashion (Jiminez *et al.*, 1993) while recent work has been more analytical.

Pethick (1998) has used surface models of estuarine morphology in attempts to predict the likely geometry of future estuaries under sea level rise scenarios. Hardisty *et al.* (1998) used a surface model of the Humber estuary in conjunction with real-time monitoring of discharge and sediment concentrations in order to integrate a sediment budget for the estuary.

9.3.5 Coastal Geomorphological Modelling and GIS

Surface models are primary forms of input and output for coastal geomorphology process models. As such GIS have been widely used in association with process models as supporting applications.

Process modelling has been used in geomorphology primarily at site rather than regional scale and specifically in the nearshore zone. Detailed accounts of nearshore modelling using the SEDSIM and WAVE programs can be found in Martinez and Harbaugh (1993). Liebig (1996) assesses the alternative ways that coastal process models can be linked with a GIS.

9.4 TECHNIQUES FOR SURFACE AND SOLID MODELLING

From the review above it can be seen that the techniques of surface and solid modelling play a central role in many forms of coastal geomorphological monitoring, modelling and analysis. As such the techniques to produce these surface and solid models take on greater significance since their characteristics can influence coastal geomorphological data and hypotheses. In this section some of the key issues for the use of these techniques in the coastal environment will be discussed. Lane *et al.* (1998) contains an extended discussion of the generic issues for surface modelling in geomorphology generally.

9.4.1 Surface Modelling

Surface modelling for coastal geomorphology can be produced from many forms of input ranging from contour lines on charts through points collected by intertidal ground survey to bathymetric sonar traces. In each case the data input must be digitised and imported into a GIS in point (e.g., elevation value), line (e.g., cross section) or polygon (e.g., contour) form. At this point the decision must be taken to "structure" or to "interpolate" the original data into a surface model. This key step determines the characteristics of the surface model. Whatever method chosen, this step should never be taken by default ("the tools were available") but be subject to an explicit search for the right tools for the approach needed in the application.

Structuring elevation points, cross sections or contour polygons involves linking the data elements together to form a coherent model of local and model-wide surface form without adding any additional information. The archetypal method is the use of a TIN to connect the elevation points or the constituent points of lines/polygons together using a Delauney triangulation. The value of this approach for coastal geomorphology is that it permits the creation of very conservative surface models based upon linear connections between known data points. When the data points have been obtained by purposive sampling at high density this approach allows excellent surface models to be created (Raper *et al.* 1999). The disadvantages are that the linear connection may result in planar facet representations without natural geomorphological curvature and too many flat triangles may be formed either side of crucial elevation values such as LAT or mean high tide. To alleviate these problems the triangular facets can be smoothed using splining without loss of the exact fit to the data points.

Interpolation of elevation points or the constituent points of lines/polygons involves the use of a model to predict values at unknown locations. Many surface models are the raster grids produced by an interpolation algorithm. While interpolation may be the only way to create a surface model if a raster is required (for example as an input to a process model), the many parameters to an interpolation make the output highly sensitive to automated or semi-automated decisions. Key issues for interpolation of coastal data points include the identification of any anisotropy in the surface model among the coastal landforms and the adaptation of the interpolation through the use of breaklines where necessary (e.g., at the high tide line or along channel floors). Also most coastal geomorphological models have open boundaries to land and sea offshore and alongshore. Such boundaries may need to be constrained to maximum depths or to be linked to processes e.g. "depth of closure" for the operation of the Bruun Rule. Such boundaries should also be located where process "divides" occur such as at inlets alongshore or beyond the breaker zone offshore. In many studies these boundaries need to be placed so as to create a geomorphological spatial unit in the coastal zone analogous to a drainage basin on land.

9.4.2 Solid Modelling

Solid models of coastal geomorphological landforms and sedimentary structures can now be made using techniques and tools reviewed in Raper (1989, 1999). Solid modelling poses few problems specific to coastal geomorphology except the handling of z co-ordinates which are both positive and negative and the need to ensure that variation of any property through the solid is oriented as appropriate to offshore or alongshore gradients. These techniques are as yet under exploited since there are few geomorphological data sets describing the internal structure of landforms in the coastal zone.

9.4.1 Visualisation Techniques

Surface and solid modelling is usually best understood through powerful visualisations allowing interactive exploration. GIS currently provide good visualisation tools with interactive "fly-through" capabilities and the means to drape any appropriate "texture" from imagery or analysis over the surface. One powerful form of visualisation is the draping of a net elevation change texture over a current surface model. This allows the viewer to see where change has occurred and how the surface has evolved to its current form.

Such visualisations can be made even more effective using the virtual reality modelling language (VRML) format that can be viewed by most World Wide Web browsers. VRML browsers allow true interactive exploration of surfaces and solids either locally or across the Internet.

9.5 CONCLUSION

Surface and solid modelling offer considerable potential to the coastal geomorphologist if the techniques are used knowledgeably. As such techniques are being progressively built-in to GIS surface and solid models will become widely used. The key question that is likely to arise is how to provide metadata outlining how such models have been produced so that the researcher can assess the model's potential and limitations. Otherwise surface and solid models may proliferate without the means to evaluate their quality.

9.6 REFERENCES

Allen, J.R., Shaw, B. and Lange, A.F., 1995, Modelling shoreline dynamics at Sandy Hook, New Jersey, USA using GIS/GPS. In *Proceedings of Coast GIS '95* (Cork, Ireland: University College, Cork), pp. 193-206.

Andersen, R., 1997, MIKE INFO Coast: A GIS-based tool for coastal zone management. In *Proceedings of Coast GIS '97* (Aberdeen: University of Aberdeen).

Brown, N.J., Cox, R., Pakeman, R. Thomson, A.G., Wadsworth, R.A. and Yates, M., 1997, Initial attempts to assess the importance of the distribution of salt marsh communities on the sediment budget of the N. Norfolk coast. In *Proceedings of Coast GIS '97* (Aberdeen: University of Aberdeen).

Cornelius, S. C., Sear, D. A., Carver, S. J., and Heywood, D. I., 1994, GPS, GIS and geomorphological fieldwork. *Earth Surface Processes and Landforms*, **19**, pp. 777-787.

De Vriend, H.J., 1991, G6 coastal morphodynamics. In *Coastal Sediments '91*, edited by Kraus, N.C., Gingerich, K.J. and Kriebel, D.L. (American Society of Civil Engineers), pp. 356-370.

Gould, R.W. and Arnone, R.A., 1998, Three dimensional modelling of inherent optical properties in a coastal environment: coupling ocean colour imagery and in situ measurements. *International Journal of Remote Sensing*, **19**, pp. 2141-2159.

Hardisty, J., Middleton, R., Whyatt, D. and Rouse, H., 1998, Geomorphological and hydrodynamic results from DTMs of the Humber Estuary. In *Landform Monitoring, Modelling and Analysis*, edited by Lane, S., Richards, K. and Chandler, J. (Chichester: John Wiley & Sons), pp. 421-434.

Humphries, L.P. and Ligdas, C.N., 1997, A GIS application for the study of beach morphodynamics. In *Proceedings of Coast GIS '97* (Aberdeen: University of Aberdeen).

Jiminez, J.A., Valdemoro, H.I., Sanchez-Arcilla, A. and Stive, J.F., 1993, Erosion and accretion of the Ebro delta coast: a large scale reshaping process. In *Abstract Volume, Large Scale Coastal Behavior '93*, U.S. Geological Survey Open File Report 93-381, edited by List, J. (St. Petersburg, Florida: U.S. Geological Survey), pp. 88-91.

Lane, S., Richards, K. and Chandler, J., 1998, *Landform Monitoring, Modelling and Analysis* (Chichester: JohnWiley & Sons).

Leick, A., 1995, *GPS Satellite Surveying* (New York: John Wiley & Sons).

Liebig, W., 1996, Mathematical models and GIS. In *Proceedings of the Joint European Conference on Geographical Information* (Barcelona, Spain: European GIS Foundation), pp. 527-536.

Livingstone, D.L, Raper, J.F. and McCarthy, T., in press, Integrating aerial videography and digital photography with terrain modelling: an application for coastal geomorphology. *Geomorphology* .

Martinez, P.A. and Harbaugh, J.W., 1993, *Simulating Nearshore Sedimentation* (Oxford: Pergamon).

Mason, D.C., Davenport, I.J., Flather, R.A. and Gurney, C., 1998, A digital elevation model of the intertidal areas of the Wash, England, produced by the waterline method. *International Journal of Remote Sensing*, **19**, pp. 1455-60.

Newsham, R., Balsam, P.S., Tragheim, D.G. and Denniss, A.M., 1997, Determination and prediction of sediment yields from recession of the Holderness coast. In *Proceedings of Coast GIS '97* (Aberdeen: University of Aberdeen).

Pater, C.I.S., 1997, Collating the past for assessing the future: analysis of the subtidal and intertidal data records within GIS. In *Proceedings of Coast GIS '97* (Aberdeen: University of Aberdeen).

Pethick, J.S., 1998, Coastal management and sea level rise: a morphological approach. In *Landform Monitoring, Modelling and Analysis*, edited by Lane, S., Richards, K. and Chandler, J. (Chichester: John Wiley & Sons), pp. 405-420.

Pugh, D.T. , 1987, *Tides, Surges and Mean Sea Level* (Chichester: John Wiley & Sons).

Raper, J.F., editor, 1989, Three Dimensional Applications in Geographical Information Systems (London: Taylor & Francis).

Raper, J.F., 1999, *Multidimensional Geographic Information Science* (London: Taylor & Francis).

Raper, J.F., Livingstone, D.L., Bristow, C.S. and Horn, D.P., 1999, Developing coastal process response models for spits. In *Proceedings, Coastal Sediments '99*, edited by Kraus, N.F. and MacDougal, W.G. (Long Island, New York: ASCE).

Riddell, K.J. and Fuller T.W., 1995, The Spey Bay geomorphological study. *Earth Surface Processes and Landforms*, **20**, pp. 671-686.

Sims, P.C., Weaver, R.E. and Redfearn, H.M., 1995, Assessing coastline change: a GIS model for Dawlish Warrant, Devon, UK. In *Proceedings of Coast GIS '95* (Cork, Ireland: University College, Cork), pp. 285-301.

Wang, Z.B., Louters, T. and de Vriend, H.J., 1995, Morphodynamic modelling for a tidal inlet in the Wadden Sea. *Marine Geology*, **126**, pp. 289-300.

Welch, R., Remillard, M. and Alberts, J., 1992, Integration of GPS, remote sensing, and GIS techniques for coastal resource management. *Photogrammetric Engineering and Remote Sensing*, **58**, pp. 1571-1578.

Wijnberg, K.M. and Terwindt, J.H.J., 1993, The analysis of coastal profiles for large-scale coastal behaviour. In *Abstract Volume, Large Scale Coastal Behavior '93*, U.S. Geological Survey Open File Report 93-381, edited by List, J. (St. Petersburg, Florida: U.S. Geological Survey), pp. 224-227.

Zujar, J.O., Parrilla, E., Perez, J.M. and Loder, J., 1997, Tracing the recent evolution of the littoral spit at El Rompido, Helva, Spain using remote sensing and GIS. In *Proceedings of Coast GIS '97* (Aberdeen: University of Aberdeen).

CHAPTER TEN

Real-time GIS for Marine Applications

Gerald A. Hatcher, Jr. and Norman Maher

10.1 INTRODUCTION

This chapter describes GIS technology (software hardware and personnel applied to the management and analysis of spatially referenced data) used in the field during the collection of oceanographic data. This is referred to as "real-time GIS".

The mention of real-time GIS to many may conjure up visions of tracking radio tagged moose across a map of the tundra or routing emergency vehicles as they move through a city street network. To the oceanographer however, it often means the tracking of a research vessel and possibly a submarine, remotely operated vehicle (ROV), drifter, or other instrument as they move through the ocean and collect data.

Although the GIS display is an abstraction of reality it can provide much more information to the oceanographer about their surroundings than a look out of a ship's porthole across a featureless sea. This is especially true when the GIS is used to display data as it is collected along with previously existing data. As described by Hatcher *et al.* (1997) the real-time GIS provides a spatial context for the oceanographer at sea which is otherwise unavailable in the absence of terrestrial boundaries such as roads, mountains, and rivers.

10.1.1 Real-time GIS: Concept One

The first method of using a GIS in real-time occurs when data are immediately entered into the GIS as soon as they are received from a sensor. For example, the navigation data stream is often collected from the research ship directly to a GIS. This allows for the creation of a representation of the ship that moves dynamically on the GIS display in correct geographical relation to other data. As mentioned by Langran (1992) a rewarding pay-off of adding this temporal aspect to a GIS is that it allows the conditions and experiences of the observer to be placed into the context of an animated map of their study area.

10.1.2 Real-time GIS: Concept Two

"Real-time" GIS also involves a situation where new data, possibly with preliminary interpretations from an earlier part of a cruise, are added to a GIS database at sea, which is then is used to guide and update the remainder of the cruise. The GIS can be a handy means of monitoring cruise progress and deciding among the logistical trade-offs that inevitably are made. The science of oceanography is uniquely suited to this type of real-time GIS because an entire GIS lab complete with technicians, PCs, workstations, plotters, scanners etc. can be taken to the field during the oceanographic expedition.

10.2 CRUISE PREPARATION

Even without base maps the GIS is useful for visualising the cruise progress and data collection at sea. However, additional information such as bathymetry or a recent satellite image will often provide a valuable "big picture" frame of reference that can be highly informative. Therefore, in preparation for getting underway, if previously collected data exists for an area of intended research, effort should be made to import as much of it into the GIS as time and space allow. These data layers then act as the "base-maps" which place the cruise plan and events into a geographic context. It has been found on Monterey Bay Aquarium Research Institute (MBARI) cruises that certain results or events may create an unexpected need for data not originally considered. Therefore, including data thought only mildly interesting or loosely related to the mission may also be desirable.

10.2.1 Data Compilation and Base-map Preparation

One of the real strengths of the GIS for cruise planning is its ability to seamlessly combine a multitude of data in a way that provides a great deal of usable information, and within a single software package. Because of this, the GIS can save a great deal of time and energy by enabling many aspects of cruise planning to be considered simultaneously. For example, combining scientific data with information about shipping lanes and navigational hazards will help avoid cruise planning mistakes such as locating sampling stations in a field of lobster traps or in a high traffic shipping lane. Viewshed analysis could be used to determine areas where radio or cellular communication with shore support may be unavailable due to terrain obstructions. With bathymetric data, it can be used to estimate the optimal placement of an acoustic transponder network for subsurface navigation with a submersible.

The capability of a GIS to create high quality hardcopy maps should not be overlooked either. In some situations such as inside a manned submersible where there is no room for a laptop or where the out-gassing of batteries is a safety concern, the printed map may be the only option. Alternatively, producing a set of printed maps customised for the mission at hand is cheap insurance against the unthinkable hard disk or other complete computer failure. Contrary to what the computer-oriented GIS enthusiast might believe, the usefulness of a printed map is not diminished by this technology. It may in fact be enhanced. It is now possible to easily create high quality map sheets with specialised data of critical interest, which are far more convenient than a computer screen for a group of scientists to gather around during discussions. This is true both during mission planning and while underway.

Humphreys (1989) mentions several critical issues related to accuracy when compiling data sets for use as map overlays in a GIS. Specifically it is important to ensure spatial agreement among the individual data layers by converting them all to a common geographic datum and map projection. The datum and projection should be chosen to minimise the effort required to integrate newly collected data at sea. The transformation of existing data may require considerable time and expense. Without this initial investment however, merging data sets in the GIS will be less accurate or impossible, severely limiting some of the most powerful capabilities of the GIS.

10.2.2 Point-and-Click Logistics

Once existing data have been compiled into the GIS, it can be used to plan cruise logistics. Track lines, dive sites, fuelling stops, contingency plans etc., can be efficiently prioritised and organised using an iterative approach with this technology. After the plan has been roughed out, the GIS is used to fill in the details and insure its feasibility. This is analogous to a mechanical engineer using a computer aided drafting (CAD) program to create a workable design from a conceptual sketch. Time estimates, co-ordinate measurements, and distance calculations can be done accurately and quickly with a few mouse clicks, and the results "pasted" into a word processing program or emailed to participants. The alternative of a paper chart, dividers, parallel rulers, notebook and eraser begins to look very labour intensive, error prone, and time consuming once one has a GIS that is tailored to the user, along with accompanying digital data.

10.3 REAL-TIME CUSTOMISATION

The GIS at sea needs to be set up so that it is very robust and easy to use. Most serious computer users have learned that highly capable software is unavoidably quite complex with a deep user interface. The software vendor may boast of its "user-friendly software" but that user-friendliness often ends as soon as one tries to perform a function that is not described in the "getting started" tutorial. The challenge for a sea-going GIS specialist is to anticipate routinely performed functions from a convoluted series of operations and then to streamline them into a simple mouse click or two through customisation. Such effort will be quickly rewarded once the cruise is under way.

10.3.1 Software Development

Much of the functionality typically required of the GIS by the oceanographer is already included in the typical "off the shelf" GIS, however there will most likely be some functions that are missing or in need of streamlining. When creating modifications or enhancements sound software engineering principles that improve the maintainability and reusability of the development should be followed as much as possible.

As no two research vessels are alike, a good modular design that can be easily maintained (even remotely via email) is important. Being able to quickly modify custom software to dovetail into an existing seagoing set-up, with only minor modifications, is critical. Most shipboard computing personnel are (justifiably) reluctant to change well working critical systems, so modifications should have as little impact as possible on the pre-existing shipboard configuration.

The returns on GIS customisation result both from making the software easier to use and through improvements to the quality of the data collected. The quality improvements are gained from simplified data entry into the GIS allowing new data to be used before the cruise is completed. Enabling the scientist to use their new data while they are still at sea gives it a preliminary quality check while there is still time to fix problems. In addition, data processing at sea reduces the difficulty associated with trying to sort out the order of cruise events and data collection after the cruise is complete.

With this in mind, it is apparent that when choosing GIS software an important factor to consider is its extensibility (Woodcock, 1990). This is especially true when considering real-time GIS for oceanography. If using a public domain software package that includes the source code as part of the distribution, unlimited extensibility is present by default (usually, however, at the expense of documentation and support). If deciding among commercial GIS packages, two important factors to consider are its capacity to incorporate existing code, and the malleability of its user interface. The purchase of a GIS package that can incorporate existing "field tested" software routines and with a customisable user interface will save time and aggravation.

When developing software for use on an oceanographic cruise there are several factors to consider that are easily overlooked. One is that the software may be expected to run continuously for days or even weeks. Therefore, during development, a critical test to perform is to let the software run overnight or, better yet, for a day or more to ensure annoying memory leaks or other problems will not occur. Another is the "bullet proof" simplicity of the graphical user interface (GUI) that is required. If a series of obscure mouse clicks can cause the program to crash, then the chances are that someone will find just the right combination at just the wrong time. When making choices for the GUI design, the operating environment should be considered: e.g., low/bright light, sophistication of the users, the number of simultaneous systems the users will be monitoring, etc. If the cruise is to encounter the International Date Line, the Equator, or be in operation near the poles, modifications must be tested for these special circumstances.

10.3.2 Real-time Data

Navigation is probably the most useful real-time data stream to access from within the GIS. Among other things, it provides the basis to attach geographic co-ordinates to many other types of data. Only recently have commercial GIS vendors started to provide the software to make this possible for most users. Until now it has been up to end users to either create the software for real-time tracking on their own or to resort to simpler means, such as updating a marker on a map by hand.

Digital tracking provides features such as instantaneous event marking, precise navigation logging, automatic execution of functions based on location, and more.

Although real-time navigation tracking is now being offered in some commercial GIS software packages (e.g., ArcView Tracking Analyst), often there is still the necessity to convert the data from shipboard to GIS format. When creating this software "glue", the ability to simulate the shipboard data gathering process before the actual cruise, with duplicate hardware or other means, has been found to be invaluable. The simulation is necessary because documented data specifications and the actual data provided are often different, due to misprints, updates or other reasons. This is especially important if the developer will not be on the cruise.

10.3.3 Near Real-time Data

Some data may be difficult or impossible to store in the GIS database at the moment of its collection. The data may be too voluminous, require complex processing by a dedicated system, or, only available after an instrument is recovered from a deployment. In these cases, it is useful to create software that converts the data, once available, into GIS layers quickly and easily. These tools are typically for reformatting, and can be created in advance of the cruise from a data description and a sample data file.

The GIS itself can be used to generate information in addition to that collected by oceanographic instruments. This information is typically in the form of cruise plan updates based on any number of factors such as weather conditions, cruise progress, or scientific findings. For example, a seafloor survey cruise might use a GIS "tool kit" which helps to generate survey track lines and estimates of the time it will take to run them. Or a "toolkit" may be created which can be used to relocate study areas based on earlier cruise findings in order to optimise scientific gains. In these situations, it is useful to be able to create output formatted in a way that can easily be used by other members of the science party and the ship's officers.

10.3.4 Development at Sea

If the intention is to complete a series of cruises where GIS is important to the success of the mission, it is invaluable to send developers to sea on at least one leg (preferably the first). This provides them with first hand knowledge of working conditions at sea and of exactly how their software will be used. It is also inevitable that bugs or problems will occur on the first cruise due to various oversights. Having the actual software developers aboard can save the day.

Placing the primary users and developers together at sea can also create an interaction whereby the people who know what is possible have a productive interaction with those who know what they require for their science. This creates a situation where fruitful compromises and ideas abound, enabling software tools to be quickly conceived, created and evaluated. It will often be found that once a software tool is created it will immediately prompt a new idea previously not conceived. The software evolves in this situation far more rapidly than if the development were done between cruises based on reports brought back from sea. From the MBARI experience, it is estimated that a month of at-sea development is nearly as productive as an entire year on land, due to the synergism mentioned previously along with the fact that the developer is immersed in the problem with little interruption or distraction.

10.4 SHIPBOARD SET UP

10.4.1 Scalable and Networked

A shipboard system configuration should be designed to the specific science requirements. Factors to be considered are the physical space available, the length of the cruise, the sophistication of the GIS operators, how distracted they will be with other tasks, and the number of people needing simultaneous access to the GIS.

The minimal system used by MBARI is a single scientist familiar with GIS on a standalone laptop. The laptop is connected directly to the navigation data that are then fed to the real-time GIS using custom software. The data layers are stored on the laptop's hard drive or CDROM. This configuration works well for short trips on any size ship, is simple, quick to set up, and provides a great deal of functionality.

At the other end of the spectrum (Figure 10.1) is the multiple platform, client/server deployment, with several users accessing real-time data simultaneously via the ship's network. This set-up is usually needed only on longer cruises with several scientists and a multifaceted science mission.

Today's network and computer technology make it possible to design a system to be both modular and scalable. Scalability does not come automatically, however, and should be kept in mind during initial design. The MBARI experience has shown that it is easier to design for the more complex system from the outset, simplifying it as needed for the stand-alone setting, rather than trying to add complexity to a simpler system. With the proper design, the stand-alone system can be inserted or removed from the more complex set-up with only minor configuration changes (Figure 10.1).

Two important issues to consider are the method used to distribute the real-time navigation to multiple clients and the potential of multiple clients to make conflicting changes to the same data file at the same time. To solve these problems for the multiple platform GIS lab, MBARI incorporates a client/server model. The server is used as the "data concentrator" by running a program that processes all the real-time data feeds into a standard format and then provides that information to clients as-needed over the ship's network (Figure 10.1). The server is also used to set access permissions to safeguard against simultaneous file changes.

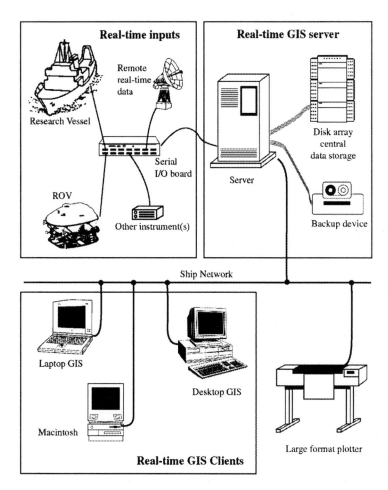

Figure 10.1 A typical real-time GIS lab set-up with multiple clients and a single server. The server acts as a data concentrator, which processes all real-time data into a standard format. The real time data are then distributed as required to multiple clients via the ship's network.

The standalone mode is created from the client/server model in three steps. First, the required data are copied to the laptop's hard drive. Second, the navigation data source is connected directly to the laptop. Third, a local copy of the real-time data distribution program is run on the laptop. The GIS then communicates to the local copy of the distribution program directly.

10.4.2 GIS Specialist

An important component of the ocean going GIS is the "GIS specialist", the person who knows and is responsible for the GIS. The GIS specialist in the case of the standalone laptop set-up can simply be the primary user or scientist. However, with the full GIS lab at sea there is much more of a system administration role to be provided and that responsibility is non-trivial. It will include setting up the GIS software and network configuration, running communication lines for the real-time data feeds and physically securing the systems against possible damage from the pitch and roll of the ship. Once underway, the GIS specialist will be responsible for monitoring systems, performing data entry, and solving the inevitable problems that arise. Finally, at the completion of the cruise, the specialist needs to make data backups, print maps, and dismantle the GIS before it is reassembled back on land.

10.5 AT-SEA OPERATIONS

10.5.1 Scenario One: Day Trips

When the cruise is of short duration, the main functions provided by the GIS are mission status display and digital recording of cruise events. The cruise plan entered into the system prior to leaving the dock consists of data layers such as base maps, track-lines, sampling sites, or other "targets". Once the cruise is underway and real-time navigation tracking is functioning, the GIS display is used to monitor the progress of the cruise to ensure that the plan is followed as intended.

The navigation track of the cruise, as well as events such as instrument deployments or sample collections, may be stored for later reference and cruise reports. This information is stored digitally and accurately to create a digital logbook of the cruise.

"Off-the-shelf" features typically available with a commercial GIS such as on-screen measurement, point and click interrogation of data, and feature symbolisation based on attributes will usually be used more than complex analysis tools. In this mode, development of specialised GIS tools while at sea will not be practical because of the limited time available.

10.5.2 Scenario Two: Longer Cruises

On a longer duration cruise the GIS is used in several modes. It is used on a daily basis as described in the previous section, and with more time available the analysis capabilities can play a larger role.

The cruise progress and data collected are used "off line" to modify or update the working cruise plan. The modified plan then becomes the new working plan. In this way the GIS is used in an iterative fashion to maximise the efficiency and results of the mission. The updates may be based on any number of factors such as the analysis of collected data, weather conditions, equipment performance or failures. This cycle of "sailing the plan," updating the plan, and then "sailing the update" may be repeated many times during the course of the cruise.

It is also possible that on a longer cruise one will need or want to develop specialised GIS software tools that perform functions not anticipated prior to sailing. Prototypes can be created and tested in the field, and, if required, "cleaned up" at a later time after the cruise is completed. This can be a very fertile environment for software development as previously mentioned.

10.6 RETURNING TO LAND

On day trips, there may not be time for any significant GIS work to be done during the transit back to the harbour. However, on a longer duration cruise transit time can be put to good use organising newly collected data and creating data backups. Additionally, it has been found that transit time is also an excellent opportunity to produce printed maps of cruise navigation and preliminary results. This serves two purposes: (1) it provides scientists with a means of describing the mission to their colleagues while the GIS is being demobilised and reassembled back on land; and (2) printed maps produced at sea become an invaluable reference for future data retrieval, processing, and analysis. The maps form a critical portion of the cruise log. A good map of the entire cruise annotated with significant events can help avoid hours of frustration and confusion spent trying to decipher samples, data, and the order of events after a cruise has ended.

10.7 STRENGTHS AND WEAKNESS

10.7.1 Some Shortcomings and Pitfalls

This chapter has summarised many of the things that a GIS does well for the oceanographer at sea. There are, however, a few shortcomings to consider and some easy traps to fall into.

- One is the realisation that in order for the system to be valuable at sea it needs to be monitored closely, and newly collected data need to be diligently entered. Therefore a member of the science party needs to have ample time to spend on GIS tasks. For the chief scientist this may require a difficult trade off in the allocation of human resources as only a finite number of people may be able to sail on any one cruise.
- The GIS can be a powerful tool for all phases of an oceanographic cruise but with this power comes a significant level of complexity and a non-trivial learning curve. Plan to set aside adequate time to gain a good understanding of

the system and what it can/can't do well before the cruise, or be ready to experience plenty of frustration at sea!

- The GIS may perform wonderfully when the required GIS functions are anticipated and prepared for prior to sailing. However, if an unanticipated operation or analysis is needed at sea it may take some time to determine how it should be performed. This delay can be frustrating both to those asking the questions and to the GIS specialist trying to determine the solution.
- It should be realised that the GIS is a companion tool to the printed map and not a replacement for it. When split second decisions need to be made a quick look at a map and a couple of measurements with dividers can be much faster than waiting for a large data layer to load and a multitude of mouse clicks to be made.

10.7.2 Several Advantages

Shortcomings aside, the GIS at sea can be an excellent tool for the oceanographer in many ways (several of the points below are also supported by Wright, 1996).

- At a minimum, the clear picture of cruise status given by the GIS when loaded with base map layers and real-time navigation is worth the effort of set-up.
- Another powerful feature of GIS is its ability to produce publication quality thematic maps at sea (provided a high quality plotter is available). This is remarkable when compared with the limited capabilities available only several years ago.
- By using the GIS with real-time and/or near-real time data the science benefits both qualitatively by having a first order check on the data while there is still time to fix problems and quantitatively by helping insure that the intended data are collected.
- Additionally, tracking real-time navigation in relation to bathymetric data can help avoid costly subsurface equipment losses or damage due to collision with the bottom when working in areas of steep seafloor topography.
- Cruise logs can be enhanced by the rapid and accurate recording of cruise into a digital notebook, along with the plotting of those events on annotated maps.
- The ability of a GIS to seamlessly combine multiple data types with different resolutions and spatial extents make it an ideal platform for cruise planning for updating and monitoring the plan while at sea, and for reporting cruise results upon completion (e.g., Wright, 1996).

10.8 FUTURE DIRECTIONS

Opportunities to apply real-time GIS to marine science will steadily grow as a result of the rapid development occurring in several areas of related technology such as global communications, data collection systems, and object oriented software. Recently available global cellular telephone systems and the continual drop in the size and cost of electronics enable real time data to be collected economically and simultaneously from anywhere on the planet. These data can then be transmitted to a central location via the Internet where science missions requiring carefully orchestrated data collection and analysis can be managed from a single location with customised GIS technology. The GIS tailored to a particular task will become more common as Rapid Application Development (RAD) software tools provided by technologies such as Microsoft's™ OCX or Sun's™ Java Beans make it possible for individuals to easily create custom applications. This component technology allows only the GIS functionality required to be included in applications reducing their complexity and improving overall usability (Environmental Systems Research Institute, 1996).

10.9 REFERENCES

Environmental Systems Research Institute, 1996, *Building Applications with MapObjects* (Redlands, California: Environmental Systems Research Institute), pp. 7-10.

Hatcher Jr, G., Maher, N.M. and Orange, D.L., 1997, The customization of Arcview as a real-time tool for oceanographic research. In *Proceedings of the 1997 ESRI User Conference, San Diego* (Redlands, California: Environmental Systems Research Institute).

Humphreys, R., 1989, Marine information systems. *The Hydrographic Journal*, **54**, pp. 19-21

Langran, G., 1992, *Time in Geographic Information Systems* (London: Taylor & Francis).

Woodcock, C. *et al.*, 1990, Comments on selecting a geographic information system for environmental management. *Environmental Management*, **14**, pp. 307-315.

Wright, D. J., 1996, Rumblings on the ocean floor: GIS supports deep-sea research. *Geo Info Systems*, **6**, pp. 22-29.

CHAPTER ELEVEN

Electronic Chart Display and Information Systems (ECDIS): State-of-the-Art in Nautical Charting

Robert Ward, Chris Roberts, Ronald Furness

11.1 INTRODUCTION

The paper chart has been a fundamental tool for the navigation of ships for centuries. However, with the advent of satellite position fixing and powerful cheap computers, a potent additional tool is now available to the mariner. Electronic Chart Display and Information Systems (ECDIS) are destined to replace paper chart based navigation in many vessels, providing increased benefits for safety and efficiency. The development of ECDIS is at least as significant to mariners and to ship safety as was the introduction of radio or radar to the bridges of ships earlier this century.

Why is a chapter which overviews ECDIS included in a book such as this? Much is written about the "power" and "potential" of GIS for coastal zone monitoring and management, as well as how GIS offers a powerful tool in the armament of those charged with deciding what to do after some sort of significant negative environmental impact. Many of these contributions and significant examples deal with GIS at the regional or local level. ECDIS demonstrates the introduction of a GIS-based capability at the global level. The relevance of ECDIS thus is at least threefold:

- It is a significant contributor to safety of navigation and thus will prevent many of the groundings, which have contributed to much of the world's ocean pollution.
- It is a demonstration of the need for *global* co-operation at the government and government agency level if such "powers" of GIS are to be realised fully to the benefit of the world.
- It demonstrates the immense capability, which can be unleashed by the proper, and appropriate development of GIS-based systems.

11.2 WHAT IS ECDIS?

11.2.1 Description

ECDIS is an advanced navigation information system for use in ships. It has been developed to lighten considerably the navigation workload, freeing the mariner for other important navigation-related tasks such as maintaining a safe lookout and for collision avoidance. It is a real-time decision aid, which provides the navigator with accurate and reliable information about a ship's position and its intended movements in relation to charted navigational features. A typical ECDIS installation is shown in Figure 11.1.

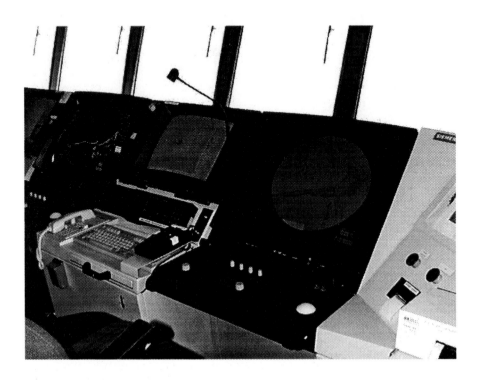

Figure 11.1 A typical ECDIS installation found aboard many commercial and scientific research vessels.

ECDIS combines satellite position fixing, ship's sensors and other data with a sophisticated electronic database containing chart information. The electronic chart database is known as an electronic navigational chart or ENC. An ENC is much more than a computer copy of a paper chart, though it can look very similar when displayed on ECDIS equipment. ENCs are sophisticated and strictly controlled vector navigational chart data bases containing high levels of textual, spatial and graphical data representing not only the material already shown on a paper chart, but also additional data and information drawn from other publications and from source survey data. ENCs are only produced by or on the authority of Government-authorised hydrographic offices or relevant government-authorised organisations.

ENCs are divided up into manageable geographic areas called "cells". Each cell is rectangular and defined by two geographic parallels (latitude) and two meridians (longitude). Individual cell boundaries are chosen by compiling authorities to best meet their requirements. Cells with the same navigational purpose cannot contain overlapping data, however cells of different purpose may overlap. Cells may not be more than 5Mb in size.

The fact that chart information is held in a data base rather than as a fixed image means that it can be analysed, manipulated and compared with other available information to provide a powerful decision making tool on the bridge of a ship. ECDIS continually analyses the ENC database and compares it with a ship's position and its manoeuvring characteristics to give timely warning of approaching dangers or notable events in the navigational plan. For example, ECDIS can display and respond to a nominated safety depth contour based on a ship's draft and give advanced and automatic warning of the approach to potentially dangerous situations.

ECDIS provides the mariner with a continuous, unified presentation of relevant navigation information, allowing him or her to decide what to do rather than spending valuable time determining whether action is required. ECDIS allows the level of chart detail and various day/night colour schemes to be selected (within strictly defined limits) by the operator. This enables the prominent display of the most important chart detail and in the manner best suited to the circumstances. For example, the display can be simplified to show only the ship's safety contour, rather than all bathymetric contours and supporting depth soundings. Meanwhile, ECDIS continues to rely on a full bathymetric data set to monitor the progress and safety of the ship and to trip any warnings or alarms.

Additional information such as scanned images of photographs and navigational notices and warnings are contained in the ENC and can be accessed when and as required through simple point and click procedures (Figure 11.2). This obviates the need to refer to many of the additional reference documents, which are normally used in conjunction with a paper chart. These publications include tide tables, sailing directions and lists of lights and radio signals.

ECDIS provides many other navigation and safety features including continuous voyage data recording and playback. The mariner can also supplement ENC data with additional information such as personal notes and warnings. These can be input manually and used to raise additional alarms and warnings.

The ship's radar signal can also be incorporated into an ECDIS and the radar image or contact data displayed on screen as an overlay. This helps provide a comprehensive and fully integrated appreciation of the navigational situation and brings together the charted information (which is essentially a record of expected conditions) with a record of the current circumstances as seen by radar. There are plans which are well advanced (1999) to incorporate additional information such as

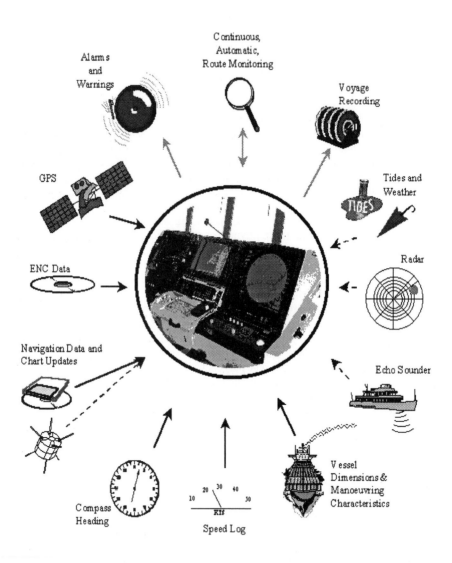

Figure 11.2 A typical ECDIS data flow diagram.

weather, ice reports, vessel position reporting and the application of observed tides. This information will be passed to ships in real-time through telemetry links. ECDIS operates on a relatively simple computer platform (486 or Pentium PC) and is interfaced to an electronic positioning system (usually GPS), as well as conventional ship's radars, speed logs and compasses.

11.3 WHY THE NEED FOR ECDIS?

It is broadly estimated that some 95 per cent of ship groundings are caused by navigational errors. In turn, ship groundings are the cause of roughly 33 per cent of major pollution incidents worldwide. Navigation errors include the incorrect plotting of a ship's position, the setting of incorrect courses, the failure to alter course when required, and the failure to fully appreciate the dangers to navigation that exist. Ironically, a major reason for these accidents, along with collisions at sea, is that the paper chart, the indispensable navigation aid, is but a passive representation of known navigation hazards and the aids (such as buoys and lighthouses) which are installed to avoid them. There is an unavoidable time lag between plotting one's position on a paper chart and knowing accurately the ship's position.

The navigation watchkeeping officer currently provides the mechanism for correctly plotting a ship's position on the chart and for integrating that fact with other information relating to the navigation hazards and aids and inputs from sources and sensors on board the vessel, notably the radar. In the 60 seconds (say) it takes a navigator to fix position on a paper chart, a ship moving at 15 knots travels 460 m. By the time the navigator looks up from his chart table to consider what he should do next, the "fixed" position is a quarter-mile astern. This procedure and an aid to assist in the subsequent decision making process are therefore obvious candidates for automation. ECDIS has been developed to provide solutions.

Not only do navigation accidents sometimes cause loss of life, there are very serious financial and environmental consequences in most cases. The grounding of the *Exxon Valdez* in Prince William Sound, Alaska in 1989, resulted in an environmental disaster estimated to have cost over $2 billion to address. The *Exxon Valdez* incident is an example often quoted as probably avoidable had the ship been fitted with ECDIS.

11.3.1 History of ECDIS

By the early 1980's advances in technology were showing the potential to combine electronically information relating to navigation hazards, ship's position and data from a ship's radar and provide a powerful navigation decision making aid and warning systems for the mariner. Development systems were demonstrated to show that the concept was possible.

In 1988 *The North Sea Project* aimed to assess the practical measures and co-operation required to establish an international electronic chart database (ECDB) to the specifications of the International Hydrographic Organisation; to test it at sea on a variety of commercial electronic charts; to evaluate methods of updating electronic charts; and to demonstrate and analyse the potential usefulness of electronic charts. Experience was also gained with early commercial systems being used in Canadian Waters.

In 1986, the International Hydrographic Organisation (IHO) and the International Maritime Organisation (IMO) set out to develop standards and specifications for what was ultimately to become known as ECDIS. The IHO is the organisation which co-ordinates world hydrographic charting policy and standards. The IMO is the UN organisation responsible for co-ordinating safety policy and standards for world shipping.

An ECDIS Performance Standard was drawn up which set out the functional requirements for ECDIS and in 1995 the International Maritime Organisation (IMO) adopted this Performance Standard as Resolution A.817(19) which endorsed ECDIS for use in ships as an alternative to carrying official paper charts.

The remaining part of the decade was taken up developing and refining technical specifications and formats for ECDIS. ECDIS and ENCs first became available commercially late in 1998.

11.4 ECDIS STANDARDS

11.4.1 Description

ECDIS is governed by four major standards. These are managed by the IHO and IMO through a number of technical committees and working groups and the International Electro-technical Commission (IEC). The standards are:

- IMO Performance Standards for ECDIS (IMO Resolution A.817(19))
- IHO S-57: Transfer Standard for Digital Hydrographic Data
- IHO S-52: Specifications for Chart Content and Display Aspects of ECDIS
- IEC 61174: ECDIS Operational and Performance Requirements: Methods of Testing and Required Test Results.

11.4.1.1 The ECDIS Performance Standard

The ECDIS Performance Standard is the over-arching standard, which provides the basic performance framework for ECDIS. It describes what functions and capabilities are required from ECDIS. The other standards listed above then provide the detail on how the requirements of the ECDIS Performance Standard are to be achieved.

11.4.1.2 IHO S-57

IHO S-57 is the common standard for the exchange of digital hydrographic data which enables the transfer of digital hydrographic data between national hydrographic offices, and to manufacturers, mariners and other data users while ensuring that the integrity of the data is preserved. The exchange format was formerly known as DX-90. The standard provides the format and structure for encapsulating the information, which can be found in a navigational chart and associated publications. The exchange language must be English, however other languages may be used as a supplementary option.

S-57 is structured to include product specifications. These describe the governing arrangement and content of individual hydrographic products and list in detail how the various S-57 constructs should be put together to form the particular product. The first product specification to be included in S5-7 is the electronic navigation chart (ENC) Product Specification. This specifies the hydrographic content of an ENC, how the data is to be formatted and how it is to be attributed through the extensive object catalogue provided in S-57. The product specification for ENC calls specifically for a sophisticated vector database.

The ENC specification includes many mandatory data items including quality attributes and source codes. Quality attributes cover such things as the positional and depth accuracy of bathmetric data. The inferred quality of the resultant seafloor model is designated through a "zone of confidence" attribute. An optional attribute called "pictorial representation" is available for all geo objects. This provides a reference to a file embedded in the data and may be a graphics file containing an image of such things as prominent land features or navigational marks such as buoys or lighthouses. All co-ordinates in the database must be referenced to the common geodetic datum WGS84 and all depth and height units must be in metres. Future editions of the standard will, amongst other things, include product specifications for raster and matrix data.

11.4.1.3 IHO S-52

S-52 provides specifications for the chart content and display aspects of ECDIS. In other words, how information in an ENC will be displayed and perform. It includes appendices describing the means and processes for updating, specifies the colours and symbols to be used (including an extensive presentation library), and a glossary of ECDIS related terms. S-52 also includes numerous conditional procedures which trigger warnings and alarms in ECDIS. An example is a submerged wreck with a known depth. The conditional procedure compares the depth of the wreck to the ship's specified safety depth. If the wreck is shallower than the safety depth, then it will be displayed as a dangerous symbol and will trigger warnings at a specified "look ahead" distance to warn that the vessel is approaching a hazard to navigation.

S-52 specifies a range of new symbols, for example, how buoys and navigation lights are shown, as well as allowing the use of traditional symbols. S-52 also specifies how ECDIS will show waypoint and route information on the computer screen. In addition, it covers how radar and other information can be combined and displayed.

11.4.1.4 IEC 61174

IEC 61174 is the International Electro-technical Commission (IEC) technical specification for testing ECDIS equipment for international type approval.

11.4.2 Development of the Standards

S-57/S-52 was originally published in May 1991 and adopted as an official IHO standard in May 1992. As experience was gained in using S-57, numerous proposals for changes, such as new objects and attributes, were addressed to a working group tasked to maintain the standard. In January 1993, a workshop was held with a view to harmonising S-57 and the emerging military Digital Geographic Information Exchange Standard (DIGEST). This resulted in the adoption of an explicit theoretical model for S-57 and Version 2.0 was released in November 1993. Development of the next version continued with the preparation of a product specification for ENCs, additional objects and attributes, restrictions on the size of data sets and the introduction of unique object identifiers to facilitate ENC updating. As well as including the development issues mentioned, a binary implementation of the format was added, a new cell structure concept was introduced, time varying objects such as tidal heights were included in the Object Catalogue and a description of four levels of topology for vector data was added. S-57 Edition 3.0 contained significant extensions from earlier versions and was released in November 1996. Edition 3.0 will remain in force for at least four years in order to provide a stable standard for ENC data production by Hydrographic Offices and the development of ECDIS equipment by manufacturers. Edition 3.1 will enter force in November 2000 to include minor corrections and additional values for some attributes, with the structure and main body of the standard being frozen for at least a further two years.

Development of Edition 4.0 may include additional product standards, extensions to the object catalogue to embrace ice objects, vessel traffic management systems, traffic reporting, weather and real-time tide information.

11.4.3 ENC Compilation Progress

The creation of reliable and comprehensive ENCs is complex and progress has always been slower than desired. This is because ENCs have the potential to capture more information than is currently shown on a paper chart. In addition, some of the detail shown on paper charts has necessarily been generalised or omitted for clarity or for other cartographic purposes. The paper charts may also be based on horizontal and vertical datums, which are different from the ECDIS standard or contain depths in fathoms and feet, which is similarly unacceptable in S-57. In such circumstances, the charts must be effectively recompiled from source surveys as though they were new charts.

Because of this, the use of official raster charts in ECDIS has been agreed by the IMO as an interim solution pending comprehensive world coverage by ENCs. It is referred to as "ECDIS dual fuelling". "Dual-fuelling" uses raster nautical charts (RNCs). An RNC is a digital facsimile of the official paper chart and provides an up-to-date, geographically precise, distortion-free image of the paper chart for use in an electronic chart navigation system. Like ENCs an RNC is issued only by or on the authority of the relevant government authority.

The "dual-fuelling" concept was introduced to promote and accelerate the uptake of ECDIS amongst seafarers. This is because "dual fuelling" will immediately provide mariners with the use and benefits of ECDIS for almost any voyage worldwide. Alternatively mariners would wait some years more before ECDIS could be used in many parts of the world, or they could resort to non-official data of questionable quality and currency. This is undesirable both for safety and financial liability reasons.

11.4.4 Scalability of ENC Data

The utility of chart data for navigational purposes is driven fundamentally by its positional accuracy and its veracity. Paper charts can be divided into a number of usage categories, including: series charts for route planning; series charts for coastal and offshore navigation; port approaches, critical areas which are considered to be dangerous or difficult for navigation; sets of plans and port charts. Each of the chart types is drawn at progressively larger scales and depends upon increasing amounts of detail and increasing accuracy of that detail. The fact that the data in an ENC database can easily be manipulated, analysed, simplified or combined to provide map-like presentations at almost any scale belies the fact that they are no more scale independent than their paper chart forebears. Accordingly, ENCs, like paper charts, are compiled with specific navigational purposes in mind. ENCs can be created for the following usages: overview, general, coastal, approach, harbour and berthing. In some cases, this means that more than one ENC will exist for a particular area, for example there may be a "harbour" ENC containing detailed information and the same harbour could be reflected in a "coastal" ENC, where more generalisation or different information such as coast radio reporting stations are included. There are potential pitfalls in ensuring consistency between the different databases, though none too dissimilar to paper chart production.

11.4.5 Data Fusion

Unlike land mapping, which is usually based on direct observation or measurement (for example, photogrammetry or direct field measurement and observation), chart compilation requires significant interpretation and judgement with regard to what information can be used, how to depict that information, and how much of it is relevant to the mariner. For example, seafloor topography must be inferred from whatever depth measurements are available, shapes of coastlines must consider the effect of tides and represent the view as seen from seaward. Reported dangers must be included in the chart until positively disproved. Furthermore, as a general principle, all the information shown on a chart must err on the side of safety as far as the mariner is concerned.

The information used in compiling a chart is derived from many sources, particularly where there is no authoritative overriding survey, which is usually the case. Data almost invariably comes from surveys and information of varying quality, scale and reliability. This poses its own complexities with regard to encoding meta-objects in particular. For example when the source scales are different, which value should be adopted for the compilation scale meta-object?

Similarly, how should a generalised depth contour, compiled from several overlapping surveys and biased to ensure safe depths are depicted, be treated?

11.4.6 Accuracy of Source Data/GPS

A further complication in the compilation of ENCs (and paper charts for that matter) is the increasing mismatch between the inferior positional accuracy of the source data and the increasingly precise capability of mariners to fix their position using such things as differential GPS (dGPS). Many charts rely on information collected some time ago when methods of fixing were poorer, geodetic datums may have been ill defined and technology limited or non-existent. Some of the world's charts still rely on information collected in the 19th century. The "zones of confidence" attribute was developed from an Australian proposal specifically to help provide some measure of the reliability that can be placed on the quality and accuracy of information presented to the mariner.

11.4.7 Updating

11.4.7.1 Replacement

Replacement usually entails replacing a whole chart file or dividing a chart into a number of cells and updating affected cells by replacement. This is similar to issuing a new edition of a chart or pasting a block correction over part of a paper chart. The previous information is completely replaced and so the updates do not depend directly on the application of earlier updates. However, it is still important to ensure that all updates have been applied otherwise the chart may not be fully up-to-date and properly corrected. Where cells are used for updating, the up-to-date chart is made up of the base or original electronic chart with any new cells overwriting the originals.

11.4.7.2 Incremental Updating

Incremental updates operate on an object by object basis. This is similar to the way traditional Notices to Mariners corrections are carried out on paper charts – by making individual corrections to items. This saves on the amount of data involved in an update because changes are made to individual chart items rather than to blocks of detail. The up-to-date chart is made up of the base or original electronic chart with new information overwriting or replacing the original. In the case of incremental updates, it is very important that all outstanding updates have been applied. Later updates may amend earlier corrections; these will not make sense and will be applied incorrectly or cannot be applied at all if earlier adjustments are not present.

11.4.7.3 ENC Updating

S-57 and S-52 define the protocols for updating ENCs. The regime relies on incremental updating with checking procedures included in ECDIS software to ensure that all updates are applied in order and that any missing updates are identified. Updating is an automatic process achieved by loading an update file into ECDIS. The methods of providing updates to ships are developing and will include e-mail messaging, satellite communications as well as delivery on disk.

While ECDIS relies on an ENC for its charting information, the ECDIS Performance Standards forbid access to or alteration of the contents of the ENC itself. This ensures that the integrity of the ENC data is not compromised. ECDIS therefore continuously creates a System Electronic Navigational Chart (SENC) which is the active or "ready-use" database used to display information and to provide real-time warnings and monitoring. The SENC is the combination of the ENC (which cannot be altered), any updates to the ENC and any other relevant information which may be added by the mariner or is input from other equipment.

11.4.8 Security and Data Integrity

While the chart data bases used in ECDIS are created using an open format, the data is usually protected from unauthorised access. This is achieved by various combinations of encryption, password protection and licensing.

Both the user and the supplying government are protected through these security mechanisms. The user can be certain that the data is unadulterated and thereby fit for its intended purpose while the supplying hydrographic organisation ensures that the considerable public investment and intellectual property that resides in the data is safeguarded.

11.4.9 Liability

Hydrographic organisations provide official charts as part of their obligations under the UN Safety of Life at Sea (SOLAS) Convention. Under the convention contracting governments are obliged to prepare and to issue official nautical charts and nautical publications which satisfy the needs of safe, precise navigation. By publishing such charts, a government holds itself as having expertise in such matters and as such has a duty of care to supply reliable information. Thus, the production of ENCs carries with it significant potential liability.

Unlike a paper chart or written publication, the ENC data must be manipulated and processed by systems and software over which the hydrographic organisation has little or no direct control. In the event of a dispute involving ECDIS, there will inevitably be the potential for complex and involved legal argument over whether the fault lies with the underlying data or with the processing capabilities or with the use and interpretation of the results.

11.4.10 Copyright

Generally speaking, nautical charts attract copyright. There resides considerable intellectual property in the published charts and the databases which, increasingly, underpin them. Although it is fair to say that there are differing views around the world in different jurisdictions as to the extent of protection offered to the owners of such copyright it is self-evident to the authors that a chart, in any form, is much more that a simple collection of geographical facts. Indeed, the judgement of both surveyor and cartographer is evident in every feature on a paper chart and in every object in its underpinning database. There has been a degree of variance in the acceptance of the significance of copyright but, suffice to say here, it is a critical factor in assuring that only quality charts of the highest integrity actually get to the mariner to support navigation tasks.

11.5 THE FUTURE

11.5.1 Commercial Benefits of ECDIS

ECDIS has already provided considerable commercial benefits to ship operators and will continue to do so in the future. The most compelling evidence so far has come from Canada where the use of ECDIS-like systems has enabled vessels to operate in conditions which were previously impossible. For instance night time, winter and reduced visibility navigation in The Great Lakes. Installation of systems has dramatically reduced the incidence of berthing and manoeuvring incidents. Owners' costs associated with groundings and collisions ranged from $60,000 per year to $2,000,000 per incident, depending on the vessel. Operators have been overwhelmingly supportive of ECDIS and feel that safety and operating has been improved through its use. By 1998, no collisions or groundings had been reported from vessels fitted with ECDIS type systems. Reduced marine fees and insurance rates have been further consequential benefits in Canada.

11.5.2 Improvements in Hydrographic Data Management

ECDIS and particularly the requirement to create and publish ENCs is leading to major changes in the way hydrographic organisations will handle their data holdings in future. More and more data is being captured and processed digitally and exchange formats are being standardised.

Within some hydrographic organisations' sophisticated digital data assessment, processing and production facilities are also under development. These rely on GIS technology and offer the capability not only to more easily identify what hydrographic data is available, but also to provide tailored and sorted data sets in appropriate formats according to a user's requirements. Such a system is the major Digital Hydrographic DataBase project due for delivery to the Australian Hydrographic Office by 2001.

11.5.3 Availability of Hydrographic Data for Other Uses

As well as being responsible for publishing a country's official navigation charts, most government hydrographic offices are also custodians of their national hydrographic archive. As such they maintain the most extensive, up-to-date and comprehensive single collection of hydrographic material covering their waters. This hydrographic information forms a substantial and fundamental spatial data set for effective coastal and ocean management, for example. It is also an essential data set in assessing the potential for offshore resource exploitation.

Unfortunately for scientists the information shown on a published chart is optimised for navigation, and information is relatively sparse. However, beyond the chart lies a rich bathymetric data set, which may be of immense value to engineers, scientists, modellers and administrators who are interested in coastal management. Until recently, it has often been difficult for resource managers both to access this information and to discover what information may be available. The advent of electronic charting and ECDIS and the improvements in hydrographic data management generally, are now beginning to turn this situation around.

11.5.4 Marine Information Objects, VTM, Tides, AIS, Weather, etc.

Having established ECDIS and the ENC, further consideration is now being given to enhancing and advancing S-57 to provide the mariner with an even more sophisticated interactive navigation information system and decision aid. This will be achieved by building on the S-57 format and constructs or developing additional product specifications, which are compatible with ECDIS where this is more appropriate.

It is envisaged that during the next decade ECDIS will be able to access not only the traditionally available chart information, but will incorporate and account for information about such things as the actual state of the tide, weather and ice conditions, and vessel traffic management and reporting schemes (including automatic vessel identification).

CHAPTER TWELVE

The National Oceanic and Atmospheric Administration's Vents Program GIS: Integration, Analysis, and Distribution of Multidisciplinary Oceanographic Data

Christopher G. Fox and Andra M. Bobbitt

12.1 INTRODUCTION

The Vents Program of the National Oceanic and Atmospheric Administration's (NOAA) Pacific Marine Environmental Laboratory (PMEL) is an interdisciplinary research initiative that brings together scientists from a wide range of disciplines, including geophysics, geology, physical oceanography, chemistry, acoustic monitoring, whale research and deep-sea biology. Research within this program is primarily directed toward establishing relationships between the geology and tectonics of seafloor-spreading centres and patterns and impact of the expulsion and regional transport of hydrothermal emissions (Hammond *et al.*, 1991). Data generated from the various research disciplines span scales from the entire Pacific Ocean to the precise location of a particular submarine seafloor marker or hydrothermal vent. Comparison of these diverse data sets provides insight into the processes at work on mid-ocean ridge systems that could not be discerned by examining individual data sets in isolation.

In 1991, Vents researchers began digitally recording acoustic signals from the U.S. Navy's SOund SUrveillance System (SOSUS), an array of underwater hydrophones designed to monitor Soviet submarine movements during the Cold War (Fox and Hammond, 1994). The acoustic signals are used by Vents Program analysts to monitor microearthquakes associated with seafloor spreading events (Fox *et al.*, 1995). The derived earthquake locations have been used to direct field observation efforts to study the hydrothermal activity associated with these volcanic episodes (Fox, 1995). The data sets collected by the Vents Program through acoustic monitoring and field observation techniques have become increasingly large and complex. The acoustic data alone constitutes over 10 gigabytes of information to be analysed each week. Besides significantly increasing the overall data volume to be managed by the Vents Program, the ability to acoustically detect and rapidly respond to seafloor spreading events increased the need for ready access to available data sets for any area in the Northeast Pacific Ocean.

Although the Vents Program's primary research goal is the understanding of seafloor-spreading centre processes, technology developed within the program has led to other research endeavours. Methodologies developed for seismic studies formed the foundation of a parallel monitoring effort to study large marine mammals through their low-frequency vocalisations (Stafford *et al.*, 1998). Mammal location data can be combined with bathymetric data (initially collected for seafloor-spreading research), climate and oceanographic data from numerical models, and real-time satellite imagery for analysis. The addition of marine mammal research to the Vents Program adds a completely different interdisciplinary aspect to an already diverse research initiative.

The interdisciplinary nature of studying hydrothermal ridge processes requires data integration of marine geological and geophysical, geochemistry, physical oceanographic, and biological data sets. Much of the data are collected in digital form. Non-digital data sets are either converted to digital formats (for example, the entry of laboratory chemical analyses into spreadsheets), or the data are interpreted and the interpretation stored in digital form, as in the case of bottom photographs (Fox *et al.*, 1988). The result is diverse data sets collected and archived in multiple formats (binary, ASCII, vector, and raster) on various computer platforms (VMS, UNIX, and personal computers). Despite these variations, all the data share the common characteristic of physical location. This geo-referencing feature of the data allows the integration of data from all sources and types under a single platform, a Geographic Information Systems (GIS) (Bobbitt *et al.*, 1997). Figure 12.1 summarises graphically some of the complexities of the VENTS Program data sets.

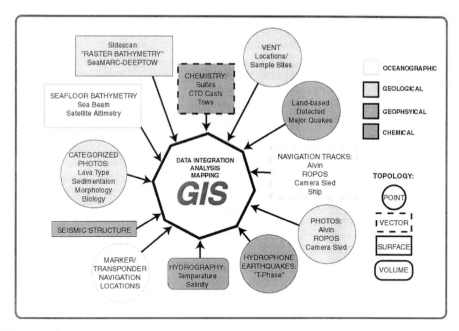

Figure 12.1 Graphical summary of the NOAA Vents Program data environment.

12.2 REQUIREMENTS

Before a GIS was chosen as a solution to the increasing data volume and complexity, the various Vents data sets had to be categorised according to their topology and ultimate usage. This is a critical step in determining which type of GIS system (vector-based, raster-based, or hybrid), if any, would best serve the end users. After interviewing the various Principal Investigators within the program, a table similar to Table 12.1 was created, categorising the Vents data sets by topological representation and application within the given disciplines.

12.2.1.1 Point Data

Many of the data sets are "simple" geographic points. These data include locations of vents, markers, sample sites, moorings, instruments, earthquakes, and whales. These data varied in the number of features (attributes) associated with the particular point. The location of an instrument may be the only attribute, besides the name, that is really used by researchers. In contrast, acoustically derived earthquake epicentre records are also point data but include numerous attributes such as location and origin time of the event, statistical accuracy of the location and time, and earthquake intensity values with corresponding statistics. Hydrothermal vent fluid sample data may include relatively large tables of various chemical compound concentrations. One of the most useful data sets for Vents Program geologists is the categorised photo data from camera tows. Each frame from the camera is categorised for lava type, sediment thickness, hydrothermal indicators, morphology and biology. These attributes are attached to the inferred location, orientation and areal coverage of each photo and the resulting data set can be used to create geological maps of features such as lava flow morphology boundaries or apparent sediment cover (Fox *et al.*, 1988).

12.2.1.2 Line Data

Line topology data sets include tracklines, latitude/longitude gridlines, and bathymetric contours. Within this topological realm, only the bathymetry contours have a relative simple attribute association: the bathymetric value of the contour line. For all the navigation tracklines, the data are more complex. Not only do researchers use the line representation for the data to assess overall data coverage, they also require direct access to given points on the line that may be associated with a specific photograph, sample site or geological feature. In fact, many of the point data sets above derive their geographic location from the navigation point of the vehicle used to collect the data. For all these types of data, therefore, both point and line topologies are required for analysis.

12.2.1.3 Polygon Data

Polygon topology data are the least-used data type within the Vents data sets, but still must be accommodated. Bathymetric data can be represented by enclosed polygons for a various bathymetry range (such as 2000-2100 m), however, most of the bathymetric data available are not complete in coverage because very little of the seafloor has been completely mapped. Data gaps present problems with polygon topology generation so bathymetry, represented as contours, are better suited for vector-based systems. The only major polygon topological data set is that representing the boundaries of individual lava flows or lava morphology provinces. These boundaries are derived from other data sets such as categorised photos from camera tows, sidescan sonar interpretation (Embley *et al.*, 1990), and bathymetric mapping depth differencing from repeat surveys of different years (Fox *et al.*, 1992).

12.2.1.4 Raster Data

Raster data types used in Vents research include climatological, bathymetric and sidescan and other imagery data. Some of these data comes as pre-processed images that only require geo-referencing in order to use the image as another interpretative data layer. An example of such a data set is satellite infrared imagery processed for sea surface temperature. Other data sets, such as sidescan sonar and multibeam bathymetry, must be combined with vehicle navigation and orientation information to georeference each pixel on the seafloor. In some cases, these data sets must be used in combination to derive a usable product, for example combining digital terrain models derived from multibeam sonar systems with towed sidescan sonar data to remove distortion of the images by bathymetric features (Lau and Fox, 1991). Video cameras and electronic still cameras can produce massive amounts of raster data, however, these data are generally reduced by classification of the image to point data sets.

12.2.1.5 Multidimensional Data

Besides two-dimensional raster images and grids, physical oceanographic data are often collected in three dimensions using CTD "tow-yos", in which a Conductivity-Temperature-Depth (CTD) sensor is moved vertically through the water column as the ship moves providing a three-dimensional view of the water properties. In some cases, these three-dimensional surveys are conducted in the same area at different times (Baker, 1994) resulting in a four-dimensional data set. Broad-scale circulation models and time-dependent ocean climate models also take this form. Representing these time-varying fields is difficult graphically and challenging from the GIS perspective as well.

12.2.2 Summary

Based on the various data types in use within the Vents program at the inception of GIS development, a vector-based system was chosen to be the primary GIS development tool. The limited raster data could still be incorporated into the vector system, however the image processing capabilities of the raster systems would not be initially available within the GIS. Since most of the raster data sets were already acquired as interpreted images this did not present many problems. Raster images, in particular bathymetric grids, were frequently used as a primary under-layer for the GIS displays. As the requirement for raster-based information eventually increased, a hybrid system combining vector and raster implementations in a single platform was developed. The raster system was carefully chosen to be compatible with the vector system to limit the need for data conversion.

12.3 VENTS GIS DEVELOPMENT AND IMPLEMENTATION

The need for establishing a GIS arose due to the growing complexity of data, the expense of personnel dedicated to programming specific software to analyse data, and the increasing reliance upon the few individuals experienced using this specialised software. Data collected since the early 1980s, including data from multiple research cruises each year along the entire Juan de Fuca Ridge with different and technologically advancing equipment, required spatial co-registration and integration for scientific analysis in a timely manner for decision making and planning purposes (Bobbitt *et al.*, 1997). The adequacy of a GIS-based solution was determined through a progression of steps. The objective of the GIS evaluation process was to find a single-platform (both hardware and software) that any scientist could easily use to analyse data. The system had to be able to overlay various types of data, produce high-quality graphics, and provide analysis and query capabilities for this complex multi-disciplinary data set. Finally, the system had to work with available hardware and require minimal personnel and data processing maintenance.

12.3.1 Previous Efforts

A single prior example of GIS used in conjunction with deep-sea hydrothermal research (Wright *et al.*, 1995) began the GIS evaluation process. In that implementation, a GIS was employed in the lab and in the field to analyse submersible, towed sidescan and camera data collected over a two-year period from several research cruises. This GIS application integrated data sets similar to those produced by the Vents Program including raster sidescan images, multibeam bathymetry contours, submersible and towed instrument navigated tracklines, sample sites and interpreted geological features such as faults and fissures. However, compared to data archived at NOAA/PMEL and planned for GIS implementation, this earlier oceanographic application was limited in data volume. In addition, the effort described in Wright *et al.* (1995) was undertaken before good

Table 12.1 Summary of data types collected by the Vents Program with corresponding topology and disciplinary focus (modified from Wright *et al.*, 1997).

DATA	Topology				Function			
	Point	Line	Polygon	Grid	Geology	Geophys	Chem	Biol
Bathymetry (Multibeam)		X	X	X	X	X	X	X
Bathymetry (Mesotech)		X		X	X	X	X	X
Bathymetry (Imagenex)		X		X	X	X	X	X
Bottom Pressure Recorder	X				X	X		
Camera Tows	X	X			X			X
Categorised photos	X				X			X
CTD casts	X						X	
CTD tows	X	X					X	
Current Meter Locations	X						X	
EQ focal Mechanisms	X					X		
EQ locations Land seismometers	X					X		
EQ locations Hydrophones	X					X		
Extensometer	X				X	X		
Geologic features		X	X		X	X	X	
Gridlines (lat/long)	X	X			X	X	X	X
Images (35mm)	X			X	X		X	X
Images (satellite)				X	X	X		X
Markers	X				X	X	X	X
Moorings	X				X	X	X	X
MTR (miniature temperature recorder)	X				X	X	X	X
Rock cores	X				X			
Samples	X				X	X	X	X
Ship tracks (bathymetry)	X	X			X	X	X	X
Sidescan sonar	X	X		X	X	X	X	X
Submersible tracks	X	X			X		X	X
Transponders	X				X	X	X	X
Vent locations	X				X		X	X
Video footage	X	X		X	X	X	X	X
Whale locations	X							X

programming and interface tools were commercially available and therefore depended heavily on custom programming. Advances within vendor-provided software providing graphical user interfaces (GUIs) which minimise custom programming for data conversion and improved user accessibility alleviated these concerns.

12.3.2 Scientific Information Model

Following the initial evaluation of a GIS implementation, a more formal scientific information model was constructed (Wright *et al.*, 1997). This approach rigorously integrates the data structure, GIS, and the interaction and feedback from the various researchers using the system. The results of this study indicated that although very complex in architecture, the Vents GIS model could potentially succeed in accomplishing the research goals of the project with GIS technology.

12.3.3 GIS Platform

The Vents GIS was implemented with commercially developed software, without purchasing additional dedicated hardware platforms and utilising less than one full-time employee per year for development and maintenance. Although initially more expensive, commercial GIS packages offer long-term stability and continuous development, while shareware/freeware or native solutions may terminate with the interests of their developers. Also by choosing the commercially developed GIS, some programs and technical advice have been obtained and shared freely through inquiries to a product-specific Internet discussion list. The Vents GIS today consists of three software and six GUI licenses residing on a Digital Unix platform as well as a single GUI for a sea-going, portable PC-based GIS. Additionally, a raster-based GIS, fully compatible with the vector-based systems, resides on a networked Unix workstation for easy exchange between the two platforms. Users access the system via a network of personal computers or through the Internet. The primary Digital Unix platform supports multiple processors, allowing for ready expansion as demand grows. Graphics can be printed on colour plotters and laser printers. The GIS was developed within the existing, stored data formats requiring minimal and simplistic data conversion routines.

12.3.4 Access by the Broader Community

The final technological issue to be addressed was how to best serve the data to the user community outside of the immediate Vents Program scientific staff. This community includes several classes of users. Scientific research colleagues require full access to the Vents GIS and the ability to integrate their own data sets into the system. Other scientific researchers may not require integration of their own data, but may still desire custom products from the GIS. The general scientific community may find value in examining the methods used in this research program (or this GIS implementation) for application to their own efforts. Finally, the general public has a right to access the data for education, amateur scientific studies or just native curiosity.

Sharing data with collaborative researchers has been facilitated using GIS and the World Wide Web (WWW). Providing integrated data sets on specifically established web sites allows world-wide easy access to the data. Researchers can download and examine the data using their own platforms (GIS or other types) for their specific analytical research or for data integration. The Vents Program GIS also provides user interfaces resident on PMEL platforms, which allow interactive access to the data over the WWW. These interfaces are often too slow to provide a long-term analysis platform, but can easily handle a quick look at the data or the generation of a custom product. Finally, the WWW interface is used to provide simplified GIS views and customised presentations for public education and access.

12.4 APPLICATIONS OF THE VENTS GIS

The true test of any GIS is its value as an analysis tool as reflected in its use in solving real research problems. The Vents GIS has proven to be useful in a number of applications, both in the laboratory and in the field, and has been successfully used as the basis of data distribution and integration with other efforts.

12.4.1 Data Analysis

The ability to overlay related data has enhanced the capability of Vents Program researchers to analyse their complex data sets within the laboratory environment. Several examples are given below.

12.4.1.1 Seismic Analysis

The analysis of acoustic data collected from SOSUS and autonomous hydrophone instruments for seismicity has revolutionised the study of low-level seismicity in the ocean basins. The ability to detect small ($m_b > 1.8$) earthquakes at long range has led to a fundamental increase understanding of active oceanic plate tectonic processes and has enabled the detection of submarine volcanism for the first time (Fox *et al.*, 1994). The interpretation of water-borne seismic phases is a relatively new endeavour and much of the interpretation of a given seismic event depends upon viewing the distribution of the earthquake patterns in time and space in the context of seafloor geology, in particular bathymetry. In addition, information from the land seismic networks provides details of the mechanics of the large magnitude events that can not be derived from hydroacoustic phases. The Vents GIS has provided a tool to easily import acoustically derived epicentre information and compare it to pre-existing mapping data and historical large earthquake properties.

12.4.1.2 Field Mapping

Geological mapping of the seafloor requires the use of numerous tools, collecting data at different scales, with different navigational constraints (Embley *et al.*,

1990). Large-scale multibeam bathymetry often forms the base, followed by towed sidescan sonar, bottom photography, and rock sampling and analysis. Added to this geological data set are biological and chemical sampling, often from manned submersibles. Each set of observations is influenced by the other processes. For example, the distribution of seafloor biota follows the discharge of hydrothermal fluids, which in turn are controlled by the seafloor geology. The most valuable contribution of the Vents GIS may be in allowing rapid, interactive access to these various data sets. It has been used extensively by geologists, chemists, and biologists from the Vents Program and outside universities in the laboratory and in the field.

12.4.1.3 Marine Mammal Habitat

The distribution of marine mammals in the open ocean is very poorly known, and the historical distributions of various species have been altered catastrophically by commercial whaling in the last one hundred years. Traditional methods of assessing whale populations have depended on visual observations from surface vessels. This method is quite expensive and produces a fairly limited data set in the open ocean. Low-frequency acoustics offers a complementary monitoring technique that can detect vocalising whales at long range and during periods when visual observations are impractical (Stafford *et al.*, 1998). To properly apply the acoustic techniques, however, the two methods must be inter-calibrated. The Vents GIS provides an ideal platform for this analysis and a major effort is currently underway to address this question by several NOAA and university laboratories. Besides the inter-calibration effort, the Vents GIS can be used to analyse whale distributions in relation to oceanographic features derived from historical, surface ship, or satellite observations. This analysis will lead to a better understanding of the ecology of large whales in the open ocean.

12.4.2 Field Operations Support

A major benefit of establishing the Vents GIS came in 1995 during at-sea research expeditions using submersibles. A sea-going version of the Vents GIS (Figure 12.2) was loaded with previously collected data pertinent to the planned submersible dive area. Data from each dive during the cruise was easily added to this existing database within hours after dive completion. Scientists were then able to analyse the daily dive plan with all the data information available quickly and graphically. This allowed decisions to be made as to which areas needed to be revisited and which new areas needed to be explored, thus more efficiently using the limited dives and bottom time of the submersible (Bobbitt *et al.*, 1996; Wright, 1996).

The GIS also provided a pre-dive experience for many of the submersible passengers who would use the GIS to become familiar with the area prior the dive. Only two passengers can accompany the pilot in the submersible to the bottom and surface communication is very limited. Most dives have several disciplinary objectives however it is impossible to have a representative from each research discipline aboard. By using the GIS prior to the dive, researchers could familiarise themselves with known seafloor features and markers and view video from prior submersible dives with tracklines displayed using the GIS graphics for analysis. This process also reduced the time spent searching for dive targets allowing more time for research work. Since 1995, Vents researchers have continued to use the sea-going version of the GIS for data analysis at sea.

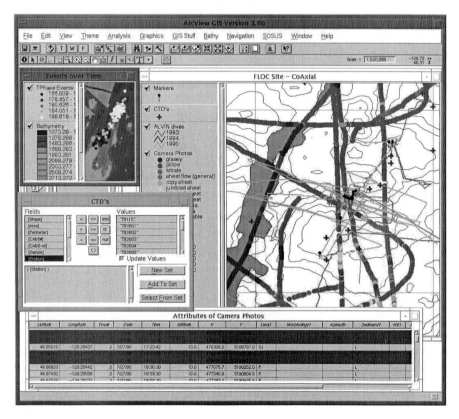

Figure 12.2 Example of a typical graphic interface for Vents Program scientists working in the field. Display includes a detailed map of the study area with previous submersible tracklines, CTD stations, interpreted camera images, bathymetry and lava flow boundaries. Insets include a time-dependent epicentre map with bathymetry of the broader study area, metadata for CTD casts, and an attribute table of interpreted camera images with fresh lava flows highlighted. This data base was continuously updated as new observations were collected.

The interactive, graphical displays of the data provided by the Vents GIS have also facilitated navigation data correction at-sea and during the post-cruise data processing effort. Working with instruments several miles underwater, precise location information on the seafloor is not directly available but is acquired using acoustic transponder beacons. The transponders are deployed from the ship, free-falling to the seafloor, and then calibrated by acoustic signals sent from each beacon back to the ship. Using statistical algorithms, positions are determined relative to each transponder and then linked geodetically to the ship's satellite-derived position. The calibration process is not precise and can have errors on the order of ten meters or more, and large biases from one year top the next. When previously explored markers or other features are re-visited, the offsets in the navigation can be conveniently measured and viewed using the GIS. Analysing these combined offsets enhances dive planning by reducing search times and data re-processing.

12.4.3 Detection and Response

Over 80% of the planet's volcanic activity occurs in the oceans, but until recently there had been no method to detect the activity to allow its study. The initiation of acoustic monitoring in the Northeast Pacific Ocean in 1991 has already resulted in the successful detection of three major and several minor volcanic episodes. In each case field parties were able to investigate the sites within days or weeks of the activity. Many of the phenomena associated with this fundamental earth process are transient, for example the generation of large hydrothermal plumes, and must be studied on-site during or shortly after the eruption. The Vents GIS offers a convenient mechanism to provide historical data to field parties investigating volcanic events. Data sets are provided on a PC platform and have been pre-configured to contain all historical observations for all areas of the Northeast Pacific spreading centres. This approach was used successfully during the 1996 eruption of the northern Gorda Ridge (Bobbitt and Fox, 1996) and is being maintained for future rapid response cruises.

12.4.4 Data Distribution

Providing data to the community beyond the local laboratory has been significantly enhanced by the use of the Vents GIS. Formerly, requests for data or custom data products required a significant effort to generate and often required several iterations before the desired product was produced. The ability to produce custom products easily using the Vents GIS has made this effort much more efficient. In addition, advances by commercial GIS vendors have led to the ability to serve GIS users over the internet via the WWW. Using this technology, the Vents GIS (in reduced form) is now served to the WWW through the Vents Program web site, allowing users to produce their own custom products with no interaction with PMEL personnel. This has led to a significant reduction in the time spent responding to user requests. It is anticipated that this technology will eventually be used in-house via an intranet implementation.

12.4.5 Data Integration

The Vents Program is not the sole community performing geological studies in the Northeast Pacific. A major effort by the National Science Foundation's Ridge Interdisciplinary Global Experiments initiative (NSF-RIDGE) has led to the collection of large data sets by university laboratories, particularly in the Endeavour Ridge region. An NSF-RIDGE project is currently underway to produce a system similar to the Vents GIS for this area and is being constructed on a similar architecture to allow for easy integration (Wright and McDuff, 1998). It is anticipated that a large, distributed GIS will eventually result, where the data reside with the investigators that collected them, but are available to the entire community in a seamless form.

12.5 PROBLEMS

Problems associated with the development of the Vents GIS can be sorted into two categories: cultural and technological. Cultural problems refer to the difficulty in inspiring researchers to abandon their present analysis tools in favour of the more integrated GIS tool. Experience has shown that this reluctance can only be overcome by providing a GIS platform that is easy to use and of significantly greater power than the existing systems. This necessarily means that there will always be a lag between the development of the GIS tool and its acceptance by the researchers. Information managers must invest for the long term in this technology and not become impatient when the initial response to such a major investment is not overly enthusiastic.

Technological problems encountered during this development were many and many shortcomings of the system still exist. Initially, the available user interfaces and programming tools were primitive and difficult to use. These deficiencies are being continuously addressed by the commercial developers but require active feedback from the community as to its needs. In many cases, initial releases of the GIS software were not capable of performing many of the basic functions required for scientific research. The primary reason for this deficiency seemed to lie in the historical development of the GIS as a land-use planning tool, rather than as a scientific analysis tool. This is a deficiency that must be constantly addressed and active feedback from the user community is critical to achieving a powerful scientific-research GIS. Finally, the initial tools for interfacing local GIS to the community via the WWW are very cumbersome and inefficient to the point of severely impacting their usefulness. Only minimal data sets can currently be served in this manner. It is anticipated that continuing efforts by commercial developers will improve this shortcoming.

12.6 FUTURE PLANS

Future efforts within the Vents Program for GIS development continue along the lines described here. New data and new data types are being collected each year and new regions of the global ocean are being investigated that require continuous updating of the Vents GIS technology. Increased use of raster and multidimensional data sets will continue to require expansion from a simple vector-based GIS to a more complex, hybrid system. When problems with Internet access are resolved, amore concerted effort is planned to distribute data internally and externally via automated access. Finally, there are numerous other marine GIS efforts currently being undertaken around the world, and it is anticipated that a large-scale, distributed GIS will eventually result that incorporates large segments of the oceanographic research community.

12.7 CONCLUSION

The GIS developed by the Vents Program is now an integral component of the scientific analysis process within this oceanographic environmental research program. The GIS is used both in the laboratory for scientific analysis, at sea to provide baseline information and support for field programs, and as a tool for data distribution and integration. Future developments call for better access through the WWW, integration with raster GIS technology, and collaboration with other efforts to ultimately provide a seamless, distributed GIS to the broader oceanographic research community.

12.8 ACKNOWLEDGEMENTS

Thanks to Dawn Wright, Andy Lau, and John Graham for their efforts in the development of the Vents GIS. The project was funded by the NOAA Vents Program, the NOAA ESDIM Program, and the Office of Naval Research's Marine Mammal Science Program. PMEL Contribution #2048.

12.9 REFERENCES

Baker, E.T., 1994, A 6-year time series of hydrothermal plumes over the Cleft segment of the Juan de Fuca Ridge. *Journal of Geophysical Research*, **99**, pp. 4889-4904.

Bobbitt, A.M., Dziak, R.P., Stafford, K.M. and Fox, C.G., 1997, GIS analysis of oceanographic remotely-sensed and field observation data, *Marine Geodesy*, **20**, pp. 153-161.

Bobbitt, A.M. and Fox, C.G., 1996, Response co-ordination and data dissemination during the 1996 Gorda Ridge eruption using WWW and GIS, *Eos, Transactions of the American Geophysical Union*, **77**, p. F3.

Embley, R.W., K.M. Murphy and Fox, C.G., 1990, High resolution studies of the summit of Axial Volcano, *Journal of Geophysical Research*, **95,** pp. 12785-12812.

Fox, C.G., 1995, Special collection on the June 1993 volcanic eruption on the CoAxial segment, Juan de Fuca Ridge, *Geophysical Research Letters*, **22**, pp. 129-130.

Fox, C.G. and Hammond, S.R., 1994, The VENTS Program T-phase project and NOAA's role in ocean environmental research, *Marine Technology Society Journal,* **27**, pp. 70-74.

Fox, C.G., Radford, W.E., Dziak, R.P., Lau, T-K., Matsumoto, H. and Schreiner, A.E., 1995, Acoustic detection of a seafloor spreading episode on the Juan de Fuca Ridge using military hydrophone arrays, *Geophysical Research Letters*, **22**, pp. 131-134.

Fox, C.G., Dziak, R.P., Matsumoto, H. and Schreiner, A. E., 1994, Potential for monitoring low-level seismicity on the Juan de Fuca Ridge using military hydrophone arrays, *Marine Technology Society Journal,* **27**, pp. 22-30.

Fox, C.G., Chadwick, Jr., W.W. and Embley, R.W., 1992, Detection of changes in ridge crest morphology using repeated multibeam sonar surveys, *Journal of Geophysical Research,* **97**, pp. 11149-11162.

Fox, C.G., K.M. Murphy and Embley, R.W., 1988, Automated display and statistical analysis of interpreted deep-sea bottom photographs, *Marine Geology*, **78**, pp. 199-216.

Hammond, S., Baker, E. Bernard, E., Massoth, G., Fox, C., Feely, R., Embley, R., Rona, P. and Cannon, G., 1991, The NOAA VENTS Program: Understanding chemical and thermal oceanic effects of hydrothermal activity along the mid-ocean ridge, *Eos, Transactions of the American Geophysical Union*, **72**, pp 561-566.

Lau, T-K. and Fox, C.G., 1991, A technique for combining SeaMARC I sidescan sonar and gridded bathymetric data to display undistorted seafloor images, *Oceans 91: Proceedings*, **2**, pp. 1140-1145.

Stafford, K.M., Fox, C.G. and Clark, D.S., 1998, Long-range detection and localization of blue whale called in the Northeast Pacific Ocean using military hydrophone arrays, *Journal of the Acoustical Society of America*, **104**, pp. 3616-3624.

Wright, D.J., 1996, Rumblings on the ocean floor: GIS supports deep-sea research. *Geo Info Systems*, **6**, pp. 22-29.

Wright, D.J. and McDuff, R.E., 1998, A geographic information system for the Endeavour Segment. *RIDGE Events*, **9**, pp. 11-15.

Wright, D.J., Fox, C.G. and Bobbitt, A.M., 1997, A scientific information model for deep sea mapping and sampling, *Marine Geodesy*, **20**, pp. 367-379.

Wright, D.J., Haymon, R.M. and Fornari, D.J., 1995, Crustal fissuring and its relationship to hydrothermal and magmatic processes at the East Pacific Rise crest (9°12-54'N), *Journal of Geophysical Research*, **100**, pp. 6097-6120.

CHAPTER THIRTEEN

Integrated Geographical and Environmental Remotely-sensed Data on Marginal and Enclosed Basins: The Mediterranean Case

Vittorio Barale

13.1 INTRODUCTION

The development of remote sensing (RS) techniques using orbital sensors has shown that a novel view of marginal and enclosed basins, combining both terrestrial and maritime elements, can be obtained from space. In the marine realm, RS offers a wide range of capabilities, complementing conventional *in situ* data gathering techniques, for the synoptic and systematic assessment of interacting bio-geo-chemical and physical processes at the regional, as well as the global, scale. However, single remotely sensed pictures of the sea surface, although spectacular, are seldom enough for a sound approach to the exploitation of the technique's information potential. The real advantage of marine remote sensing is to be found in the long-term, large-scale monitoring of entire basins (Barale *et al.*, 1997). The main problem to be solved, in order to fully exploit such capabilities, is that RS techniques produce huge amounts of data, which must undergo several levels of processing, requiring special facilities and expertise, before reaching the end user. Moreover, the highly dynamic nature of many coastal and marine environmental processes requires that this kind of information be analysed on a statistical basis, and hence starting from historical time series, for the assessment of environmental trends over suitable periods of time. New data can then be used for monitoring anomalies, which diverge from the statistical conditions described by the climatologies.

These requirements have led to the adoption of concepts analogous to those of classical geographic information systems (GIS), even though with a number of caveats due to the peculiarities of typical maritime space–time scales, for the integration of RS-derived geographical and environmental value-added data (Barale and Murray, 1992). In the following, an application of these concepts will be presented, using the Mediterranean basin as an example. Historical time series of satellite-based observations, collected by a suite of sensors over the last 20 years, will be introduced, and used to differentiate between geographical provinces shaped by coastal patterns and plumes, mesoscale features such as permanent eddies, and large-scale structures. Further, the main trends observed for the same parameters in these provinces will be considered, providing insights on the seasonal characteristics of the various geographical components of the basin.

Provinces and trends can be used to explore the relationships between the patterns of surface indices appearing in the RS-derived data. This, in turn, can shed some light on the role played by geographical setting and atmospheric forcing in establishing the observed space–time distribution of surface parameters (Barale and Zin, 1998). Based on the analysis of integrated geographical and environmental RS data, the potential impact of morphologic and climatic traits of the basin and of its continental margins will be addressed, leading to the hypothesis that such factors may have a major influence on the bio-geo-chemistry and dynamics of the entire Mediterranean Sea.

13.2 REMOTE SENSING OF THE SEA SURFACE

Although seemingly uniform to the uninitiated, the marine environment is characterised by pronounced space and time heterogeneity. Traditional measurements collected at sea, from moored sensor arrays, by oceanographic campaigns or along shipping lanes, cannot generate a uniform global coverage. They can thus provide only a sketchy scientific description of the world's oceans. Moored sensors can provide continuous, long-term data at given locations, as a function of depth in the water column, and excel in the assessment of the temporal domain. Drifters, as well as ship campaigns, are most useful to extend the investigation into the spatial domain, although on limited scales. However, ships in particular move too slowly to permit a clear separation between time dependent changes of the environment and spatial changes, whether geographical or induced by water dynamics. Given this speed limitation, and the prohibitive costs involved in covering a large area with *in situ* sensors, it is in the combination of extended spatial coverage and repetition capabilities that RS can play a major role for environmental surveying.

Typical environmental phenomena can affect entire marine regions over large spatial scales (from hundreds to thousands of km) and short time scales (from hours to days), and such evolutionary processes are difficult to follow with *in situ* techniques. On the other hand, most variations in environmental conditions have specific signatures on various sea surface properties, which can be best assessed by orbital sensors. This qualifies RS as a powerful, if not exclusive, tool for studying and monitoring various processes on the sea surface, provided that its basic concepts, including observation mechanisms and limitations, are well understood.

Using classical RS techniques (i.e., observations with either passive or active sensors such as imaging radiometers, spectrometers, or imaging radars), the main features accessible when observing the sea are essentially surface colour, temperature and structure, including, within this term, both roughness and elevation (which can be obtained also by means of non-imaging radars called altimeters). Such features summarise the whole range of surface properties, optical, thermal and structural, which can be inferred from the interaction of electromagnetic radiation with the air-water interface.

A whole suite of environmental parameters can be assessed and quantified using remote sensors operating in suitable regions of the electromagnetic spectrum. In general, different methodologies may be applied, depending on the objectives and boundary conditions of the observations. Passive remote sensing techniques use reflected visible and near infrared sunlight, or thermal emissions in the infrared spectral region, or again emitted microwave radiation, to assess various surface parameters. Active remote sensing techniques use transmitted impulses of visible or microwave radiation, for a subsequent evaluation of the nature, delay and shape of the returned signal.

Each of the spectral regions available for passive or active radiometry (corresponding to suitable atmospheric windows of propagation) has specific merits and drawbacks. For example, passive observations of optical or thermal properties at the visible and infrared frequencies are limited to a cloudless, clear atmosphere, but take advantage of a stronger natural signal and allow greater spatial resolution. Analogous observations of the same properties at microwave frequencies can be performed through clouds, but have lower spatial resolution.

It must be stressed that the environmental parameters, which can be derived from remote assessments of sea surface properties, cannot be measured directly by remote sensors. Rather, the sensors measure some other, spectral, quantity (e.g., reflected or emitted natural radiance, or the reflection of an artificial electromagnetic signal), which is only indirectly related to the parameters of interest (e.g., water constituents concentration, or sea surface temperature, or wind speed, etc.). Further, none of these, spectral, quantities has a unequivocal relationship with a certain parameter, so that no sensor, or single spectral band, is sensitive to just one phenomenon. Rather, each responds to a combination of atmospheric and marine environmental conditions. This complicates the interpretation of data, and usually requires that a set of multi-spectral observations, each sensitive to somewhat different combination of phenomena, be performed simultaneously to provide unambiguous information.

13.2.1 Optical Properties

Optical properties can be measured using both passive and active techniques, observing the reflection of natural sunlight or transmitted laser infrared digital airborne radar (LIDAR) signals respectively, in the visible and near infrared spectral regions. An evaluation of optical properties can be used to estimate the concentration of various water constituents (e.g., plankton pigments, dissolved organic materials, suspended inorganic sediments), to a certain depth in the water column, or some optical characteristic of the marine surface layer (such as diffuse attenuation coefficient, at fixed wavelengths). Direct use of these parameters, or their interpretation as natural tracers, allows the analysis of phenomena like water quality and plankton blooms, primary productivity, bio-geo-chemical cycles, coastal and fluvial runoff, sediment transport, water dynamics, depth variations in shallow waters, exchanges between coastal zone and open sea, climatic events, and so on.

13.2.2 Thermal Properties

Thermal properties can be assessed by means of passive techniques sensible to Earth emissions in the thermal infrared or microwave parts of the electromagnetic spectrum. Similarly, through the evaluation of thermal properties, (i.e., sea surface temperature), one might derive information of physical, dynamical or climatic nature. Currents, fronts, eddies, upwelling and vertical mixing events, as well as surface slicks of certain kinds, are some of the features which can be quantified by means of this parameter. Synoptic assessments of sea surface temperature are also important for water circulation modelling, and for the assessment of energy exchanges at the interface between sea and atmosphere.

13.2.3 Structural Properties

Structural or dynamical properties can be measured with both passive and active methods using reflected visible light or, primarily, through microwave active techniques. The surface roughness parameters derived from measurements of structural properties can further contribute to the quantitative identification of dynamical phenomena such as winds, waves, wakes, and alterations of the water surface texture due to circulation features, to bottom profiles, or to the presence of surface films (e.g., hydrocarbons). Finally, dynamical properties (i.e., essentially the marine surface elevation with respect to the geoid) provide information on water motion and circulation at large, planetary, scales, or on deep geological features.

13.3 HISTORICAL ARCHIVES OF MEDITERRANEAN RS DATA

The integration of geographical and environmental data, derived from RS campaigns over the Mediterranean Sea, and used here for the assessment of provinces and trends in the basin, was done using long-term data sets collected by the visible/infrared radiometers Coastal Zone Colour Scanner (CZCS; 1978-1986) and Advanced Very High-Resolution Radiometer (AVHRR; 1982-1991), and by the radar altimeters on Geosat (1986-1989), the European Space Agency Remote Sensing satellite (ERS-1; 1992-1995) and Topex (1993-1997). These data sets originate from the activities of various independent projects, which generated the historical archives used to derive climatologies of surface parameters (see Barale *et al.*, 1998, and references therein). The time series of satellite data were originally processed to derive composite images of parameters such as plankton concentration, surface temperature and wind speed, covering the entire Mediterranean basin. A GIS-like integration of the composite images was performed so as to obtain geographically coherent data sets, remapped over a standard geographical grid, covering an area of 4000 x 2000 km^2, using an equal-area (Alber's) projection, and constant resolution, with a 5 x 5 km^2 pixel size (Figure 13.1).

The raw CZCS and AVHRR data (respectively 2465 and 9396 original images, collected on a daily basis when favourable meteorological conditions occurred) were processed to apply sensor calibration, to correct for atmospheric contamination, and to derive chlorophyll-like pigment concentration and surface

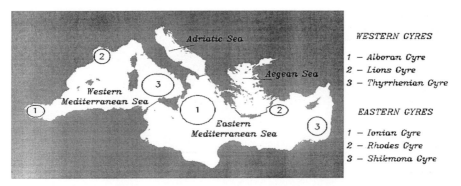

Figure 13.1 Standard geographical grid for the Mediterranean Sea, covering an area of 4000 x 2000 km², using an equal-area (Alber's) projection, and constant resolution, with a 5 x 5 km² pixel size. Geographical features and names correspond to those reported in the text.

temperature (Sturm, 1993; Nykjær 1995). Individual images were generated for each available day, co-registered using the same geographic equal-area projection and resolution, and then averaged pixel by pixel, to compute monthly and annual composites.

As for the altimeter data, wind speed at 10 m above the sea surface, at 1 second intervals, corresponding to 6-7 km intervals along the satellite ground-track, was obtained from the strength of radar returns (Witter and Chelton, 1991). The altimeter is a narrow-swath instrument, with a footprint of 5-10 km, which gives good coverage along-track, but leaves wide gaps between tracks. The distance between tracks depends upon their repeat pattern and the latitude. For Geosat, in a 17-day repeat, over the Mediterranean Sea, this is about 120 km; for ERS-1, in a 35-day repeat, 50 km; and for Topex, in a 10-day repeat, 220 km. Hence, monthly mean wind speeds were obtained by analysing data from each calendar month from areas covering 2° latitude by 2° longitude. Because of the high correlation along-track and to remove dubious outliers not discarded by quality checks, the median wind speeds from each satellite pass across each area were calculated and these values averaged. There were about 50 to 100 transects *per* month in each area. Results were checked by visual inspection of the 12 monthly means and of the first few eigenvectors from an empirical orthogonal functions (EOF) analysis. Anomalous means in a few areas were changed using values obtained by fitting an annual cycle to the 12 monthly means (the monthly mean values from the medians were also checked against means from all the data, but no significant differences were found). Means over the 5-km grid were derived by replication and smoothing algorithms.

The statistical results obtained should be considered with caution, due to the sensors limitations in retrieving quantitative assessments of surface pigments, temperatures and winds, and to the poor spatial and temporal resolution of the observations. However, an analysis of recurrent patterns was derived from the long-term composites, showing the estimated mean conditions of the Mediterranean, which can be interpreted in terms of the main environmental traits of the basin.

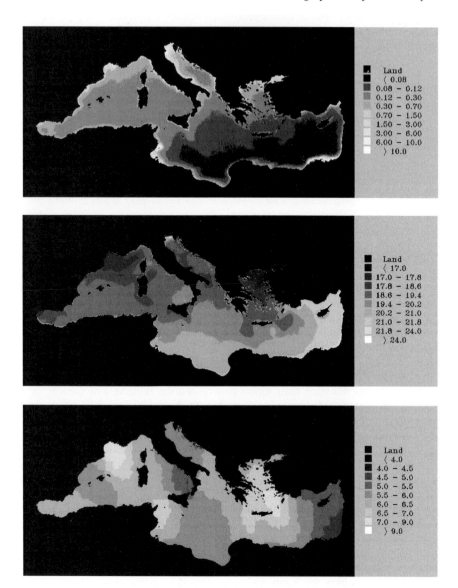

Figure 13.2 Integrated geographical and environmental data from RS archives for the Mediterranean Sea. Top: mean chlorophyll-like pigment concentration [mg/m3] derived from the CZCS (1979-1985) data set (upper plate). Centre: mean sea surface temperature [°C] derived from the AVHRR (1982-1991) data set (middle plate). Bottom: mean wind speed [m/s] derived from the Geosat (1986-1989), ERS-1 (1992-1995) and Topex (1993-1997) data sets (lower plate). Original images in colour.

13.4 PROVINCES AND SEASONS OF THE MEDITERRANEAN SEA

The Mediterranean Sea mean pigment concentration, surface temperature and wind speed, derived from historical satellite-based observations, and integrated in a GIS-like structure, are shown in Figure 13.2. In broad terms, the images support the classical geographical subdivision between western and eastern sub-basins, inshore and offshore domains, northern and southern near-coastal areas. The northwestern sub-basins seem to be dominated by a stronger pigment signal, while the southeastern sub-basins by a more pronounced temperature signal. The wind speed signal has maxima corresponding to the main, known, wind patterns of the basin.

The geographical provinces emerging from the analysis of the integrated imagery include coastal areas under the direct influence of major river plumes (i.e., those of the Ebro, Rhone, Po, and Nile), or other intense coastal interactions. Also included are minor river discharges, and non-point sources of runoff, along the Italian coast in the Tyrrhenian Sea, along both the Italian and Albanian coastlines in the Adriatic Sea, along the northern shores of the Aegean Sea and the Marmara Sea. On the other hand, in the coastal area off southern Tunisia, the enhanced pigment signal is due to direct bottom reflection in an area of shallow clear waters around Kerkenna Island, with variable temperatures and wind patterns.

A distinct type of province covers the Ligurian–Provençal–Balearic sub-basins, and the Adriatic–Aegean Seas, and includes near-coastal areas with high pigments, low temperatures, and high wind speed. Possibly, the rim of enhanced pigment values and lower temperatures around most of the northern Mediterranean is associated with the impact of runoff from continental margins (i.e., both a direct impact due to the sediment load and one induced on the plankton flora by the associated nutrient load), as well as with the vertical mixing due to the prevailing winds (i.e., the Mistral over the north-western Mediterranean, the Bora over the northern Adriatic and the Etesians over the Aegean), as suggested by the high wind speed of this region. The gyres formed by the incoming Atlantic jet, in the Alboran Sea, and the giant filament of Capo Passero, at the southern tip of Sicily are examples of dynamical features related to mixing processes in the water column. Hence, they are linked to higher pigments and lower temperatures, but not to a corresponding wind speed signal.

Finally, the remainder of the basin is occupied by open sea provinces, with increasingly oligotrophic traits, separated (in the western basin) from the coastal areas by intermediate frontal zones, and characterised (in the eastern basin) by a permanent eddy field, i.e., a string of mesoscale gyres, such as the Cretan and Rodhes (cyclonic) gyres, and the (anticyclonic) gyre between the two.

The integration of monthly averaged values of pigment concentration, temperature and wind speed (not shown here) has allowed also the evaluation of trends in the time domain. A pattern of higher pigments, lower temperatures and higher wind speed develops in winter, i.e., from late fall to early spring, and then reverses in summer, i.e., from late spring to early fall. The mean parameter values computed for the entire basin, on a monthly basis, and their regression analysis (Figure 13.3), point to a high correlation between pigment and wind, and to an inverse correlation between both variables and temperature. While the basin seems to follow such a model, specific areas present a different seasonality. Notably, the

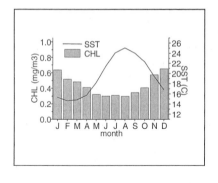

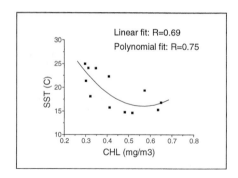

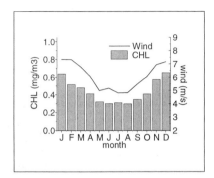

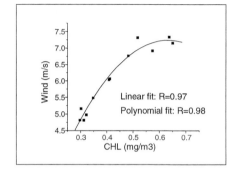

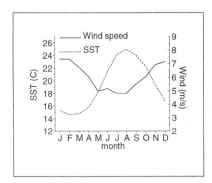

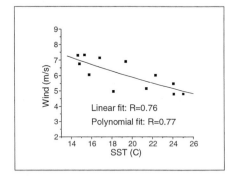

Figure 13.3 Seasonal evolution (left column) and second order polynomial regression analysis (right column), computed over the whole Mediterranean basin, on a monthly basis, for mean pigment concentration and mean surface temperature (upper graphs); mean pigment concentration and mean wind speed (middle graphs); mean surface temperature and mean wind speed (lower graphs).

Adriatic Sea and the Aegean Sea display spring enhancements in pigments, superimposed to the general annual trend. Local conditions describing an analogous behaviour can be recognised also elsewhere, in both the western and eastern Mediterranean. In the western basin, while the Alboran Gyre does not appear to have a pronounced seasonal signal, except for the higher pigments in fall, the Tyrrhenian Gyre and the Lions Gyre also display a combination of seasonal signatures. In particular the northwest, i.e., the Ligurian-Provençal basin, appears to have low pigments in winter, a pronounced spring bloom followed by a summer minimum, and another (quite large) bloom in fall. In this area, the lower pigment values of the cold season (i.e., the so-called "blue hole" corresponding in the pigment field to the Lyons gyre) are coupled to very high wind speeds and to very low surface temperatures, presumably originated by deep convection processes, occurring systematically when the Mistral wind blows, mixing waters down to 1500-2000 m (THETIS Group, 1994). In the eastern basin, the Ionian Gyre, but primarily the Rhodes Gyre, and even the Shikmona Gyre, show the usual temperature seasonal cycle, but also late-winter, early-spring (February) and fall enhancements in the pigment cycle.

13.5 CONCLUSIONS

The application of RS to assessments of coastal and marine environmental parameters, in enclosed and marginal basins, is a well-established technique (Barale and Folving, 1996). Its major contribution lies in the ability to extend observations beyond the *in situ* platforms' domain. To a large degree, the lasting uncertainties in (global) estimates of several marine environmental parameters can be attributed to the fact that direct measurements seriously undersample the ocean for the purpose of characterising its actual scales of spatial and temporal variability. In order to overcome such problems, RS of the sea surface must be used to complement *in situ* measurements.

The information derived from the application of remote sensing techniques can be of paramount importance in the development of mathematical simulations (and predictive capabilities) of marine and coastal environments. At present, these developments are limited by inaccuracies in the knowledge of the system initial state, as well as of how and at what rate the system components interact. Marginal and enclosed seas, let alone the open oceans, are still quite poorly sampled, and a synoptic GIS of environmental parameters is critical to specify the initial conditions for the systems to be modelled. Further, model updating is required, as the state of the system must be re-specified as it changes.

The exploitation of RS data for the above tasks is limited by a certain number of complexities, which hamper somewhat its potential benefits. These complexities can actually prevent the average user (who, in practice, usually works with limited operational resources) from taking full advantage of the wealth of available RS information. Remote sensors produce not only a new kind of data, but also far more data than have been normally handled by environmental scientists or managers. Large quantities of data are generated in short periods, sometimes exceeding the capacity of available computer systems. Moreover, these data must pass through several levels of processing before reaching the end user. Their interpretation and integration with other kinds of data further requires dedicated facilities and special expertise. Therefore, as testified by the consistent amount of unused data residing in many RS data banks, there is a continuing need for the editing, formatting, and processing of existing data sets not yet in accessible, or usable, form, and this need shall be perpetuated in the foreseeable future by a growing demand for, and collection of, remotely sensed data.

These kinds of information management problems can be eased through the application of GIS concepts, as shown here for the case of the Mediterranean Sea. Experience has shown that RS images by themselves, spectacular as they may be, are not enough for an assessment of marine environmental conditions, since the observations must be turned into time series and relevant statistics of quantitative, navigated values of geophysical parameters. On the other hand, the Mediterranean Sea example shows that the integration of environmental data, based on historical time series of satellite-based observations, allows the assessment of the main geographical provinces in the basin. Further, the derived statistical information points to the existence of a specific relationship between remotely sensed indices and geographic/climatic features of the region. The provinces can be defined in terms of coastal processes and open sea conditions, as well as in terms of transition conditions between these two extremes.

The seasonal trends, derived from the integrated mean monthly parameters, point to a behaviour of the Mediterranean Sea similar to that of a subtropical basin, where light is never a limiting factor, but nutrients always are (Yoder *et al.*, 1993). In such a scenario, the maximum concentration of surface pigments, coupled to minimum values for surface temperature, would occur in the cold (windy and rainy) winter season, and would be related to surface cooling, vertical mixing and runoff; whereas one would find a minimum in surface pigments, coupled to a maximum in surface temperature, in the warm (calm and dry) summer season, when the water column is strongly stratified and no nutrient supply, from coastal zones or deeper layers, is readily available.

Some provinces display a seasonality closer to that of a subpolar basin, with lower pigment concentration in winter, because of reduced light and, more important, because of the vertical turbulence and the deep vertical mixing due to the prevailing wind field, which prevents algae to be stabilised in the upper well-lit layers. This seems to be particularly true for the northern part of the western basin, where the lack of high pigments, and the very low temperatures, in winter might be linked to the extreme conditions generated by the overturning of the entire water column. The ensuing spring bloom should be triggered by the relaxation of these conditions, when the wind field relaxes, the water column becomes sufficiently stable, and stratification occurs in the basin.

The approach adopted here suggests the hypothesis of a relationship between geographic and climatic factors and bio-geo-chemistry of the Mediterranean Sea. The fertilisation of the basin would be mainly due to coastal interactions linked to atmospheric forcing, and then to its thermohaline dynamics. Therefore, the bio-geo-chemistry of the system would seem to be driven primarily by morphological and meteorological key features of the basin.

13.6 REFERENCES

Barale, V. and Folving, S., 1996, Remote sensing of coastal interactions in the Mediterranean region. *Ocean and Coastal Management*, **30**, pp. 217-233.

Barale, V., Meyer-Roux, J., Schmuck, G. and Churchill, P., 1997, Time series of remote sensing data in the Mediterranean basin: complementary tools for environmental applications. In *Remote Sensing '96: Integrated Applications for Risk Assessment and Disaster Prevention for the Mediterranean*, edited by Spiteri, A. (Rotterdam: A.A. Balkema), pp. 11-17.

Barale, V., Zin, I. and Carter, J.T., 1998, Mediterranean Sea trends in the surface colour, temperature and wind fields. In *Proceedings of the International Symposium on Satellite-based Observation: A tool for the Study of the Mediterranean Basin*, Tunis (Toulouse: CNES).

Nykjær, L., 1995, Seasonal variability of coastal upwelling of Northwest Africa and Portugal, 1981 to 1991. *Geo Observateur*, **6**, pp. 5-15.

Sturm, B., 1993, CZCS Data processing algorithms. In *Ocean Colour: Theory and Applications in a Decade of CZCS Experience*, edited by Barale, V., and Schlittenhardt, P. (Dordrecht: Kluwer Academic Publishers), p. 95-116.

THETIS Group, 1994, Open ocean deep convection explored in the Mediterranean. *EOS, Transactions, American Geophysical Union*, **75**, pp. 217-221.

Witter, D.L. and Chelton, D.B., 1991, A GEOSAT altimeter wind speed algorithm and a method for wind speed algorithm development. *Journal of Geophysical Research*, **96**, pp. 18853-18860.

Yoder, J.A., McClain, C.R., Feldman, G.C. and Esaias, W.E., 1993, Annual cycles of phytoplankton chlorophyll concentrations in the global ccean: A Satellite View. *Global Biogeochemical Cycles*, **7**, pp. 181-193.

CHAPTER FOURTEEN

Mapping Submarine Slope Failures

Brian G. McAdoo

14.1 ABSTRACT

Geographic information systems (GISs) are ideally suited for mapping submarine landslides. Deep-water landslides, though seldom affecting humans, are responsible for moving vast quantities of sediment from continental slopes to the deep ocean. Earthquakes are believed to be responsible for many mid- to lower-slope slides, which in turn can be used as indicators of past earthquakes. The increasing amount of oil exploration and production in deepwater environments raises concerns over slope instability. Multibeam bathymetric data are used, with or without seafloor reflectivity data from side-scan sonar, to locate and quantitatively characterise very large submarine mass movements. The data acquired on size, water depth, thickness, slope gradient on which the slide occurred, gradient of the runout zone, gradient of the headscarp, and slope within the slide scar help in assessing slope stability. Once the conditions that triggered existing slides are defined, they can be applied to present-day conditions (sedimentary, seismic), and landslide hazard may be assessed.

14.2 INTRODUCTION

Submarine landslides are not only an important process in shaping the seafloor on continental margins and transporting sediment from the continental shelf break to the deep-water environments, but can have significant influence on humans as well. A submarine landslide triggered by an earthquake on the Grand Banks in 1929 created a turbidity current that severed trans-Atlantic communication cables almost 600 km away (Heezen and Ewing, 1952; Hampton *et al.*, 1996). In 1964, the great earthquake in Prince William Sound, Alaska triggered submarine slides that eventually retrogressed past the coastal zone, swallowing the towns of Seward and Valdez, and creating tsunamis that repeatedly washed over low-lying coastal areas (Coulter and Migliaccio, 1966; Lemke, 1967; Hampton *et al.*, 1996). Twenty-meter high waves associated with Hurricane Camille caused slope failures which damaged petroleum platforms on the Texas and Louisiana Gulf of Mexico shelf in August of 1969 (Bea *et al.*, 1983). Although many submarine slides occur in the deepwater, far offshore, their influences on humans can be substantial, and mapping them can prove beneficial to many concerns.

The primary challenge in studying submarine landslides is the relative inaccessibility and invisibility of the submarine environment. Slides undoubtedly occur without our knowledge. The only way to know where and when a submarine slide has occurred is to have data before and after a slide. Presently, governments and industry are in the process of acquiring high-resolution side-scan sonar and bathymetric data for ongoing studies of the seafloor for academic, economic, and military purposes. Data from side-scan sonar systems are similar to aerial photographs that show acoustic reflectivity (analogous to visual reflectivity in aerial photographs). Multibeam bathymetry data yield spatially referenced depths to the seafloor that can be morphometrically analysed in three dimensions. In digital format, these data are ideally suited for compilation and spatial analysis in a GIS framework.

McAdoo *et al.* (in prep.) created a GIS database of almost 100 geomorphically expressed submarine landslides on the continental slopes offshore Oregon, California, the Texas/Louisiana Gulf Coast, and New Jersey/Maryland ranging in size from 1 km^2 to over 5,500 km^2. The database has morphometric measurements including area, scar slope gradient, runout length and slope, headscarp height and slope, and slope gradient of the adjacent unfailed slope, which is used as a proxy for the pre-failure slope angle. Collected data can be used to assess and mitigate potential landslide hazards in offshore regions of similar geology and geomorphology, and perhaps start investigating the triggers for deepwater mass movements.

14.3 DATA

A ship collects "multibeam" or "swath" bathymetric data by broadcasting an acoustic signal in a fan pattern beneath the ship while surveying at speeds generally around 10 knots (Figure 14.1). Sixteen or more beams collect depth data, corrected for beam angle, geographically referenced to the ship's global positioning system. The bathymetric data used to identify the submarine landslides in the McAdoo *et al.* (in prep.) study were collected using the Sea Beam swath-mapping system (deMoustier, 1988). The data have a vertical resolution of $\leq$ 1% of water depth and a position accuracy $\leq$ 50 m (Grim, 1992). Pratson and Coakley (1996) created gridded data sets (100 m) from the Northeast Consortium for Oceanographic Research and the National Oceanic and Atmospheric Administration (NOAA) surveys of the Oregon, California, Texas/Louisiana, and New Jersey/Maryland margins. Existing data sets can be acquired from a number of sources, often with only minimal charge. The National Geophysical Data Center (NGDC) is a clearinghouse for numerous types of geophysical data. Their web site, http://www.ngdc.gov/mgg/bathymetry/multibeam.html is the gateway to these data.

The formats of the data are as varied as the software packages that create and analyse them. For mapping even the largest submarine slides, it is best to use data sets with grid size < 0.04 km^2 (< 200 m or 0.002° on a side for a square grid cell), but these particular data sets are often difficult to acquire. Much of the gridded

Figure 14.1 Cartoon of multibeam bathymetric data collection from ship. Surveying at 10 knots, a ship can cover between 100 and 4000 km² per day in water depths between 200 and 6000 m.

multibeam bathymetric data come in either a binary Generic Mapping Tools (GMT) format or Arc interchange (.e00) format for those with access to the Environmental Systems Research Institute's (ESRI) Arc/INFO or ArcView software. GMT is collection of Unix tools designed for spatial data (x, y, and z), and can display data in many different map projections (Wessel and Smith, 1991). Graphical output is in the form of Encapsulated PostScript (EPS) files. It can be downloaded, free of charge, from http://imina.soest.hawaii.edu/gmt. Once a gridded data set is acquired, it is possible to convert GMT binary grid (.grd) files to Arc/INFO™ ASCII format. A public domain converter can be downloaded from http://dusk.geo.orst.edu/arcgmt (Wright *et al.*, 1998).

To make morphometric measurements using ArcView's Spatial and 3-D analyst tools, it is helpful to have the same x, y, and z units. In Arc/INFO, re-project geographic projections with x (longitude) and y (latitude) decimal degrees, and z (usually) in meters, to a Universal Transverse Mercator (UTM) projection where all units are in meters. This way, the analysis programs that calculate slope gradients and aspects, and artificial hillshading will operate properly.

14.4 DATA ANALYSIS

When mapping any geomorphic feature, be it incised steam beds, drumlins, or landslides, it is useful to view the data in a format that accentuates rapid changes in surface elevation. The artificial hillshading and slope gradient algorithms available in many software packages will cast "shadows" on or illuminate regions of steep local slope. This is very helpful in identifying landslide scars.

The first step in identifying slides is to make slope and artificial hillshade maps. Regions with rapid changes in slope gradient help point out the landslides (McAdoo *et al.*, in prep). Figure 14.2 shows how clearly the diagnostic arcuate headscarps are illuminated by the slope gradient maps. The slope gradients of the landslides' headscarps and sidewalls can be measured at any point in the grid. By casting various artificial sun angles and altitudes, headscarps, sidewalls, and occasionally material at the slide base can be alternately lit up or darkened (Figure 14.3). Aspect maps are helpful in identifying regions of anomalous direction, and highlight regions such as canyons and slides' sidewalls where the erosive walls are directed ~90° to the headscarp and local downslope trend (Figure 14.4). Again, to get a statistically significant value of slope, it is helpful to average as many points within the area of interest as possible.

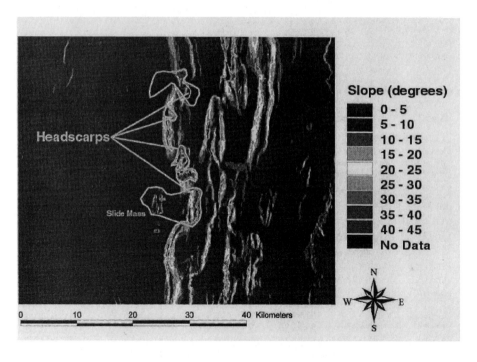

Figure 14.2 Slope gradient map of the base of the Oregon continental slope. In this figure, lighter shading indicates higher gradients and darker, lower. Arcuate headscarps have anomalously high slope gradients. Smooth, continuous regions of high slope gradient are the flanks of anticlinal ridges caused by the subduction of the Juan de Fuca plate beneath the North American plate.

Figure 14.3 Artificial hillshade maps of the base of the Oregon continental slope. Same view as Figure 14.2. Top figure has a sun illumination angle from the northwest (315°), and the bottom figure is with the sun angle at the same altitude, but from the southeast (225°).

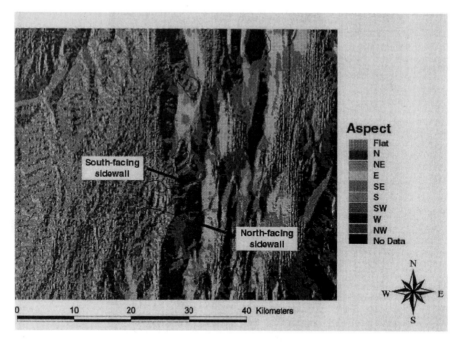

Figure 14.4 Aspect map of the base of the Oregon continental slope. N, NE, E, etc. represent compass directions north, northeast, east, etc. The anticlinal ridges at the base of the slope are directed dominantly to the east and west. The sidewalls of the landslides show with as north and south aspects.

Cross-sections or profiles through a slide help verify of potential landslide surfaces. From downslope profiles, headscarps are easily identified, and their height can be measured in various sections (Figure 14.5). Profiles taken across slides, can be helpful in identifying the lateral extent of erosion or deposition, and help constrain the geometry of the slide deposit (Figure 14.5). In addition, they can assist in identifying the characteristics within the slide scar. For instance, slump blocks, which have detached from a headscarp, and do not exceed the downslope extent of the sidewalls (Figure 14.6). Profiles through the slope maps identify headscarps as the region of steepest slopes, and yield the point of the maximum slope within a given profile.

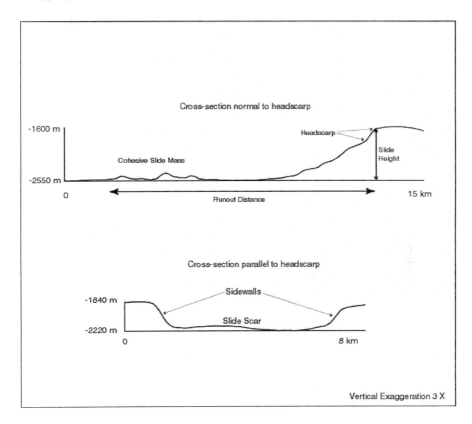

Figure 14.5 Profile normal and parallel to a slide's headscarp. Vertical exaggeration is 3 times. The cross-sections are from the southernmost slide illustrated in Figures 14.2 and 14.3. Note the slide mass at the base of the slide, with a 10 km runout distance from the headscarp. The height of the slide is measured from the top of the headscarp to the level at which the slide mass came to rest, or the base of the slide scar if slide mass is not apparent. The headscarp is the steepest portion at the head (top) of the slide.

14.5 DATA INTERPRETATION

The data acquired by morphometric mapping in a GIS framework can be interpreted to provide critical information on the location, distribution, and hazards associated with submarine slides. Bathymetry data provides information on the water depths where failures occur, slide thickness, heights from top of the headscarp to base of the scar, runout maximum, and height of the headscarp and slide thickness. Slide area assessment can tell us how important landslides are as sediment transport and landscape shaping processes in particular regions. Furthermore, defining the slide area helps with hazard mitigation by showing us how large slides tend to be in a given region.

Slope gradient measurements are critical in slope stability analyses and subsequent hazard recognition. There are numerous methods of calculating slope stability, most of which include some combination of material properties (strength, degree of consolidation or previous burial, ability to build up pore-pressures with cyclic earthquake loading, etc.) and slide morphology, including slope and thickness of slide. Assuming a slope with similar physical properties (which is not unlikely over relatively small areas in the marine environment), the only variables in calculating a factor of safety for a particular region are slope angle and slide thickness. By looking at the morphology of existing slides, we can assess slope stability based on the slide thickness and slopes where previous slides have occurred.

Charts and tables of the measurements can be compiled directly into the spatial framework of the GIS. Outlining the extent of the slide with a polygon, then tying in each of the variables on the attribute table for that polygon is an effective means of cataloguing the data on each of the slides in a particular area. Location, type of failure (i.e., cohesive or disintegrative), depth to headscarp, area, runout distance, headscarp height, and slope gradients of the headscarp, local unfailed slope, scar, and runout zone are added to the attribute table of the theme. Individual slides can be interrogated for their statistical measurements, and the statistical data can be viewed as a whole in the theme's attribute table.

14.5.1 Bathymetry

One of the more fundamental measurements made on submarine slides is water depth of the headscarp. Some continental margins, especially active convergent margins such as offshore Oregon, tend to have continental slopes that get steeper with depth (Pratson and Haxby, 1996). Working on the assumption that slides tend to occur where slopes are the steepest, we might expect to see more slides on the lower slope of active margins. A simple histogram of a number of slides in given water depths can be used to assess this hypothesis. If there are more slides in shallow water, on or near the shelf break, perhaps slides are related to rapid sediment deposition and fluid overpressuring during times when shelf break deposition is occurring. On the other hand, perhaps erosion and deposition cycles erased past landslide activity, therefore more slides are preserved in quieter, deeper water. Interpretations may vary based on how much is known about depositional history of a region, and its present-day geotechnical properties.

Headscarp height is a good proxy of material strength (McAdoo *et al.*, in prep.). Regions with overconsolidated, high strength material tend to produce failures with higher headscarps, resulting in deeper-seated slides (McAdoo *et al.*, in prep.). These deep-seated slides expose material that is even more overconsolidated than the original unfailed, but eroded, surface. These slides often maintain a degree of post-failure cohesion, leaving blocky masses at the base of the slide scar (Figure 14.2). Shallow seated slides, like those at the base of Mississippi canyon in the Gulf of Mexico (Figure 14.7), suggest weaker (perhaps high fluid pressure) sediment. These slides are quite often very large, despite occurring on gentle slopes (McAdoo *et al.*, in prep.).

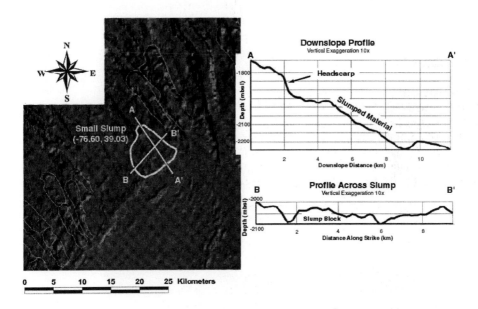

Figure 14.6 Cross-sections through a slump offshore New Jersey. The top section is approximately normal to the headscarp and the lower section is taken across the slide. Note the steep headscarp at the top of the slide followed by a downslope decrease in gradient at the top of the slump block. The cross-section from across the slumped mass shows separation of the mass from the sidewalls.

Mississippi Canyon Region
Gulf of Mexico

Figure 14.7 Very large slides in the Gulf of Mexico, east of Mississippi canyon, offshore Louisiana. These slides combined encompass over 5,000 km² of seafloor. Slides of this magnitude could pose substantial hazard to offshore drilling and/or production infrastructure.

The height of the failure's headscarp is also important when considering landslides as a sediment transport mechanism, and when assessing hazard based on runout distance potential (Figure 14.7). Runout length is one way that submarine landslides differ substantially from their subaerial counterparts. Runout for a given submarine slide tends to be on the order of 10^2 to 10^3 times higher than a subaerial slide with the same slide height (Keefer, 1984; McAdoo *et al.*, in prep). Runout distances (and slide area) in submarine environments are higher due in part to hydroplaning of submarine slides (Mohrig *et al.*, 1998).

14.5.2 Area

The larger the area of a submarine landslide, the more sediment is moved, the greater the area of seafloor affected, and hazards potential. Slide area can be divided into two regions: the scar (erosive) and slide mass (depositional). It is often difficult to find evidence of the slide mass, either in the bathymetry data, because the slide mass is too thin to be resolved, or in a side-scan system such as GLORIA, because the failed material does not sufficiently contrast with the surrounding seafloor.

The minimum slide area that can be reasonably measured in a 100-m gridded data set is ~1 km². McAdoo *et al.* (in prep.) report a slide at the base of Mississippi canyon that exceeds 5,500 km² (Figure 14.7). Slides such as this are responsible for transporting large quantities of sediment. A rough estimate of slide volume (V) can be calculated by using the slide area (A), headscarp height (h), and a shape coefficient (k) related to the slide geometry.

$$V = k\,A\,h \tag{14.1}$$

Figure 14.8 shows the two end member models of the volume calculation. If the sidewalls of the slide are of similar height as the headscarp, then $k = 1$. If the sidewalls taper to nothing at the slide base, the geometry is approximately wedge-shaped, and $k = 1/2$.

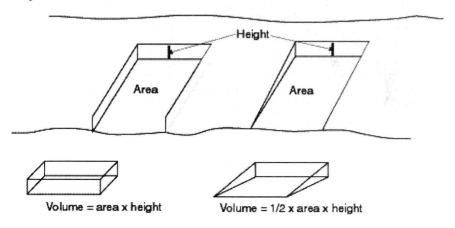

Figure 14.8 3-D cartoon of two submarine slides. The slide on the left has sidewalls that do not diminish in height downslope. The right slide has sidewalls that taper to nothing at the base of the slide.
End-member volume calculations are approximated by a shape with a given surface area and uniform heights around that surface area (volume is the area of the base multiplied by the thickness of the slide), and one where heights taper off at the base of the slide, where volume is approximately half of the volume calculated using uniform thickness.

Erosive processes below ~200 m are limited primarily to deepwater ocean currents, sediment laden hyperpycnal flows, and mass movements of rock/sediment. Over time, multiple landslides in the same location will have a significant effect on the morphology of the continental slope. Similarly, regions that lack evidence of repeated mass movements will have a different appearance (Figure 14.9). By comparing the landscape of margins in different geologic and tectonic environments, we can gain some insight into sediment transport and landslide triggering mechanisms.

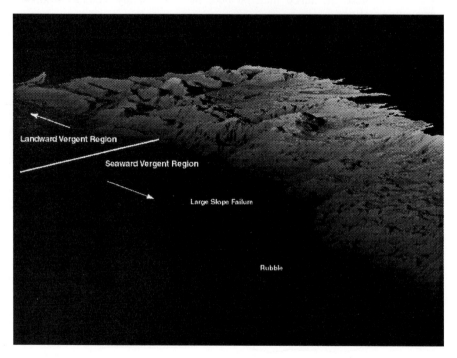

Figure 14.9 Three-dimensional perspective of the lower Oregon continental slope comparing eroded and non-eroded regions. View is looking northeast, vertical exaggeration ~3 times. In the southern (lower) portion of the image, the slope has a rough, blocky appearance. North, in the regions where the anticlinal ridges formed by accretion verge landward, the slopes are equally steep, but smooth and non-eroded.

Measurements of the area and volume of submarine landslides can also directly influence hazard determination. Submarine deltas are often regions of offshore oil production (i.e. Mississippi and Niger deltas). The search for oil is moving exploration into deep-water slope environments, where slope stability is an issue. Large runout slides that often occur in overpressured sediment are capable of wiping out entire production platforms (Bea, 1971; Prior and Coleman, 1982; McAdoo *et al.*, in prep.). By analysing runout pathways in a GIS, it is possible to illustrate where landslides have occurred in the past, what path they followed, and where they are likely to occur in the future, and what path they are likely to follow.

Landslides triggered by earthquakes may add to tsunami hazard if a large volume slide displaces the seafloor creating a wave in phase with the wave created by the uplift or subsidence caused by fault displacement. For example, the M_s = 7.2, 1992 Nicaragua earthquake displaced a water mass of ~30 km^3, creating a 9.7 m wave run-up and was responsible for ~170 deaths (Sataki, 1994). A 300-km^2 landslide with a 100-m high headscarp (which is large, but reasonable, McAdoo *et al.*, in prep.) could produce a similar size tsunami. Mapping submarine landslides will disclose margins with potential for landslide-generated or landslide-enhanced tsunamis.

14.5.3 Slope

Measurements of slope gradient are critical for assessing slope stability and aid in examining submarine landslides as a landscape altering process. By knowing the slope gradient, the stability of a region can be assessed assuming reasonable ranges of physical properties. Many GIS software packages have algorithms that automatically calculate slope by averaging the gradient between the eight neighbouring cells. Resulting maps (Figure 14.2) can be interrogated for the average slope at individual cells. To find the slope of a particular feature such as a headscarp or runout region, it is important to average the values of numerous cells within the given area. Small-scale topographic roughness can produce both flat and very steep regions that may not be representative of the feature in question.

The majority of submarine slopes on continental margins are very low (< 6°; Pratson and Haxby, 1996) when compared to their subaerial counterparts. With this in mind, unconsolidated marine sediment will not fail on a slope less than the angle of repose (ϕ, which ranges between 25° for silt to 36° for some sands and gravel (Lambe and Whitman, 1969) unless disturbed or affected by an increase in fluid pressure.

A simple factor of safety (F) is calculated by dividing the sediment strength by the stress imposed on the slope.

$$F = [C_0 + \mu(\gamma' \, h \, \cos^2\theta - P_f)] \, / \, (\gamma' \, h \, \sin\theta \, \cos\theta) \qquad (14.2)$$

where C_0 is the material cohesive strength, $\mu = \tan \phi$, γ' is the buoyant unit weight of the sediment ($\gamma' = \gamma_{sed} - \gamma_{water}$), h is the depth to a potential failure plane, θ is the slope angle, and P_f is the existing pore-fluid pressure. An $F \leq 1$ suggests that the slope is statically metastable. A slope with $F > 1$ is most likely stable under the present conditions, and increasing F values indicate increasing stability.

The above equation suggests that an increase in fluid pressure or stress could trigger a slope failure for a gradient below the angle of repose. Rapid sedimentation and loading by large storm waves can lead to increased fluid pressures, and possibly slope failure (Seed and Rahman, 1978; Prior and Colemen, 1982), but are unlikely in deepwater continental slope environments. Earthquakes, however, act independent of water depth and can both increase the stress on the sediment and increase fluid pressure (Lee and Edwards, 1986). Lee *et al.* (in prep.) have proposed a model for slope stability that requires only slope angle, sediment unit weight (γ), and an empirical relationship between water content and sediment type, and strength degradation associated with cyclic loading (CSR_{10}). The critical acceleration (k_c as a fraction of gravity) that would cause a failure on slope of θ with γ unit weight and γ_- buoyant unit weight is

$$k_c = (\gamma/\gamma_-)\ (CSR_{10} - \sin\theta) \tag{14.3}$$

Lee *et al.* (in prep.) implement this model using measurements of sediment total density from numerous cores in the Eel River basin (northern California). Using the core location and slope data in a GIS database, they create hazard maps that show where a range of critical accelerations would cause failure.

14.5.4   Statistical   Analysis

Relationships between measured features yield insight into submarine landslide behaviour. McAdoo *et al.* (in prep.) present measurements of the slide area, headscarp height, water depth, runout distance, along with slopes gradients of the headscarp, scar, runout zone and the local unfailed slope (LUS) which is used as a proxy for the pre-failure condition.

Histograms often elucidate interesting relationships. For example, where are slides occurring? More slides in shallow water near sediment depocenters (upper slope) might suggest sedimentation-related failures. What is the most common mode of failure size? Is there a difference in the areas of slides that maintain post failure cohesion and those that disintegrate? A distribution of headscarp heights skewed towards the smaller end might indicate weaker sediment.

Cross-correlations show trends (or lack thereof) between variables. For example, McAdoo *et al.* (in prep.) found that runout distance tends to decrease with the LUS and runout slope (perhaps because steeper runout slopes may distribute the material so that there is no remaining bathymetric or reflectivity signal). Curiously, McAdoo *et al.* (in prep.) found that the shallowest slopes in the Gulf of Mexico produce the largest slides. In addition, steep slopes tend to produce steep scars and steep headscarps, reinforcing the idea that steep slopes are stronger than less steep slopes. The relationships between various variables may vary based on the local prevailing conditions, giving us insight into process in different geologic/tectonic settings.

14.6   IMPLICATIONS   AND   CONCLUSIONS

Submarine landslides are important erosional and depositional processes. GIS gives us the power to examine the spatial relationships of landslides in submarine, subaerial, and even extraterrestrial environments. The analysis provides insight into many critical processes, including hazards, paleoseismology, sediment transport and landscape evolution. The field is in its infancy, and data acquisition technology along with the volume of existing databases are increasing at a rate that will make the deepsea environment the exploration and research frontier for the 21st century.

There has been increasing concern about hazard mitigation, especially with steadily increasing costs associated with replacing damaged property (van der Vink *et al.*, 1998). Oil companies are increasingly looking towards the deepwater as shelf exploration is becoming less economic. Extraordinary costs are associated with continental slope exploration and production, and the chance of a catastrophic landslide destroying production/exploration platforms increases on steeper slopes. Advances in directional drilling in the oil industry make moving a platform (for the sake of possibly saving it) an economically wise decision. Increasing coastal development not only means far more loss due to hurricanes (van der Vink *et al.*, 1998), but also increases the potential for significant tsunami hazard, especially on the Pacific rim. Maps of existing submarine slides and predictions of possible landslide locations can help mitigate property loss by clearly identifying the most hazardous regions.

From an academic point of view, mapping submarine landslides is critical in understanding erosion, deposition, and landscape forming processes on the continental slope. Are landslides on the continental shelf break responsible in part for slope deposition? What triggers submarine slides? There are numerous examples of seismically active regions where landslides are scarce including the Eel River basin and the Cascadia margin (McAdoo *et al.*, in prep.; Lee *et al.*, in prep.). Repeated landslides over time may armour the continental slope by exposing overconsolidated material. The resulting slope will able to withstand steeper gradients because it is stronger than slopes with normally consolidated or overpressured sediment. On accretionary and convergent margins, landslides may play a role in maintaining the critical taper of the slope (Davis *et al.*, 1983).

All of these issues can be addressed by a well-organised and thorough GIS. With the increasing technology and data library, GIS will be able to integrate multibeam bathymetry, side-scan sonar, slope stability and core data with bottom photographs and samples, along with sub-bottom profiles and structure location of man-made structures. These databases will lead us to a better understanding of the marine environment.

14.7 FUTURE WORK

GISs are ideally suited for mapping submarine landslides. Future efforts need to be concentrated on acquiring more offshore data (bathymetry and reflectivity) to investigate how important submarine landslides are as sediment transport mechanisms, and paleoseismic indicators. In addition, landslides can be an import geodynamic process, maintaining critical taper in both submarine accretionary prisms and subaerial mountain ranges. Furthermore, large mass wasting events can aid in unroofing of both compressional and extensional orogenic belts.

14.8 REFERENCES

Bea, R. G., 1971, How sea floor slides affect offshore structures, *Oil and Gas Journal*, **69**, pp. 88-92.

Coulter, H. W. and Migliaccio, R.R., 1966, Effects of the earthquake of March 27, 1964, at Valdez, Alaska, *U.S. Geological Survey Professional Paper 542-C*, 36 pp.

Davis, D., Suppe, J. and Dahlen, F.A., 1983, Mechanics of fold-and-thrust belts and accretionary wedges, *Journal of Geophysical Research*, **88**, pp. 1153-1172.

deMoustier, C., 1988, State of the art in swath bathymetry systems, *International Hydrographic Review*, **65**, pp. 25-54.

Grim, P., 1992, Dissemination of NOAA/NOS EEZ multibeam bathymetric data. In *1991 Exclusive Economic Zone Symposium: Working Together in the Pacific EEZ Proceedings*, edited by Lockwood, M., and McGregor, B.A. (U.S. Geological Survey Circular 1092), pp. 102-109.

Hampton, M.A., Lee, H.J. and Locat, J., 1996, Submarine landslides, *Reviews of Geophysics*, **34**, pp. 33-59.

Heezen, B.C. and Ewing, M., 1952, Turbidity currents and submarine slumps, and the 1929 Grand Banks earthquake, *American Journal of Science*, **250**, pp. 849-873.

Lee, H.J. and Edwards, B.D., 1986, Regional method to assess offshore slope stability, *Journal of Geotechnical Engineering*, **112**, pp. 489-509.

Keefer, D.K., 1984, Landslides caused by earthquakes, *Geological Society of America Bulletin*, **95**, 406-421.

Mohrig, D., Whipple, K.X., Hondzo, M., Ellis, C. and Parker, G., 1998, Hydroplaning of subaqueous debris flows, *Geological Society of America Bulletin*, **110**, pp. 387-394.

Pratson, L.F and Haxby, W.F., 1996, What is the slope of the U.S. continental slope, *Geology*, **24**, pp. 3-6.

Prior, D.B. and Coleman, J.M., 1982, Active slides and flows in underconsolidated marine sediments on the slopes of the Mississippi Delta. In *NATO Workshop on Marine Slides and Other Mass Movements*, NATO conference series IV, edited by Saxov, S. and Nieuwenhuis, J.M. (London: Plenum Press), pp. 21-49.

Seed, H.B. and Rahman, M.S., 1978, Wave-induced pore pressure in relation to ocean floor stability of cohesionless soils, *Marine Geotechnology*, **3**, 123-150.

Van der Vink, R.M. and 14 others, 1998, Why the United States is becoming more vulnerable to natural disasters, *Eos, Transactions of the American Geophysical Union*, **79**, pp. 533, 537.

Wessel, P. and Smith, W.H.F., 1991, Free software helps map and display data, *Eos, Transactions of the American Geophysical Union*, **72**, pp. 441.

Wright, D.J., Wood, R. and Sylvander, B., 1998, ArcGMT: A suite of tools for conversion between Arc/INFO and Generic Mapping Tools (GMT). *Computers and Geosciences*, **24**, pp. 737-744.

CHAPTER FIFTEEN

Applications of GIS to Fisheries Management

Geoff J. Meaden

15.1 INTRODUCTION

Some 20 per cent of the Earth's marine area is conducive to a natural biological productivity rate such that it is attractive to intensive fish production. Thus, most coastal shelf areas in both tropical and temperate regions, plus the major near-shore upwelling areas, have long been the scenes for the high biomass yields, which have encouraged fishery activity. More recently, this activity has turned towards some Arctic shelf waters, which are also proving to be highly productive. The other 80 per cent of the marine area, which may be considered as open oceanic, is less biologically productive but it is able to support extensive fisheries for highly migratory pelagic species such as tuna and shark.

Total fishery production grew significantly until the late 1980's, since which time it has levelled off, with yields now increasingly being maintained from aquacultural sources (Grainger and Garcia, 1996). A situation exists whereby many individual fish stocks have been, or are being, fished beyond their sustainable yield level, with the consequence that some stocks have "crashed" alarmingly. Allied to this, over-exploitation of higher trophic level species has resulted in ecosystem changes such that short lived, lower level species now predominate in many areas, and these are of little use to humankind (Pauly *et al.*, 1998). It would not be too great an exaggeration to say that a significant proportion of fishery marine areas, or of particular stocks, are in crisis, and that a range of problems, e.g., coastal zone degradation, pollution, over exploitation, habitat destruction, are increasing in an exponential way. Since the results of the crisis are variably manifest in the spatio-temporal domain, then decisions on their management and remediation can best be made using tools that are efficient at functioning in this domain, i.e., geographical information systems.

The concern of this chapter is specifically directed at fisheries "management" rather than the wider area of fisheries "science". This is because the content of much of the rest of this volume includes facets of the subject that both directly and indirectly broach the periphery between fisheries management and science. Here, then, confinement is to the imposition of authoritative controls on, support to and measurement of, fishery activities, plus those less direct management concerns that have an immediate relevance to fisheries success, such as those relating to natural fish habitats and environments. It is also the intention to cover the area of mariculture (marine fish farming), both as a marine-based activity, and as an activity which takes place in brackish waters at the marine-coastal interface. Another concern of the chapter is a consideration of the conceptual field of the subject, i.e., the nature of the considerations that must be confronted when applying GIS to fisheries management, including the complexity of the functioning domain into which we are entering. A review of the progress that has been made to date in the overall field will be included, and the main concern will be to examine the present status and fields of interest in this subject. The conclusion will largely examine the degree to which GIS penetration is now occurring in the fisheries field, and at some of the areas into which it still has to move.

15.2 PERCEPTIONS OF THE ADOPTION TASKS FOR GIS IN FISHERIES MANAGEMENT

GIS applications to terrestrial-based management concerns are widespread and diverse. The increasing rate of all kinds of GIS implementations over the past decade indicates at the least a large degree of interest, and at the best the likelihood of many successful GIS adoptions. Given this proliferation, then it is likely that GIS could hold the key to better management and control of the numerous spatially related problems that face the fisheries sector, and indeed there is recognition in this sector that this might be the case (Hinds, 1992; Loayza and Sprague, 1992). However, as will be shown in Section 15.4, the rate of GIS adoption in supervisory management by fishery authorities has been very slow. Why is this? In order to answer this question, a three-stage conceptual model will be proposed, which relates to the actual task of GIS adoption in a complex innovations milieu.

15.2.1 Components in a GIS Fisheries Task Conceptual Model

Any GIS related task must consider its range of "task components". Thus, at its very basic level, a mapping task must include three components: (i) an object, located (ii) in space, and (iii) in time, and a GIS task would normally (for example simply using GIS as a spatially-referenced database for search and retrieval does not require space-time changes) require that additionally some space/time change be examined and/or that extra objects also be considered. It is possible to conceive of the main task components relating to the use of GIS in fisheries management as being those shown in Figure 15.1. Although the components will be inter-related, for conceptual purposes they are displayed here as separate inputs to the GIS, with each having a quantifiable range for any given task in terms of quantity, range, speed, scale, etc. The actual quantitative value chosen for the task will relate to its degree of complexity and the output precision required. The acquisition of data for each task component will incur considerations relating to data collection and data entry methods, cartographic representation, object and process classification, etc., i.e., those factors which allow the real fisheries world to be suitably modelled in a digital environment.

Although any GIS task must relate to at least task components 1, 2 and 4 (see Figure 15.1), it is clear that GIS tasks for fisheries management will mainly require a consideration of all seven components. Thus, as well as the basic components, activities associated with fisheries management take place in 3-D space, thereby requiring a consideration of the vertical plane. Also, the major elements of the fishery environment are dynamic, meaning that processes must be incorporated. Both the processes and the objects themselves will exhibit varying manifestations of "dynamism", i.e., regarding speed or regularity of movement. Given this complex yet essential range of task components, then we can begin to see why fisheries management adoptions of GIS might be slow to materialise.

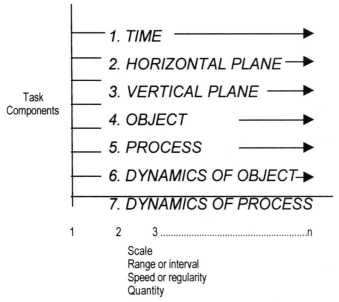

Task
Components

1. *TIME*

2. *HORIZONTAL PLANE*

3. *VERTICAL PLANE*

4. *OBJECT*

5. *PROCESS*

6. *DYNAMICS OF OBJECT*

7. *DYNAMICS OF PROCESS*

1   2   3 ....n

Scale
Range or interval
Speed or regularity
Quantity

Figure 15.1 Task components in Stage 1 of a conceptual model for a fisheries management GIS.

15.2.2 Areas in a GIS Task Complexity Model

As well as viewing the GIS task in terms of its individual input "task components", it is instructive to form a concept of the overall task. The core of any GIS task can be usefully divided into two main "task areas" of consideration, i.e., GIS functionality and spatial data. So, for a GIS to produce results, then its two basic requirements are the ability to perform a range of GIS functions (data input, manipulation, searching, analysis, display, etc.), and the supply of a quantity of at least 2-D geo-referenced (or mappable) data on which the GIS functions will be performed. Within a GIS task, both of the task areas can be shown to have x and y variables which relate to quantity and complexity, and the task areas can be modelled as shown in Figure 15.2. Here the spatial (or mapped) task area on the right has been joined to the GIS functionality task area on the left. The x-axis records both the number of maps (layers) and the number of GIS functions, with quantity increasing from the centre. The y-axis for both task areas records increasing complexity. Complexity in the GIS task area might be viewed in conceptual terms relating to a hierarchy of difficulty in programming or understanding, hard-ware requirements, the acquisition of data and the range of sources and data formatting requirements, etc. In the mapping task area it might relate to increasing intricacies associated with more complex surfaces or accuracy levels required, the periodicity of mapping required, the ease of data acquisition, or it might simply relate to a wider geographic area being considered.

Radiating around the centre (zero point) of the *x*-axis are lines which show increasing total task complexity. In the case of a fisheries management GIS, total task complexity can also be thought of as consisting of a composite of the seven task components which were outlined in Figure 15.1, i.e., each task component increases in quantity and/or complexity as it moves outwards from the central (zero) point.

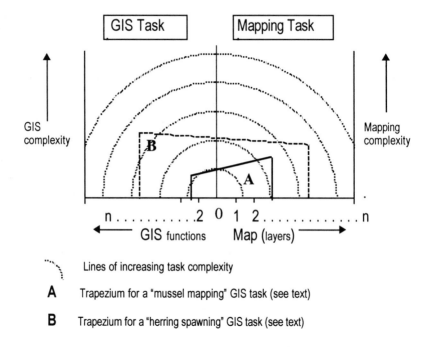

Figure 15.2 Stage 2 of a conceptual model showing "task areas" in a fisheries management GIS.

For any single GIS task, it is possible to draw a trapezium whose shape, position and area within the model reveals facets of the complexity of the task. In Figure 15.2, the trapezia for two hypothetical fisheries related GIS tasks have been drawn. "A" indicates a low complexity task such as "mapping mussel distribution in a rock pool". Here the GIS is simply required to perform basic mapping which might comprise of the overlaying of perhaps three layers, i.e., depth, rock distribution and mussel quantity. Conversely, the "B" trapezium indicates a higher complexity task such as "analysing the spatial relationship between herring spawning and the marine ecology in the River Thames estuary". Here the number of mapped layers will be larger, indicating high quantitative data inputs, the actual area being considered is also larger, and the range and complexity of GIS functionality goes well beyond simple map overlays. Clearly, the relative trapezium area indicates the overall complexity of the GIS/mapping task being undertaken, and the shape indicates the balance between mapping and GIS inputs. Relative to other mainly terrestrial-based GIS tasks, it would be true to acknowledge that present GIS applications to fishery tasks would be oriented more towards the centre (zero point) of the Stage 2 model. This "retardation" relates to data demands when working in a mobile, dynamic marine environment having both an infinity of spatio-temporal scales and of process combinations and fluctuations.

15.2.3 A Total Fisheries Management GIS Conceptual Model

Thus far, the conceptual model has been developed in an idealised or theoretical environment, with little allowance having been made for the real world. Figure 15.3 illustrates how a greater degree of reality may be incorporated. Adoption or implementation of the GIS must take place within a set of socio-economic constraints. Here, the x axis plots diminishing economic constraints, i.e., the levels of funding for, or inputs to, the GIS task, and the y axis illustrates those constraints which may be considered as social, i.e., cultural, developmental, political, institutional, etc. It can be conceived that the task complexity model (Figure 15.2) can now be variably superimposed upon the diagram, according to the prevailing socio-economic conditions, to complete a total fisheries management GIS conceptual model.

The purpose of exhibiting the conceptual models has been to reveal something of the complex milieu in which a fisheries management GIS must operate. It also illustrates that there is likely to be a huge range of different "fisheries management GISs", both in their levels of sophistication and in their management objectives. The model is useful in that it hints at the possible causes for implementational and GIS functioning delays and problems, although other problems such as:

- the functional design of 4-D databases;
- defining boundaries in a transitory (fuzzy) environment;
- the selection of appropriate temporal and spatial scales for data collection and mapping;
- allowance for statistical variance in data collected;
- the diversity, fragmentation and widely scattered nature of fishery activities, have scarcely been considered. The various levels of "task component" and "task area" complexities, plus the fact that fisheries management per se functions in a wide range of socio-economic and operational milieu, clearly reveals the likely array of factors accounting for the slow rate of GIS adoption in the fisheries management arena.

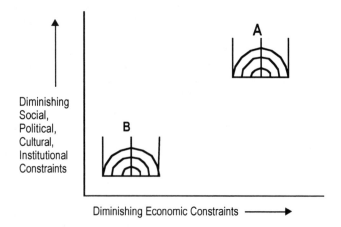

A   GIS fisheries management task being performed in an advanced
    social economy
B   GIS fisheries management task being performed in a more primitive
    social economy

Figure 15.3 Stage 3 of a conceptual model showing GIS tasks relative to socio-economic
development constraints in which the fisheries management GIS is operating.

15.3 EARLY PROGRESS IN THE ADOPTION OF GIS FOR
FISHERIES MANAGEMENT

Given the extraordinary range of barriers to GIS adoption described above, then it
was not until the mid-1980s that any form of fisheries management GISs appeared.
Although the impetus for adoptions would obviously have come from
implementations in various terrestrial based GIS applications, there are a number of
key initiators which might be identified as landmarks in propelling GIS into the
fisheries arena. In 1984 the Food and Agriculture Organisation (FAO) of the United
Nations (UN) held a workshop in Rome (Food and Agriculture Organisation, 1985)
to demonstrate ways in which remote sensing could be of help in aquaculture or
inland fisheries. Among the contributions were a number of sophisticated mapping
and/or remote sensing (RS) applications showing, for instance, how shrimp farm
locations could be planned (Mooneyhan, 1985), how algal growth could be
monitored (Travaglia and Lorenzini, 1985) and the establishing of inventories of
inter-tidal zones (Loubersac, 1985). Each of the contributions had a
mapping/analytical emphasis. In 1985, a paper was presented which provided the
first recognition of the importance of using a range of mapping techniques as

management tools in solving fisheries related problems (Caddy and Garcia, 1986). Here the authors stressed the need to apply computing power and suitable software for map construction and manipulation. At about the same time, Warner (1987) produced the first paper specifically applied to GIS in fisheries management, and part of its title gave a hint of the problems involved, i.e., "Mapping the Unmappable...". Although all of these potential initiators discussed GIS-related techniques, these were not yet actually being applied.

When GIS applications to fisheries management did arrive, all of the first were made in the sphere of aquaculture location. It is easy to see that this field offered the most ready opportunity since its inputs (production variables) are basically static and terrestrial-based, with the GIS functions largely being confined to overlaying, buffering, reclassification and Boolean operations, all of which are basic operations for both the GIS software and the GIS personnel involved. Examples of studies during this period (mid-1980s to early 1990s) include Kapetsky's (1989) work on developing a GIS for seeking further opportunities for mariculture in Johore state; studies which specifically sought sites for shrimp farming (Kapetsky *et al.* 1987; Paw *et al.*, 1993; Aguilar-Manjarrez and Ross, 1994) or for oyster culture (Durand *et al.*, 1994), and the use of GIS in site selection determination for salmon cages (Ross *et al.*, 1993). Where raw input data for these studies was difficult to acquire, as it often was, then the authors showed how an ingenious array of "proxy" data had sometimes to be assembled.

During the period that these early GIS studies were being undertaken, RS (usually satellite) data sources were increasingly being used as a means of overcoming the dearth of requisite data. As well as RS data being used in a variety of aquaculture location projects (e.g., Kapetsky, 1987; Cordell and Nolte, 1988), it was used also used in various other fisheries related tasks:

- Monitoring of fishing effort.
- Measuring and monitoring of seaweed resources.
- Tracking of algal blooms.
- Tracking pollutant or turbidity plumes.
- Measuring sea surface temperatures in conjunction with fish concentrations.
- Estimating the size and availability of various marine ecosystem or habitat types.
- Establishing shallow water bathymetry.

Examples of these uses have been referred to in Simpson (1992), Meaden and Kapetsky (1991) and Johannessen *et al.*, (1989).

As noted above, all the early uses of fisheries GIS had been as a tool to help in aquaculture location. This restrictive usage had been recognised in Simpson's (1992) seminal paper which noted that "Today, RS/GIS systems are not commonplace in the operational fisheries oceanography community" (p. 266). Simpson was instructive in recognising both the importance of ascertaining the needs of the fisheries industry before applying GIS, and in suggesting that it was likely that GISs would each need to be specifically fine-tuned to individual fishery requirements. These were important considerations in a technological sector where the tail had often wagged the dog! Thus, too frequently GIS was being seen as having itself imposed upon an unready audience. Although Simpson could say very little about the fishery applications of GIS *per se*, he did offer useful advice on implementation designs.

During the earlier part of this decade (1992 to 1996), GIS slowly expanded its range of fishery applications. By far the most important area into which it moved was that of marine habitat mapping and analysis. Much of this work consisted of basic data gathering and mapping of various important habitat types, e.g., mangroves (Long and Skewes, 1996), sea grasses (Long *et al.*, 1994), sediments and hydrology (Somers and Long, 1994), plus the detailed mapping of whole littoral environments (Liebig, 1994). Some of this data has been used for subsequent GIS based analyses. There was a miscellany of other GIS work undertaken, including the use of GIS to map and estimate marine productivity (Caddy *et al.*, 1995), plus the setting up of various integrated GIS/management information systems (MIS) for specialised activities such as fishery protection (Pollitt, 1994), or simply as MISs which could be utilised for a range of marine and fishery purposes (e.g., Li and Saxena, 1993). Other GIS work had analysed the impact of anthropogenic activities on fish habitats (e.g., Wood and Ferguson, 1995), and Gordon (1994) looked at using GIS in planning for artificial aquatic habitats.

15.4 THE PRESENT FIELDS AND STATUS OF GIS APPLICATIONS IN FISHERIES MANAGEMENT

It would be fair to state that, by the mid-1990's, GIS had become established, if still on rather an exploratory basis, as a tool which could be utilised in a variety of fisheries management functions. Not only was there a range of studies, which exhibited a comparatively high degree of sophistication, but the avenues of usage, were growing exponentially. However, if we examine the institutional array of this GIS activity, it is clear that it is restricted to a number of very specialised small "enclaves". These include a few international organisations such as the UN's FAO and International Center for Living Aquatic Resources Management (ICLARM), national or state governmental departments such as the National Marine Fisheries Service (NMFS) or the United States Geological Survey (USGS), some national agencies or institutions such as IFREMER in France, IMR in Norway, or the Commonwealth Scientific and Industrial Research Organisation (CSIRO) in Australia, plus a number of specialised programs in universities such as Stirling in Scotland and Oregon State in the US. Undoubtedly, each of these enclaves would either have had working within them individuals who had a vision of the technology's potential, and who had access to development funding, or the in-house GIS progression would have been spawned from developments in the remote sensing field. What stands out sharply is that there is little evidence, except in the USA, that the institutions utilising GIS for fisheries management are the fishery management authorities, either regionally, nationally or internationally. Even the work being carried out by the fishery authority in the US (the NMFS within National Oceanic and Atmospheric Administration or NOAA) was almost all unrelated to direct fisheries management in its authoritative support and control sense. By the mid-1990s GIS had failed to move into the applied management field, though at the very end of this decade there are some indications that this situation is beginning to change (see below).

What then is the present situation regarding GIS applications to fisheries management? Initially, it is clear that the various fields of work described in Section 15.3 are still thriving, but newly-emerging fields are now also receiving much attention. Basic work on marine habitat mapping is spreading rapidly (Meaden, 1996), and increasingly traditional cartographic mapping techniques, or mapping software, is being replaced by dedicated GIS packages in the realisation that this will allow for a greater variety of analyses and output. Further work is also being pursued into applications of GIS to aquaculture location. What is especially important here, is that some of this work is now refined to the stage whereby GIS Units are being established by government institutions or departments to plan for or to manage their coastal aquaculture. An example of this "applied GIS" is a project being funded by the FAO to set up a GIS for the zoning of coastal aquaculture in Sri Lanka (FAO Project No. TCP/SL/6712).

One of the two significant advances in the application of GIS to fisheries management, is in the mapping of catch or fishing effort distributions, and in the matching of these to basic environmental or habitat parameters. Insufficient data, or data of very low statistical significance, had precluded earlier efforts at mapping catches in anything other than for very limited areas. An example of this recent interest in catch and effort mapping is seen in work carried out under European Union (EU) funding, which reveals how cuttlefish (*Sepia officinalis*) catches made by the collective French fleet can be mapped (Denis and Robin, 1998). Thus, Figure 15.4 shows, by located proportional histograms, landed catches of cuttlefish made per month per ICES rectangle. Under this same project, attempts have been successfully made to match various cephalopod species with sea surface temperatures in the north east Atlantic, though the results from the GIS analyses indicate that other environmental controls tend to dominate during summer months. What these controls are has yet to be determined (Koutsoubas *et al.*, 1998; Pierce *et al.*, 1998). Sakurai *et al.,* (1998) have taken this type of work further by developing a GIS-based model which claims to be able to predict cephalopod (*Todarodes pacificus*) stock fluctuations according to shifts in the local climatic regime. Both Brown *et al.*, (1998) and Fogarty and Murawski (1998) have recently used GIS methods to map fishing effort on the Georges Bank (off the New England coast of NE USA) utilising otter trawl data provided by both the NMFS and the Northwest Atlantic Fishery Organisation (NAFO).

Although Fogarty and Murawski infer a relationship between demersal (and benthic) species and bottom sediment type, the scale of this mapping, i.e., using 10-minute latitude by 10 minute longitude cells, possibly precludes matching of the two data sets, i.e., in an area where bottom sediments are quite diverse. Two final studies which show varying methodologies within a GIS framework, are the works of Castillo *et al.*, (1996) and Booth (1998). The former used a GIS to analyse the distribution of anchovy, sardine and jack mackerel off the coast of northern Chile, deriving their data via the use of hydroacoustic estimations. The distribution of the three species was found associated with the occurrence and intensity of thermal and haline fronts. Booth used generalised additive modelling (GAM) to model the spatial distribution of sparid fish (*Pterogymnus laniarius*) on the Agulhas Bank, South Africa (Figure 15.5). Working within a GIS environment, use of the GAM allowed for very detailed spatial and temporal relationships to be established between the fish distributions and a variety of environmental parameters.

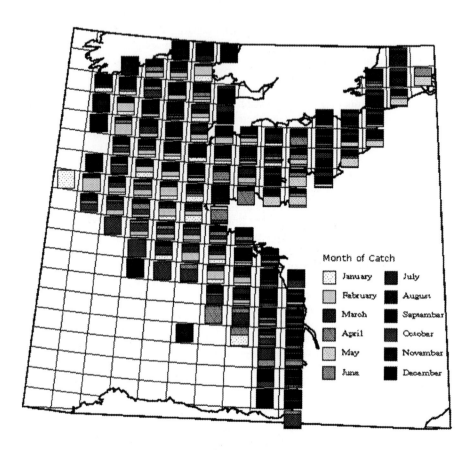

Figure 15.4 The monthly breakdown of 1996 French landed cuttlefish catches made per ICES rectangle area (adapted from Denis and Robin, 1998).

Taking the mapping and analyses of the relationships between species abundance and various environmental parameters a stage further, interesting work is now emerging which seeks to model or simulate these relationships as a means

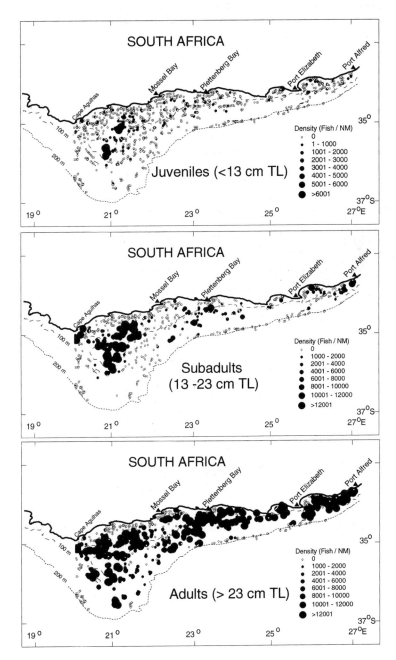

Figure 15.5 Spatial distribution of three size categories of sparid fish on the Agulhas Bank, South Africa (after Booth, 1998).

of predicting what might be happening in other marine areas. A main centre of this activity is in Florida, USA, a state where both the federal and national governments are concerned to maintain or enhance marine areas which are under considerable threat from population increases allied to preferences for water based recreation. Ault (1998) has demonstrated a sophisticated "fly-through" video animation, based on GIS/graphics, which simulates the entire life history of the pink shrimp in Biscayne Bay, incorporating likely patterns of growth, production, recruitment and migration as they are related to hydrographic variables. Also in Florida, Rubec *et al.*, (1998) have shown how a GIS is being used for habitat suitability modelling (HSM) (Figure 15.6). Indeed, "...these models, when linked to a GIS, may provide a means of mapping species distributions in estuaries not currently surveyed" (p. 22). An interesting variant on fisheries modelling is that proposed by Caddy and Carocci (in press). Using simulation methods in a GIS environment, the authors set up a model which attempts to show the amount of interaction which might occur between vessels from neighbouring ports, i.e., given ubiquitous conditions in nearshore shelf regions. They do this by creating a "dynamic Gaussian" model showing that fishing intensity gradually moves outwards from a port over time, so that eventually fishing activities from neighbouring ports overlap. The authors suggest that use of a GIS will be a valuable prediction tool in forecasting the ensuing management problems.

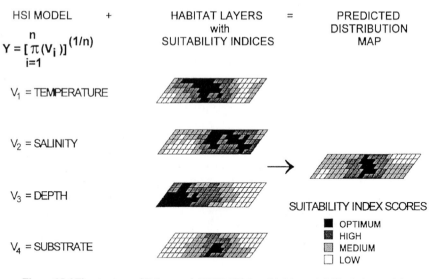

HSI MODEL   +

$$Y = \left[\prod_{i=1}^{n} (V_i) \right]^{(1/n)}$$

V_1 = TEMPERATURE

V_2 = SALINITY

V_3 = DEPTH

V_4 = SUBSTRATE

HABITAT LAYERS   =
with
SUITABILITY INDICES

PREDICTED
DISTRIBUTION
MAP

SUITABILITY INDEX SCORES

■ OPTIMUM
▨ HIGH
▦ MEDIUM
□ LOW

Figure 15.6 The structure of Rubec *et al.* (1998) GIS-based habitat suitability index model.

The second significant advance in the adoption of GIS for fisheries management is in the ability to utilise the captured data on fishing catch and effort as a means of directly monitoring and managing the fishery. Fisheries managers in most developed world fisheries have long recognised that one means of better stock control was to introduce fisheries logbooks. These have been operating in some marine areas for over a decade, and they nearly all capture fishery-related data by means of hand-written forms. It became clear to some observers that this task could be more effectively undertaken via electronic logbooks, especially given the fact that most larger vessels were already equipped with computing functionality. A number of systems have now been developed which are capable of performing a variety of catch and effort data capture tasks, and systems are adaptable for both authority stock assessment survey work and for monitoring and managing commercial fishing vessel activity.

One of the leading institutions in the applications of GIS to direct fisheries monitoring has been the French institution IFREMER. By 1995 they had published work on a prototype fisheries GIS (Le Corre, 1995), which showed the capacity to map trawling activities, in this case in the Mediterranean, and to match trawl tows against various other parameters such as simulated management zones, bathymetry or bottom sediment types. Similar software has also been developed by C-MAP Environmental in developing their "EchoBase" software (C-MAP Environmental, 1995). IFREMER's work has now made considerable advances, and their latest oceanographic research vessel, the R/V *Thalassa*, is equipped with a developmental data capture system called "FishView". Fish biomass data, based on actual catches or on acoustic estimates, are captured and entered to an electronic "logbook" database (called Casino) along with other hydrographic data captured by external instrumentation. Data from Casino can be exported to FishView (which is an adaptation of ArcView plus its associated Spatial Analyst programme), for further analyses and cartographic or graphical display (Durand *et al.*, 1998). An example of output from one of *Thalassa*'s 1998 surveys is given in Figure 15.7. Nichols (1998) has demonstrated a similar catch and effort monitoring system which has been developed in Canada. Meaden and Kemp (1998) have now produced an enhanced second version of their fisheries computer-aided management (FishCAM) software (Meaden and Kemp, 1996). FishCAM2000 operates as two modules. First, a vessel-based module provides for the real-time digital capture of fishing catch and effort by linking an on-board computer to a Global Positioning System (GPS). The captured spatio-temporal data, together with fishing vessel details, provides the substrate for the second resource management module. In this module, aggregated data from all vessels in a fleet, or in a management area, is collected by the authorities for multidimensional analyses, for long term modelling or for carrying out mapping or querying tasks. Figure 15.8 illustrates catch data which has been aggregated for cells in the eastern English Channel. The management module comprises of a purposefully-structured object-relational database management system integrated to ArcView GIS in a workstation computing environment. FishCAM2000 can be used either as a means of monitoring commercial fishing vessel activity, or for data capture and analysis by stock assessment authorities.

Given that marine fisheries takes place in a spatially extensive milieu, which itself maybe highly variable in both the 2.5-D and 3-D planes, then the quantities of data which may be required for reliable output from a GIS may be substantial. Recognition of this fact has led a number of people or groups to place major emphasis on the establishment of data sets, metadatabases, data archives, etc, as their main objective. Once sufficient data has accrued, then it is a relatively simple task to integrate this to a GIS for visual output. Work has been continuing in this direction for a number of years. Meaden and Do Chi (1996) show how the eastern seaboard of Canada is probably the best served with regard to data access, largely through the work of the Atlantic Coastal Zone Information Steering Committee (ACZISC). This committee has a workplan which itself has a number of essential components contributing to an information structure. Thus ACZISC administer:

- An ACZISC database directory containing over 600 databases.
- A coastal mapping working group who identify the major needs in terms of requisites for mapping.
- A coastal mapping project inventory listing the mapping work accomplished.
- A data exchange and standards group who facilitate the easy flow of data and who encourage uniformity in standards.

Further information and access to the ACZISC data can be made through their web site on http://is.dal.ca/aczisc/aczisc. Similar, if smaller, projects are beginning to appear elsewhere. Long (1998, pers. comm.) describes how a major project is now underway in Indonesia to create a Spatial Information Digest (SID) which will integrate the information on the marine resources of Indonesia. Other projects have been aimed at demonstrating how marine databases may best be developed (Valavanis *et al.*, 1998), and Durand (1996) has shown how IFREMER in France is developing a graphic programme (Patelle), based on ArcView GIS, which is capable of integrating and displaying a wide range of marine based data sets.

A final sphere into which GIS marine fishery applications are now moving is in the ability to provide for compendiums of data, i.e., usually in the form of atlases. Presently, most developments take the form of paper-based publications, though there are examples of marine digital atlases, e.g., the UKDMAP of the North Sea (produced by BODC at Birkenhead, UK), and the information section of the FAO in Rome (WAICENT) is presently engaged in the production of an interactive GIS-based atlas which will make use of the FAO's large array of data sets. Paper-based atlases include a mapping study of the demersal species along the Italian coast, which includes 440 separate maps (Relini, 1991), an atlas of the Gulf of Cadiz (Spain) showing various species distributions in relation to bathymetry and bottom sediment types (Ramos *et al.*, 1996), and an atlas showing tuna and billfish catches by species in the Pacific Ocean (Carocci and Majkowski, 1996). The use of GIS is particularly suited to the production of atlases since it allows for the rapid output of an almost unlimited variety of cartographic display and visualisation preferences.

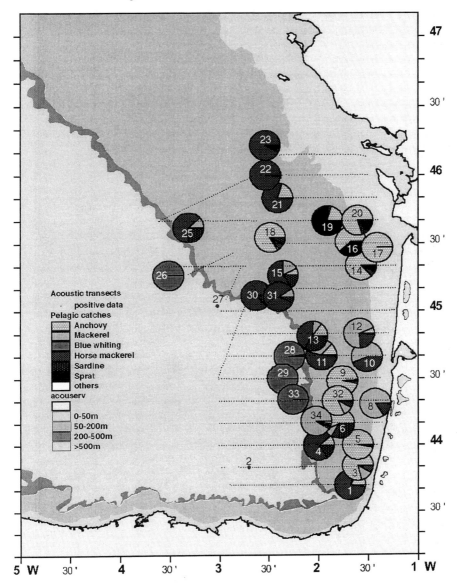

Figure 15.7 GIS output from the IFREMER survey vessel *Thalassa*, showing trawl survey tows and catches in the Bay of Biscay superimposed on stratified survey trajectories
(after Durand *et al.*, 1998).

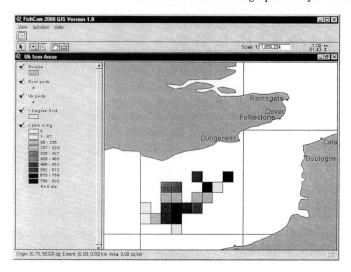

Figure 15.8 Output from FishCAM2000 illustrating aggregated fleet catches in the eastern English Channel. Here 10 km cells have been used, but resolution is multi-dimensional (after Meaden and Kemp, 1998).

15.5 CONSIDERATIONS AND CONCLUSIONS

It is possible to view the adoption of any new technology by an industrial or social sector, as occurring via a number of stages. The adoption of GIS for fisheries management might be conceived as occurring sequentially as in the stages shown in Figure 15.9. Worldwide the GIS adoption process in fisheries will clearly be in various stages, but the most advanced adoptions would be in Stage 4. What evidence is there for this? Sections 15.3 and 15.4 of this chapter have provided numerous illustrations of the work currently underway. Thus, an examination of the literature has revealed the number and variety of projects, which have proliferated over the past three or four years. In 1999, the first International Symposium of GIS in Fisheries Science was held in Seattle, Washington, USA, and here some 120 posters, oral papers or software demonstrations were presented by delegates from nearly 20 countries.

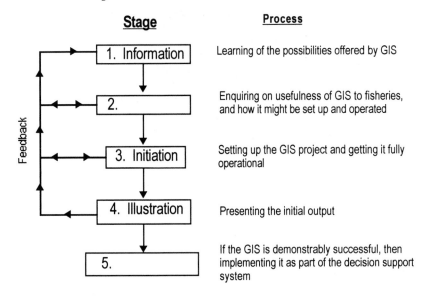

Figure 15.9 Hypothetical stages in the adoption of a GIS for fisheries management.

Although this symposium covered the wider field of fisheries science, it is indicative of the present variety of interest. Despite this enthusiasm for the promulgation of ideas, Stage 5 barely appears to have been reached. To some extent this may not be a true reflection of developments, i.e., in that what constitutes full implementation or adoption will vary according to how "fisheries management" itself is structured. For instance, there may be implementations of GIS where perhaps a strong individual initiator within a fisheries management authority is using this technology, whereas the authority as a whole has not instigated a dedicated GIS unit or section. Undoubtedly, there are not yet full working examples of the latter, though they are on the brink of inception. In the past schemes for a total fisheries management GISs have been proposed (Meaden and Reynolds, 1994; FAO-CRODT-ORSTOM, 1995; Al-A'ali and Bakiri, 1996), but these have either failed or only been partially adopted.

A factor of considerable importance to GIS in fisheries management is the degree of interchange of ideas and support which reverberates between fisheries management and associated disciplines, plus the wide breadth of these disciplines. Unfortunately, space has precluded an examination of these. Examples of important associated areas which have been excluded include fishery navigation, fisheries oceanography, database developments, system's architecture, 3-D visualisation, spatial analysis and the application of remote sensing to fishery activities. Whilst the fact that such a breadth of disciplines exists gives considerable scope for applications, it also means that developments are both hard to keep abreast of and that they occur in a somewhat fragmented way. Information dissemination flows are problematic, and much of the work fails to reach the established literature.

Despite the breadth of work in progress, there are fishery management areas into which GIS has barely penetrated. It can be safely postulated that the following areas will prove attractive for future GIS based analyses:

- Regulatory modelling : examining the repercussions of quota impositions, gear restrictions, area closures, etc.
- Stock enhancement -timing and selection of sites for artificial stocking.
- Marine reserve allocation: identifying areas according to management predefined criteria.
- Creation of economic surfaces : establishing variable profit surfaces according to stock estimations and prices obtaining in competing ports.
- Enhancement of stock assessment processes   through the capability of applying easily varied modelling algorithms.
- Fish occurrence or abundance relative to oceanographic processes, a refinement of the present examinations of the fish-environment associations.
-

Obviously, this list is far from exclusive, and largely it depends on the interpretation of "fisheries management". One factor that will spawn this enhanced range of GIS applications is the array of interested parties or groups who are legitimately engaged in fisheries management. These comprise not only of various local, national and international fishery authorities and organisations, but also of university departments, research institutions, conservation groups and private corporations. Again, this diverse list poses the dangers of fragmentation, and the difficulties, which this spawns in achieving well-directed output and the funding for necessary research and development.

Progress in applying GIS to fisheries management has been rapid, diverse and imaginative, and it will be most surprising if this does not continue in the immediate future. Although there are immense numbers of problems and difficulties which must be faced, some of which are alluded to in the final chapter of this book, possibly the greatest question of all will remain the very existence of a need to consider "The application of GIS to (in this case) fisheries management". Nobody looks at the application of "word processing", or "spreadsheets" or "statistical packages" to fisheries management, so why should we consider "GIS" as a special case? It could be said that nobody ever did look at specialist applications of word processing, etc., so GIS is a special case. But whereas it appears that all of the early work using GIS acknowledged its use, there is already research being reported which relies upon GIS for much of its analysis, but which barely recognises this fact (e.g., Fogarty and Murawski, 1998; Brown *et al.*, 1998). Will GIS have arrived when it becomes unnoticed?

15.6 REFERENCES

Aguilar-Manjarrez, J. and Ross, L.G., 1994, GIS-based environmental models for aquaculture development in Sinaloa state, Mexico. In *Proceedings of European Aquaculture Society Symposium No.21*(Bordeaux, France).
Al-A'ali, M. and Bakiri, G., 1996, A GIS for Bahrain fisheries management. In *Proceedings of the 1996 ESRI User Conference* (Redlands, California: Environmental Systems Research Institute).

Ault, J. S. and Luo, J., 1998, Coastal bays to coral reefs: Visualizations of a spatial multi-stock production model. In *ICES Annual Science Conference* Vol. S:1 (Lisbon, Portugal: ICES CM 1998).

Booth, A.J., 1998, Spatial analysis of fish distribution and abundance patterns: A GIS approach. In *Fishery Stock Assessment Models; Alaska Sea Grant College Program* (AK-SG-98-0).

Brown, R.W., Sheehan, D. and Figuerido, B., 1998, Response of cod and haddock populations to area closures on Georges Bank. In *ICES Annual Science Conference,* Vol. U:9 (Lisbon, Portugal: ICES CM 1998).

C-MAP Environmental, 1995, *EchoBase: Geographical Data Base for Management and Processing of Survey Data on Aquatic Resources* (Marina di Carrara, Italy).

Caddy, J.F. and Garcia, S., 1986, Fisheries thematic mapping: A prerequisite for intelligent management and development of fisheries. *Oceanographie Tropicale,* **21**, pp. 31–52.

Caddy, J.F. and Carocci, F., in press, GIS applications and the spatial allocation of fishing intensity from coastal ports. *ICES Journal of Marine Science.*

Caddy, J.F., Refk, R. and Do Chi, T., 1995, Productivity estimates for the Mediterranean: Evidence of accelerating ecological change. *Ocean and Coastal Management,* **26**, pp. 1–18.

Carocci, F. and Majkowski, J., 1996, *Pacific Tunas and Billfishes: Atlas of Commercial Catches* (Rome: Food and Agriculture Organisation of the United Nations).

Castillo, J., Barbieri, M.A. and Gonzalez, A., 1996, Relationships between sea surface temperature, salinity, and pelagic fish distribution off northern Chile. *ICES Journal of Marine Science,* **53**, pp. 139–146.

Denis, V. and Robin, J.P., 1998, Present status of French Atlantic fishery for cuttlefish (*Sepia officinalis*). In *ICES Annual Science Conference,* Vol. M:43 (Lisbon, Portugal: ICES CM 1998).

Durand, C., 1996, *Présentation de Patelle: Programmation Graphique de Chaînes de Traitement SIG sous ArcView,* IFREMER: Note Technique TC/04 (Plouzané: Sillage/IFREMER).

Durand, H., Guillaumont, B., Loarer, R., Loubersac, L., Prou, J. and Heral. M., 1994, An example of GIS potentiality for coastal zone management: Pre-selection of submerged oyster culture areas near Marennes Oleron (France). In *EARSEL Workshop on Remote Sensing and GIS for Coastal Zone Management* (Delft, Netherlands: EARSEL).

Durand, C., Loubersac, L. and Massé, J., 1998, Operational GIS applications at the French oceanographic research institute IFREMER. In *Proceedings of the 1996 ESRI User Conference* (Redlands, California: Environmental Systems Research Institute), p. 360.

FAO-CRODT-ORSTOM, 1995, *A Simulated GIS Exercise to Demonstrate its Usefulness in the Management of Senegalese Demersal Fisheries,* Report of the Training Course on the Application of GIS to Fisheries: FAO Project GCP/RAF/288/FRA (Rabat, Morocco: Food and Agriculture Organisation of the United Nations).

Fogarty, M.J. and Murawski, S.A., 1998, Large-scale disturbance and the structure of marine systems: Fishery impacts on Georges Bank. *Ecological Applications,* **8,** pp. S6–S22.

Food and Agriculture Organisation of the United Nations, 1985, *Report of the Ninth International Training Course on Applications of Remote Sensing to Aquaculture and Inland Fisheries,* RSC Series 27 (Rome: Remote Sensing Centre, FAO).

Gordon, W.R., 1994, A role for comprehensive planning, geographical information system (GIS), technologies and program evaluation in aquatic habitat development. *Bulletin of Marine Science,* **55**, pp. 995–1013.

Grainger, R.J.R. and Garcia, S.M., 1996, *Chronicles of Marine Fishery Landings, 1950-1994: Trend Analysis and Fisheries Potential,* FAO Fisheries Technical Paper No.359 (Rome: Food and Agriculture Organisation of the United Nations).

Hinds, L., 1992, World marine fisheries – Management and development problems. *Marine Policy,* **16**, pp. 394–403.

Johannessen, O.M., Kloster, K., Olaussen, T.I. and Samuel, P., 1989, Application of remote sensing to fisheries. In *Final Project Report to the CEC's Joint Research Centre* (Oslo, Norway: CEC), p. 111.

Kapetsky, J.M., 1989, *A Geographical Information System for Aquaculture Development in Johor State,* FAO Technical Cooperation Programme, FI:TCP/MAL/6754 (Rome: Food and Agriculture Organisation of the United Nations).

Kapetsky, J.M., McGregor, L. and Nanne, E.H., 1987, *A Geographical Information System to Plan for Aquaculture: A FAO-UNEP/GRID Study in Costa Rica,* FAO Fisheries Technical Paper No.287 (Rome: Food and Agriculture Organisation of the United Nations).

Koutsoubas, D., Valavanis, V.D. and Georgakarakos, S., 1998, A study of cephalopod resource dynamics in the Greek seas using GIS. In *Proceedings of GISPlaNET'98* (Lisbon, Portugal: GISPlaNET).

Le Corre, G., 1995, Propositions d'usages de SIG en halieutique. In *Etude et Séminaire sur les Systèmes d'Information Géographique (SIG, en Méditerranée.* Vol.2, Contributions au Séminaire, Projet TR/MED/92/013 (Plouzané: IFREMER-CEE).

Li, R. and Saxena, N.K., 1993, Development of an integrated marine geographic information system. *Marine Geodesy,* **16**, pp. 293–307.

Liebig, W., 1994, Protecting the environment: GIS and the Wadden Sea. *GIS Europe,* **3**, pp. 34–36.

Loayza, E.A. and Sprague, L.M., 1992, *A Strategy for Fisheries Development,* World Bank Discussion Paper – Fisheries Series No.135 (Washington, D.C.: The World Bank).

Long, B.G. and Skewes, T.D., 1996, A technique for mapping mangroves with Landsat TM satellite data and geographic information systems. *Estuarine, Coastal and Shelf Science,* **43**, pp. 373–381.

Long, B.G., Skewes, T.D. and Poiner, I.R., 1994, An efficient method for estimating seagrass biomass. *Aquatic Botany,* **47**, pp. 277–291.

Loubersac, L., 1985, Study of intertidal zones using simulated SPOT data: Inventorying of aquaculture sites in the intertropical zone. In *Report of the Ninth International Training Course on Applications of Remote Sensing to Aquaculture and Inland Fisheries.* RSC Series 27 (Rome: Remote Sensing Centre, FAO), pp. 261–272.

Meaden, G.J., 1996, Potential for geographical information systems (GIS), in fisheries management. In *Computers in Fisheries Research,* edited by Megrey, B.A. and Moksness, E. (Chapman and Hall, London), pp. 41–79.

Meaden, G.J. and Kapetsky, J.M., 1991, Geographical information systems and remote sensing in inland fisheries and aquaculture. In *FAO Fisheries Technical Paper No.318* (Rome: Food and Agriculture Organisation of the United Nations).

Meaden, G.J. and Reynolds, J.E., 1994, Establishing a marine fisheries GIS: The Libyan experience. In *Libfish Technical Briefing Notes No.15* (Tripoli, Libya: Food and Agriculture Organisation of the United Nations).

Meaden, G.J. and Do Chi, T., 1996, *Geographical Information Systems: Applications to Marine Fisheries,* Fisheries Technical Paper No.356 (Rome: Food and Agriculture Organisation of the United Nations).

Meaden, G.J. and Kemp, Z., 1996, Monitoring fisheries effort and catch using a geographical information system and a global positioning system. In *Developing and Sustaining World Fisheries Resources: The State of Science and Management,* edited by Hancock, D.A., Smith, D.C., Grant, A. and Beumer, J.P. (Brisbane, Australia: 2nd World Fisheries Congress), pp. 238–244.

Meaden, G.J. and Kemp, Z., 1998, Towards a comprehensive fisheries management information system. In *Proceedings of the IIFET'98 Annual Conference* (Tromsö, Norway).

Mooneyhan, W., 1985, Determining aquaculture development potential via remote sensing and spatial modelling. In *Report of the Ninth International Training Course on Applications of Remote Sensing to Aquaculture and Inland Fisheries,* RSC Series 27 (Rome: Remote Sensing Centre, FAO), pp. 217–247.

Nichols, P., 1998, Vessel monitoring systems as a fisheries management tool. In *EEZ Technology 2* (London: IGC Publishing, London), pp 167–171.

Pauly, D., Christensen, V., Dalsgaard, J., Froese, R. and Torres, F., 1998, Fishing down marine food webs. *Science,* **279,** pp. 860–863.

Paw, J.N., Robles, N.A. and Alojado, Z.N., 1993, The use of GIS for brackish water aquaculture site selection: A case study of three sites in the Philippines. In *Proceedings of the 3rd Asian Fisheries Forum* (Singapore, Malaysia).

Pierce, G.J., Wang, J., Bellido, J.M., Waluda, C.M., Robin, J.P., Denis, V., Koutsoubas, D., Valavanis, V. and Boyle, P.R., 1998, Relationships between cephalopod abundance and environmental conditions in the northeast Atlantic and Mediterranean as revealed by GIS. In *ICES Annual Science Conference,* Vol. M:20 (Lisbon, Portugal: ICES CM 1998).

Pollitt, M., 1994, Protecting Irish interests: GIS on patrol. *GIS Europe,***3,** pp. 18–20.

Ramos, F., Sobrino, I. and Jiménez, M.P., 1996, *Cartografía de Especies y Caladeros: "Golfo de Cádiz",* Informaciones Técnicas 45/96 (Junta de Andalucia, Espana: Consejeria de Agricultura y Pesca).

Relini, G., 1991, *Mapping of Italian Demersal Resources,* EU Doc. No. XIV/C/1 Med 1991/013 (Genova, Italy: Instituto di Zoologia dell'Università di Genova).

Ross, L.G., Mendoza, Q.M.E.A. and Beveridge, M.C.M., 1993, The application of geographical information systems to site selection for coastal aquaculture: An example based on salmonid cage culture. *Aquaculture*, **112**, pp. 165–178.

Rubec, P.J., Coyne, M.S., McMichael, R.H. and Monaco, M.E., 1998, Spatial methods being developed in Florida to determine essential fish habitat. *Fisheries*, **23**, pp. 21–25.

Sakurai, Y., Bower, J.R., Kiyofuji, H., Saitoh, S., Goto, T., Hiyama, Y., Mori, K. and Nakamura, Y., 1998, Changes in inferred spawning sites of *Todarodes pacificus* (*Caphalopoda ommastrephidae*), due to changing environmental conditions. In *ICES Annual Science Conference,* Vol. M:18 (Lisbon, Portugal: ICES CM 1998).

Simpson, J.J., 1992, Remote sensing and geographical information systems: Their past, present and future use in global marine fisheries. *Fisheries Oceanography*, **1**, pp. 238–280.

Somers, I.F. and Long, B.G., 1994, Note on the sediments and hydrology of the Gulf of Carpentaria, Australia. *Australian Journal of Marine and Freshwater Research,* **45**, pp. 283–291.

Travaglia, C. and Lorenzini, M., 1985, Monitoring algae growth by digital analysis of Landsat data: The Orbetello Lagoon case study. In *Report of the Ninth International Training Course on Applications of Remote Sensing to Aquaculture and Inland Fisheries.* RSC Series 27 (Rome: Remote Sensing Centre, FAO), pp. 255–259.

Valavanis, V., Georgakarakos, S. and Haralambous, J., 1998, A methodology for GIS interfacing of marine data. In *Proceedings of GISPlaNET'98* (Lisbon, Portugal: GISPlaNET).

Warner, L.S., 1987, Mapping the unmappable: Use of geographic information systems in fisheries management. In *Proceedings of Tenth National Conference: Estuarine and Coastal Management: Tools of the Trade,*Vol.2 (New Orleans, Louisiana), p. 705.

Wood, L.L. and Ferguson, R.L., 1995, Monitoring the effects of sidecast dredging on seagrass habitat using aerial photography and GIS. In *Proceedings of the 61st Annual Convention of the American Society for Photogrammetry and Remote Sensing* (Charlotte, North Carolina: American Society for Photogrammetry and Remote Sensing), pp. 403.

CHAPTER SIXTEEN

A User-friendly Marine GIS for Multi-dimensional Visualisation

Yafang Su

16.1 INTRODUCTION

Although the ocean covers almost three quarters of our planet and contains tremendous resources, geographic information systems (GIS), have drawn the interests of the oceanographic community only recently (Wright 1996; Hatcher *et al.*| 1997; Su, 1997) while terrestrial GIS applications have a history of more than 30 years. This may be explained in part by the multi-dimensional and multi-temporal nature of oceanic data and the high costs of data acquisition (Li and Saxena, 1993; Basu, 1994; Wright and Goodchild, 1997).

With the development of new technologies, more and more spatially referenced oceanographic information has been collected by many organisations with various observation platforms (Gritton and Baxter, 1992; Macaulay, 1992; McNitt, 1993; Chavez *et al.*, 1994). The Monterey Bay Aquarium Research Institute (MBARI) is dedicated to achieving scientific success through the development of state-of-the-art instruments, systems, procedures and methods for scientific research in the deep waters of the ocean (MBARI's mission by David Packard (1912-1996), the founder of MBARI). Since its establishment MBARI has collected extensive data sets in Monterey Bay from its multi-platform systems composed of remotely operated vehicles (ROV), cruising ships, buoy drifters, moorings, and a NOAA satellite data receiving station (Chavez *et al.*, 1994). These data, ranging from surface to deep water, include physical, chemical, biological, geological measurements, and numerous videotapes.

To conduct effective data management, MBARI developed the MBARI observation database system (MODB) with the support of Sybase, a major improvement in MBARI's oceanographic data management during the late 1980s (Gritton and Baxter, 1992). The data obtained from cruises and ROV dives are managed in MODB. The MODB stores all the expedition and ROV diving records, including biological, physical, chemical measurements, a portion of the data generated from laboratory analyses, and annotations of digitised videotapes. However, MODB does not incorporate spatial attributes of oceanographic data, does not support spatial analysis, does not provide suitable tools for oceanographic data visualisation, and does not have a user-friendly interface for oceanographers. In response to requirements specified by MBARI scientists, the author integrated the Arc/INFO and ArcView GISs with the Vis5D advanced visualisation system, incorporating MBARI's existing data-acquisition and database systems. The result was the Monterey Bay Marine GIS (MBMGIS), a user friendly application environment for the desktop (Su, 1997). This system allows oceanographers with little or no background in GIS to conveniently query and analyse data, as well as examine oceanographic processes, via an integrated system.

In addition to describing the MBMGIS core database system, the user interface, and the advanced visualisation tools, this chapter also shows how the MBMGIS integrates GIS, visualisation and database systems for analysing upwelling phenomena in Monterey Bay. Some new results have been produced from the different perspectives provided by this integration of GIS with scientific visualisation.

16.2 MBMGIS DATABASE

The data from various sources were stored in different formats under MBARI's previous data management scenario. In MBMGIS, the GIS database was extended to manage the data from various sources (Figure 16.1). The data from cruises and dives formerly stored in the MODB are converted from Sybase into the MBMGIS attribute INFO database to accelerate the query speed. The OASIS (Ocean Acquisition System for Interdisciplinary Science) drifter and mooring data are in ASCII formats, vector maps such as contour lines, coastlines are stored in

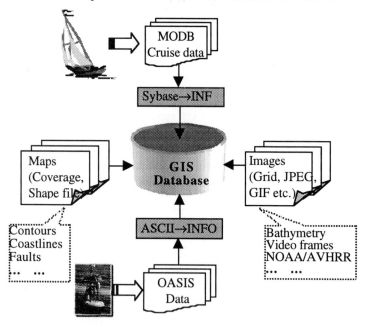

Figure 16.1 Conceptual diagram of MBARI GIS.

Arc/INFO coverages or ArcView Shape files, raster images including video frames, NOAA/AVHRR (Advanced Very High Resolution Radiometer) data, sonar beam images, and bathymetric data are stored in Arc/INFO grids, as well as the popular JPEG and GIF image formats. The database was organised according to data source,

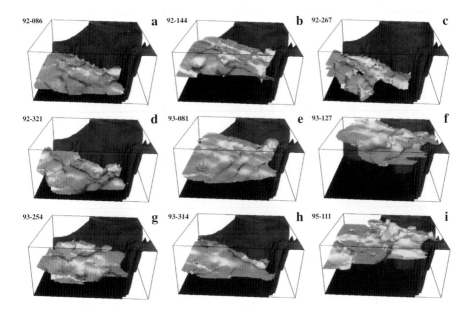

Plate 16.5   Snapshots of 10°C temperature isosurface animation from 1992 to 1995.

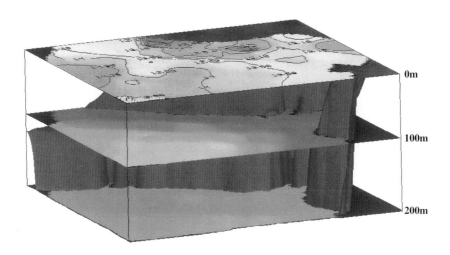

Plate 16.6   Temperature slices at different depths.

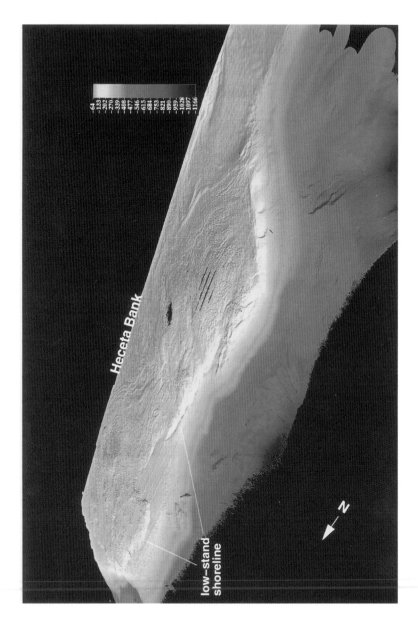

Plate 17.1 3-D perspective view of Heceta Bank, on the continental shelf off central Oregon. Data collected by C&C Technologies using a Simrad EM-300 multi-beam system and sponsored by the National Oceanic and Atmospheric Administration. Pixel size is 10m. These data clearly show detailed folding and faulting of the bank, as well as a low-stand Pleistocene shoreline and wave-cut platform, probably cut at 18ka. Data courtesy of R. W. Embley, NOAA Pacific Marine Environmental Laboratory, Newport, Oregon.

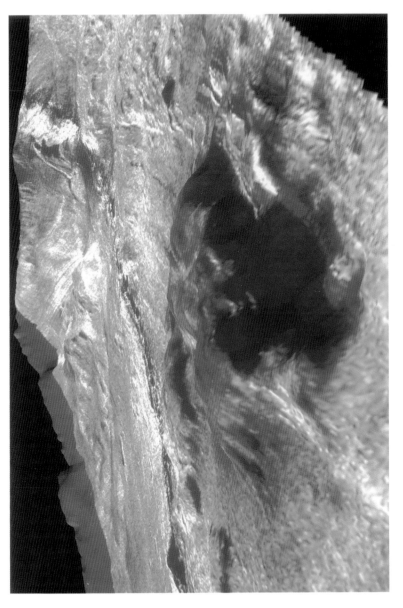

Plate 17.2 Perspective view of SeaMARC1A sidescan sonar data overlaid on Hydrosweep bathymetry. The sidescan data is colour shaded by depth derived from the bathymetry. The data came from two systems with different swath widths, hence the mismatching of coverage. The data came from two systems with different swath widths, hence the mismatching of coverage. The data reveal a large slump block, 1.5km across, that has slid onto the abyssal plain from the frontal ridge in the background. The block left "skid marks" leading back to the slump scar.

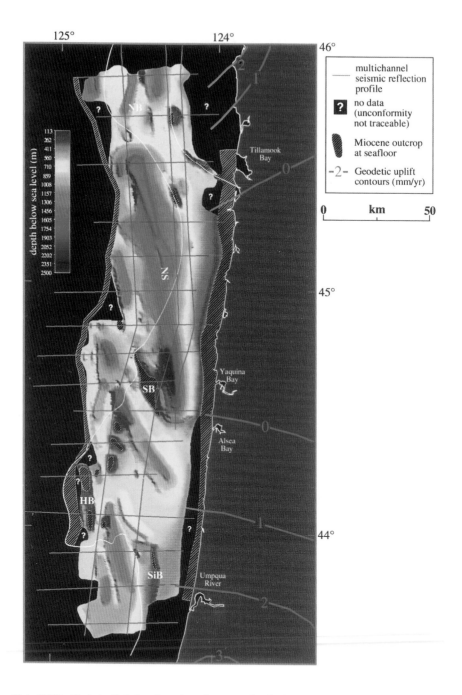

Plate 17.10 Shaded relief plot of a sub-surface unconformity beneath the Oregon continental shelf. Data were gathered from seismic reflection profiles and wells. Contours onshore are geodetic uplift rate in mm/yr from Mitchell *et al.* (1994).

data type, and collection date to optimise database operations and accelerate data query. An Arc macro language (AML) program was developed to update the database regularly. Many data in the MBMGIS are spatially- and temporally-referenced, as well as multi-dimensional. These data are essential in order for oceanographers to be able to understand the oceanographic processes in Monterey Bay.

16.3 MBMGIS INTERFACE

MBMGIS is a user-friendly system based on Unix ArcView3.0b. MBMGIS is composed of three extensions: MBARI, MODB, and OASIS. They were written in Avenue (the program language comes with ArcView). Five sets of pull-down menus in the View graphical user interface (GUI) and associated icons on the button bar and the tool bar have been developed for this three extensions. Figure 16.2 shows the View GUI when all three extensions are loaded into ArcView. Among the five sets of menu buttons (see Figure 16.3), the MODB extension is composed of the MODB menu set with an Expedition view, while the OASIS extension is composed of the OASIS menu set with a Mooring view. The other three menu sets, the View icon ⬚, the View tool icon ⬚, and the Table icon ⬚ are part of the MBARI extension. The MODB and the OASIS menu sets become available in the Table GUI and the Chart GUI when these extensions are loaded. This is to help users to conveniently access these menus from various interfaces. Users can load or unload the extensions easily when needed.

16.3.1 MBARI Extension

The MBARI extension, the general GUI for MBARI oceanographers, contains three groups of menus: Analysis, Addthemes, and Addimages (Figure 16.3), and three icons ⬚, ⬚ and ⬚.

Four commonly used analysis tools of data interpolation, contour generation, buffer zone generation and theme merging, and three data conversion tools called Point2line, Shape2gen and Shapepoint2gen are available in the Analysis menu set. The Interpolation button uses the inverse distance weighted (IDW) interpolation or the triangular irregular network (TIN) data structure in Arc/INFO to interpolate irregularly distributed cruise and ROV dive data to evenly distributed grid data. The interpolation models were pre-parameterised with appropriate values as defaults, but users can set the values of their own. The Contour button generates contour lines on gridded data such as bathymetry, temperature, salinity etc. The Buffer button determines buffer zones of sampling site points or ship track lines, coastlines and others. The Merge button merges two or more themes into one shape file. With this button, two or more expeditions can be combined into one theme for further

Figure 16.2 View GUI in MBMGIS.

MODB	Oasis	Analysis	Addthemes	Addimages
Expd_datarun	Headinfo	Interpolation	Coastline	Bathymetry
Expedition	M1_Series	Contour	Contourline	Video_frames
Physical	M2_Series	Buffer	Fault	NOAA_Sat
Chemical	M1xym	Merge	CTDsite	Sonar_Beam
Biological	M2xym	Point2line	Countymap	TM image
Video_anno	M1meas	Shp2gen		
Bio_lab	M2meas	Shppoint2gen		
Chml_lab	Drifter			
Lab_sample				

Figure 16.3 Pull-down application menus.

analysis. The locations of cruising ships, moorings and drifters are recorded as points by global positioning system (GPS) receivers periodically. The Point2line button generates track lines showing the spatial trajectory of the instruments from the GPS records. The Shp2gen button exports spatial co-ordinates of points, lines and polygons from shape files to the delimited ASCII format. The Shppoint2gen button exports point co-ordinates and a spot field (e.g., temperature, salinity, etc.) from a point shapefile to a delimited ASCII file. These delimited ASCII files can be used in Arc/INFO for more sophisticated analyses.

The Addthemes menu set adds geographic data to the active view. There are five menu buttons: Coastline, Contour line, Fault, County maps, and CTD sites (CTD is an electronic device for simultaneously measuring electrical conductivity, temperature and density). The Coastline button has two options: to add the coastline to the active view in either the Monterey Bay area (122.5°W, 36.5°N to 121.7°W, 37.1°N) or the San Francisco-Monterey Bay area (124.027°W, 35.491°N to 120.943°W, 38.017°N). The Contour line button adds contour maps to the active view with four options: at 20 or 100 meter intervals of the Monterey Bay area, or 50 meter or 100 meter intervals of the San Francisco-Monterey Bay area. The Fault button adds the known geological fault lines to the active view. The CTD site button adds the CTD sampling data in Monterey Bay to that active view. The County map button adds parcel maps to the active view with three options: Monterey county, Santa Cruz county or both.

The Addimages menu set is composed of five menu buttons, which add grid or image data including bathymetry, NOAA/AVHRR satellite images, video JPEG images and Landsat/TM (thematic mapper) images to the active view as image themes, respectively. The Bathymetry button with six options adds bathymetry or shaded relief of Monterey Bay or San Francisco/Monterey Bay at different resolutions to the active view. The Video_frames button adds video frame images captured by ROV video cameras and identified by their annotations to the active view. The NOAA_Sat button adds the NOAA/AVHRR images acquired at a specified time to the view. The Sonar_Beam and the TM image buttons add side-scan sonar images and TM images to the active view, respectively.

The ⬛ button on the View button bar starts the Vis5D visualisation module for visualising oceanographic data. With the assistance of Vis5D, oceanographers are able to inspect complex phenomena from the multi-dimensional measurements of temperature, salinity, density, light-transparency and fluorescence using isosurface, slicing, volume rendering, animation and other visualisation techniques. This module will be discussed in detail in Section 16.4.

The ⬛ icon measures surface distance in kilometres from the view map in geographic projection. The ⬛ icon in the Table GUI saves the selected records with selected fields in the active table to a new table in ASCII text, dBase or INFO formats.

16.3.2 MODB Extension

The nine menu buttons in the MODB extension are designed specifically to query data collected from ship cruises and ROV dives. The first two buttons of Expd_datarun and Expedition query all the expedition records. Expedition number, expedition date, platform used in the expedition, ROV diving number, and scientific purpose are the five optional keywords used to search expeditions from the Expedition table and the Expd_data_run table. The users can also query the expeditions spatially by clicking the mouse on a point in the Expedition view of Monterey Bay. The query results can be highlighted symbolically on a map or in a table immediately, and saved as a new table or a new map for future uses. The other seven buttons (Physical, Chemical, Biological, Video_anno, Bio_lab, Chml_lab, and Lab_sample) query the physical, chemical, biological measurement tables, the video-annotation table, the result tables from biological and chemical laboratory analyses, and a table of samples for laboratory purposes, respectively. One of the three keywords of expedition number, date range and depth range is used to query data in each category. The query results can be displayed either in a table or on a map, or both.

16.3.3 OASIS Extension

The OASIS extension consists of eight menu buttons that query and analyse MBARI mooring and drifter data. The Headinfo button displays the header information of the sensor's measurements and the definition of the fields in the tables. M1 and M2 are the names of the two MBARI mooring stations in Monterey Bay. The M1_Series and M2_Series buttons query the time series of physical and biological mooring data for a whole table, a subset within a date range, or a daily summary of selected measurements at the two stations. The M1xym and M2xym buttons query various measurements at M1 and M2 by date, and display the spatial locations of the moorings on a map. The M1meas and

M2meas buttons query the measurement tables of the M1 and M2 stations by date. The Drifter button is used to query the drifter data by drifter time. These query results can be displayed in a table, a view or a chart.

16.4 MULTI-DIMENSIONAL VISUALISATION OF MARINE DATA

Vertical profiles of CTD data (i.e., temperature, salinity, density etc.) are often collected for a variety of oceanographic research and operation purposes. These data are typically analysed only one- or two-dimensionally, although the oceans are dynamic three-dimensional entities. This is due to the difficulties in high-dimensional representation and visualisation using conventional approaches. The recent development of GIS and advanced scientific visualisation technologies provides a powerful tool for scientists to analyse and visualise marine data. Although the 3D tools recently available in GIS packages (e.g., 3-D Analyst in ArcView) are good at representing surface features, such as terrain and landscapes (Berry *et al.*, 1998), they are ineffective for representing and visualising 3-D volumes in the ocean. However, volume features are readily visualised in advanced scientific visualisation packages. Therefore, it is extremely important to integrate the database capabilities of GIS with multi-dimensional scientific visualisation for the most powerful analysis of marine data.

Vis5D was chosen as the advanced visualisation tool in MBMGIS not only because it is a readily available public domain package for visualising atmospheric and oceanographic data, but also its design and code are open for further enhancement. Vis5D (URL: http://www.ssec.wisc.edu/%7Ebillh/vis5d.html, 1996) was developed by the Space Science and Engineering Center at the University of Wisconsin, and was supported by the NASA Marshall Space Flight Center and Marie-Francoise Voidrot-Martinez of the French Meteorology Office. Vis5D provides the visualisation functions most useful to marine studies such as volume rendering, multi-dimensional visualisation, and animation. The data structure of Vis5D, as its name suggests, is simply a five-dimensional rectangle. The first three dimensions are spatial: rows, columns and levels (or latitude, longitude and depth). The fourth dimension is a singular or multiple physical variables such as temperature, salinity etc. The fifth dimension is time. The Arc/INFO grid/stack is the counterpart of this data structure.

16.4.1 Data Source and Preparation

MBARI has intensively collected CTD data for years, including temperature, conductivity, pressure, fluorescence, beam-c, etc. The nine week-long CTD data sets used in this study were collected by MBARI oceanographers between 1992 and 1995 (see Table 16.1 for detailed descriptions for each cruise). Figure 16.4 shows the CTD sampling locations and bathymetry during the week of April 1995 in Monterey Bay. The CTD data were collected up to 200 m in depth at one-meter intervals, while near the coast some samples may reach the seafloor. The 3-D boundary of the data sets is 36.48°N to 37.08°N latitude, 122.62°W to 121.82°W longitude, from surface to 200 meters in depth or to the seafloor if the seafloor is less than 200 meters in depth.

Table 16.1 CTD sampling periods.

Cruise #		Starting date	Ending date	Julian date	Note
92	1	03/27/92	04/03/92	92-086	1992 was a warm El Niño year.
	2	05/24/92	06/01/92	92-144	
	3	09/24/92	10/01/92	92-267	
	4	11/17/92	11/24/92	92-321	
93	1	03/22/93	03/30/93	93-081	1993 was a normal upwelling year.
	2	05/07/93	05/13/97	93-127	
	3	09/11/93	09/16/93	93-254	
	4	11/10/93	11/15/93	93-314	
95	1	04/18/95	05/07/95	95-111	upwelling season in 1995

Regarding the slight vertical variation of the properties in each meter, computer capacities and visualisation speed, four-meter was chosen as the vertical layer interval, and 0.010 x 0.010 as the horizontal grid size. The unevenly sampled CTD data were interpolated into a stack of 50 x 5 x 9 (2250) grid layers in Arc/INFO and then converted to the Vis5D format. The final Vis5D data set has a size of 60 x 80 x 50 x 5 x 9, describing a true 3-D space of 60 x 80 x 50 voxels with five measurements over nine time steps. The bathymetry and coastal terrain data from the USGS were imported into the Vis5D data set as the topographical boundary to assist visualisation.

16.4.3 Visualisation and Results

In the environment of MBMGIS, the processed CTD data were visualised in multiple dimensions and then animated to further understand the upwelling process in Monterey Bay. The classical understanding of coastal upwelling dynamics is a surface layer of water being driven away from the coast with the subsurface replacement of that water as a result of surface wind stress and Ekman transport (Graham and Largier, 1997). The upwelling season in Monterey Bay is usually in the early summer.

The temporal and spatial variations of water properties illustrated by the isosurface animation tool are informative in understanding upwelling. In each of the nine snapshots in a time series of 10°C temperature isosurfaces (Plate 16.5), the Monterey Bay area appears three-dimensionally with depth as the vertical axis, and the yellow isosurface shows the 3-D distribution of the 10°C isotherms in the

Figure 16.4 CTD sampling locations and bathymetry.

bay. The location and the pattern of these isosurfaces provide information about upwelling. An isosurface of a certain temperature closer to the sea surface suggests that the water is colder; therefore, a stair-shaped isosurface towards the coast indicates the presence of upwelling because the nearshore is found to be colder than offshore.

When animating the snapshot series in Plate 16.5, one can identify changing patterns of seasonal temperature in Monterey Bay for 1992 and 1993. Cold waters invaded Monterey Bay in the early summer of both years as evidenced by the rise in the corresponding isosurfaces. A warm El Niño event was the dominant atmospheric and oceanic characteristic in Monterey Bay in 1992 (Chavez, 1996). The downward trend towards the coast in each of the 1992's four snapshots (Plate 16.5a-d) implies that upwelling was not present in the El Niño year of 1992. In the normal upwelling year of 1993, the upward trend to the inner bay in the isosurface of late March 1993 records the emergence of the 1993 upwelling (Plate 16.5e). The nearly 100 m rise of the isosurface illustrates the prevalence of upwelling (Plate 16.5f). Later, the upwelling died out as evidenced by the return of the isosurface to a depth of 100 m (Plate 16.5g). The outer bay remained colder than the inner bay during the non-upwelling season (Plate 16.5h).

The visualisation technique of slicing enables the simultaneous visualisation of water properties at different depths. The rough and sharp stair-shape of the temperature isosurface in April 1995 (Plate 16.5i) represents a strong and complex upwelling pattern. The upwelling during this period was examined using the slicing technique.

The water temperature distributions at 0 m (sea surface), 100 m and 200 m depths are shown in Plate 16.6. The red, orange, yellow, cyan, and blue colours represent the waters of 14°C, 13°C, 12°C, 11°C, and 10°C, respectively. The upwelling results in a significant horizontal temperature variation at the surface: the 10°C (blue colour) nearshore water versus the 14°C (red colour) offshore water. The range of temperatures narrows down to 10-12°C as the depth goes to 100 m. The stable state of deeper water at 200 m depth is indicated by its uniform horizontal distribution. Therefore, a reasonable estimation of the maximum upwelling depth is between 100 m and 200 m.

16.5 IMPLEMENTATION AND DISCUSSIONS

The MBMGIS developed in ArcView can be accessed from any Unix workstations or PCs with Unix terminal emulation at MBARI. The documentation of this system is available on MBARI's Intranet. The user-friendly interface provides easy access for oceanographers with little or no background in GIS. They do not need to know the physical location of the data on the network. With the assistance of this system, scientists can fully use various data in the database, design new expedition track lines, deploy markers and sampling sites, and carry out various sophisticated analyses and visualisation to inspect complex processes in the ocean.

GIS and scientific visualisation packages have different strengths and weaknesses. The MBMGIS attempts to utilise the strengths of both. GIS focuses on spatial data management and analysis, while visualisation packages focus on the visual perception, enhanced by computer graphics, that provides more insight into the complex processes of nature. The GIS/visualisation combination can minimise the economic constraints of separate, redundant, closed systems, thereby enabling the user to understand and solve complex problems, each unsolvable with a solitary system. As shown by the case study, GIS supports visualisation by generating and analysing data sets, while visualisation lends itself to more powerful insights derived from the data as to the processes that are driving oceanographic phenomena. The integration is much more powerful than any one of the separate systems. This study shows that the integration of visualisation and GIS is urgently needed in marine data analyses. Hopefully the seamless integration of the two can soon be realised by collaborative efforts between the GIS and scientific visualisation communities.

16.6 REFERENCES

Basu, A., 1995, Development of a marine geographic information system for various data analysis and data integration in the Hawaiian Exclusive Economic Zone. In *Proceedings of IEEE/OCEANS'95*, pp. 146-153.

Berry, J. K., Buckley, D. J., and Ulbricht, C., 1998, Visualise realistic landscapes: 3D modeling helps GIS users envision natural resources. *GIS World*, **11**, pp. 42-47.

Chavez, F. P., Herlin, R., Thurmond G., 1994, OASIS-Acquisition system for mooring/drifters. *Sea Technology*, **35**, pp. 51-59.

Chavez, F. P., 1996, Forcing and biological impact of the onset of the 1992 El Niño in central California. *Geophysical Research Letters*, **23**, pp. 265-268.

Graham, W. M. and Largier, J.L., 1997, Upwelling shadows as nearshore retention sites: the example of northern Monterey Bay. *Continental Shelf Research*, **17**, pp. 509-532.

Gritton, B. R. and Baxter, C. H., 1992, Video database systems in the marine sciences. *Marine Technology Society Journal*, **26**, pp. 59-72.

Hatcher, G. A., Maher N., and Orange, D., 1997, The customization of ArcView as a real-time tool for oceanographic research. In *Proceedings of theESRI User Conference* (Redlands, California: Environmental Systems Research Institute), http://www.mbari.org/~gerry/esri_97paper/p676.html.

Li, R. and Saxena, N.K., 1993, Development of an integrated marine geographic information system. *Marine Geodesy*, **16**, pp. 293-307.

Li, R., 1993, 3D GIS: A simple extension in the third dimension? In *Proceedings of the ACSM/ASPRS Annual Meeting* (New Orleans, Louisiana, American Congress on Surveying and Mapping/American Society for Photogrammetry and Remote Sensing), pp. 218-227.

Macaulay, M. C., 1992, A hybrid database management system for the collection, display, and statistical analysis of large and small scale hydroacoustic survey data. In *IEEE Oceans'92 Proceedings*,Vol.1, pp. 91-96.

McNitt, A., 1993, Data modeling for distribution of oceanographic database. In *IEEE Oceans'93 Proceedings*, Vol.3, pp. III-61-65.

Su, Y. ,1997, Interfacing oceanographic database by ArcView. In *Proceedings of theESRI User Conference* (Redlands, California: Environmental Systems Research Institute), http://www.esri.com/library/userconf/proc97/PROC97/TO150/PAP136/P136.HTM.

Wright, D. J., 1996, Rumblings on the ocean floor: GIS supports deep-sea research. *Geo Info Systems*, **6**, pp. 22-29.

Wright, D. J. and Goodchild, M. F., 1997, Data from the deep: implications for the GIS community. *International Journal of Geographical Information Science*, **11**, pp. 523-528.

CHAPTER SEVENTEEN

Active Tectonics: Data Acquisition and Analysis with Marine GIS

Chris Goldfinger

17.1 INTRODUCTION

The explosion in GIS usage in marine disciplines such as active tectonics has revolutionised our ability to visualise, analyse, and interpret seafloor environments using multiple data sets and a much more "holistic" view of geologic/geophysical data than previously available. Driven by the growing capabilities of desktop workstations, we can now realistically expect to display, merge, and fly through data sets that were beyond the capabilities of all but a few specialised laboratories just a few years ago. This paper presents a brief review of some of the data gathering instrumentation available for active tectonics research, and discusses techniques for processing, combining, visualising and analysing these data in the context of active tectonics. The paper concludes with some examples of active submarine tectonic research.

17.2 SEAFLOOR IMAGING

The heart of any scientific investigation is data, and as with all studies, collection of data appropriate to the needs of the study is a critical element in a successful project. Data collection is often the most difficult and costly component, and this is particularly true of investigations involving submarine tectonics. Much of the data needed for the study of active tectonics at sea is gathered remotely, by sound and light propagation systems, and occasionally, by direct observation with submersibles.

17.2.1 Hull-mounted Multibeam and Backscatter Systems

Multibeam bathymetric sonar systems were first developed in the early 1980s. They have now evolved considerably and are capable of collecting a swath of bathymetric data up to 5-7 times the water depth at cruising speeds of 8-12 knots or more. These systems provide the fastest way to gather bathymetry data over large areas. Present deep-water systems such as the Sea Beam 2100 series, Simrad EM-12, and ATLAS Hydrosweep typically can survey swaths 3-3.4 times the water depth in deep water, 5-7 times the water depth in shallower water, and also collect backscatter data. As an example, in 3000 m of water, the maximum swath width will be about 20 km, and the nominal resolution about 70 m, thus these systems are quite good for large scale features such as plate boundaries and fracture zones.

Backscatter data from hull-mounted systems has a steep "look angle", making these data most applicable in situations where it is desirable to emphasise bottom type and "roughness" at the expense of bathymetric detail. Full ocean depth systems are low frequency to reduce attenuation, typically 12 kHz, and can penetrate many meters into the bottom, which may be desirable or undesirable depending on the application.

Recently, a newer generation of sonars has been developed for medium and shallow water depths that offer superior resolution in water depths from tens of meters to about 3500 m, and like their deep water predecessors, can collect both backscatter and co-located bathymetry at the cruising speed of the vessel. An example of this is shown in Plate 17.1.

Shallow water sonars such as the RESON Sea Bat are capable of imaging swaths of 5-7 times water depth at cruising speeds, though lower speeds are preferable. Although these systems are most often used for commercial work such as cable and pipeline surveys, they offer potential for geologic studies in water depths less than about 500 m.

17.2.2 Shallow-towed Sonars

Sidescan sonars generally fall into two classes, shallow and deep-towed. Shallow towed sonars, such as the SeaMARC II/HMR1 instrument, the GLORIA system Tamu2, and the Sys 09 vehicle operated by Seafloor Surveys International are operated at or near the vessel cruising speed (Blondel and Murton, 1997). Towing below the surface offers some advantages over hull mounting in terms of surface noise both from the sea and from the ship, while retaining the relatively high speeds needed to cover large areas. The swath width that can be mapped from a relatively high altitude above the seafloor is large, maximising areal coverage in a given survey time. These systems collect better sidescan imagery than available from early hull-mounted systems, but are now being superseded by the recent generation of hull mounted sonars such as the Simrad EM-300 except for water depths greater than 3000m. Bathymetry is calculated from phase differences received at multiple transducers.

17.2.3 Deep-towed Sidescan and Phase Interferometric Bathymetry

Deep-towed sonars offer the highest resolution sidescan imagery available. Systems range in frequency from 30-400 kHz, with swath widths of 100 m to 5 km. These systems are "flown" over the seafloor at an altitude of about 10% of the swath width. Resolution varies both along and across track in all sonar systems, complicated by the elliptical "footprint" of each beam as parcels of the seafloor are ensonified (Johnson and Helferty, 1990). These drawbacks tend to be maximised in deep-towed systems. Deep-towed sonars are applicable to detailed investigations, and can image subtle secondary kinematic features such as tension gashes, en-echelon fault strands, offset surficial features and changes in sediment type caused by tectonic activity. Although these systems offer a wide range of frequencies, they are often higher than either shallow-towed or hull-mounted sonars. The relatively higher frequency of deep-towed systems offers higher spatial resolution, but requires a shorter travel path for the signal to avoid attenuation, thus the need for deep towing. The higher frequency also means less penetration of soft sediments, which is for the most part beneficial for active faulting studies. Subtle features imaged in areas of soft sediment may actually be located in the shallow subsurface rather than at the seafloor, complicating active tectonic interpretation and causing confusion in the case of visits with submersibles or ROVs. Conversely, some penetration can be helpful in areas where sedimentation or erosion rates exceed the slip-rates of tectonic structures, burying or eroding the most recent scarps. In such cases, a lower frequency sonar can see through the soft drape and image buried scarps or carbonate alignments common along submarine active faults. Recently, the Hawaii Mapping Research Group has developed a low-frequency synthetic-aperture sonar system optimised for detection of shallowly buried objects for military applications (B. Appelgate, pers. comm., 1999). The sonar essentially collects a high-resolution swath sub-bottom image, and is being tested for use in mapping shallowly buried faults at this writing.

17.2.4 Short Range Systems

Filling a spatial gap between deep-towed and hull mounted multibeam sonars and submersible-mounted and towed cameras are several short-range scanning imaging systems. These systems can reveal morphologic features and spatial relationships at an intermediate scale, albeit with very slow collection of data. Applications thus far have consisted of surveys of up to about 0.2 km^2.

Several versions of a Mesotech scanning sonar have been mounted submersibles, including ALVIN and SeaCliff. The NOAA Pacific Marine Environmental Laboratory has used a 675 kHz, "pencil beam" scanning sonar for high-resolution bathymetric surveys of the Juan de Fuca ridge (Chadwick and Embley, 1995). The sonar data are integrated with navigation and altitude data from the submersible, and are processed as individual profiles, not unlike a single athwartships beam from a multibeam sonar. These data can then be gridded and contoured or visualised as shaded relief like other multibeam data. Similar systems are being tested on ROVs and AUVs for detailed applications (e.g., Mallinson *et al.*, 1997).

Towed camera systems have been in use in marine geology for many years. In simplest form, they take single frames with a flash unit as they are towed in an area of interest. Newer systems are typically towed video sleds that obtain a continuous imagery record along the bottom as they are dragged across a study site. Camera data, either single frames, captured video frames, or video clips, can be studied singly as "ground truth" and incorporated into the GIS as ancillary data. Although these systems cannot collect samples as can submersibles and ROVs, they provide valuable information about the nature of seafloor morphology and character at outcrop scale. These data can be used to calibrate interpretations of sidescan imagery, and can help discriminate between multiple interpretations of enigmatic imagery from other sources. Camera data from tows on mid-ocean ridges have been used to map the extent and distribution of lava flows. These data can help determine the relative ages of flows, lithologies, sediment cover, as well as vent sites and biological communities. Fox *et al.* (1988) detail an image analysis system for analysis of bottom photography. Towed-camera positions from a transponder navigation system are smoothed, and camera lens characteristics are used to calculate the field of view for each photograph. The navigation and attitude data are used to georeference the imagery, which are classified for geological and biological features.

17.2.5 Submersibles and ROVs

Submersibles and remotely operated vehicles (ROVs) are the "ground truth" for many types of investigations in the marine environment. When critical information unavailable by other means is needed, they provide direct observation of active structures on the seafloor. Even with the best high-resolution seismic reflection and sidescan sonar, it may still be difficult to determine whether a fault actually cuts Holocene sediments, or whether it is actively venting fluids. Submersibles and ROVs can provide these data and more, including the ability to do field mapping on the seafloor not unlike land studies, albeit very slowly and at great expense.

GIS can be used in several ways prior to and during submersible operations. During operations, The GIS can be used as an active navigation device to display ship and vehicle position on a background of relevant data layers such as bathymetry and sidescan sonar data. This serves several purposes: 1) eliminating navigation errors due to transposing numbers, because the vehicle is always visible relative to the target; 2) giving the observers on the ship real-time knowledge of what the vehicle is doing; 3) allowing a number of investigators to access vehicle information and discuss plans and alternatives in real-time with all available data displayed simultaneously.

For submersibles, some of this information can also be made available to the vehicle. Annotated dive maps from sonar imagery and other relevant data layers from the GIS database for each site can be taken in manned vehicles. The maps can be used by the observers to correlate their observations with the sonar imagery and to navigate to the targets of interest, with help from the surface party who are doing the same thing in parallel.

Submersible observers and ROV operators can collect high-resolution video and still images as well as visual observations and structural measurements during the dives (e.g., Wright, 1996; Hatcher and Maher, 1999). Structural attitudes (strike and dip) of faults and exposed strata can be taken frequently for field interpretation and later construction of geologic maps (Goldfinger and McNeill, 1997).

17.3 DATA INTEGRATION AND VISUALISATION

17.3.1 GIS and Supporting Tools

Ideally, a GIS would be fully three (or more)-dimensional, incorporating both three (or more)-dimensional surfaces and volumes, as well as two-dimensional data an any plane. Realistically, most available GISs are two-dimensional, with some surface rendering and overlay capability for three-dimensional perspective view plotting. Active tectonics interpretation requires more capability and flexibility than that provided by most GIS, and thus other software must be incorporated in order to address the components that the GIS cannot.

One of the most useful tools for active tectonics is a powerful and flexible visualisation tool. Since active tectonic processes generally deform the land surface or seafloor, visualisation of topography is an essential element in tectonic interpretation, and can be used alone where other data are not available. Most visualisation tools have reduced computational overhead as compared to GISs by virtue of being able to take better advantage of hardware specific rendering capabilities and eliminating complex data structures needed for GIS functionality. The relative simplicity allows for reasonable rendering speeds even with very large data sets. Tectonic geomorphologic interpretation is greatly enhanced by the speed and ease of "flying" around a scene and viewing from any desired perspective (Plates 17.1 and 17.2).

Topographic visualisation has several basic forms: shaded relief plots, contour plots, and wire-frame plots. Shaded relief plots are best for revealing detailed faulting tectonics, and overall patterns of deformation. The illumination angle is also important, and it is often necessary to make plots with several angles and combine interpretations. Such imagery emphasises features oriented at a high angle to the artificial light source, and the same topography can look very different when illuminated from several angles. As in low-sun aerial photography, low illumination angles can be used to investigate subtle topographic relationships such as faint scarps, channels, and fans. Many GISs and visualisation packages will also calculate relief shading of a topographic data set with a raster image such as sidescan sonar draped over it. The shading contrasts are applied to the image colour table, producing a combined output. This application can sometimes be useful in distinguishing sidescan sonar reflectance that is due to topography from that due to intrinsic seafloor properties such as roughness and lithology.

17.3.2 Derivative Visualisation

GIS can generate many useful derivative data sets from both raster imagery and topography. Two useful products for tectonics are slope and aspect plots. A slope map assigns colours or grey levels to the slope angle of each pixel of the bathymetric raster. This type of image may illuminate tectonic activity not apparent otherwise, and can be used to isolate ranges of topographic slope angles for statistical treatment, predictive capabilities of slope stability, or outcrop exposure (Figure 17.3). Similarly, an aspect plots assign colours or grey levels to the azimuth of the slope direction (Figure 17.4). Both of these plots can be used for either topography or subsurface horizons to illuminate trends associated with deformation.

17.4 TECTONIC INTERPRETATION

Rates of deformation on earth are almost entirely based on subaerial landforms. Seafloor imaging techniques and submersible observations now allow the mapping

Figure 17.3 Perspective view of a slope map (dark rectangle) overlaid on shaded bathymetry. Bright pixels denote steeper slopes. A channel crosses the scene diagonally from the upper left of the slope map, but is buried by a slump visible at right. Although the channel is no longer visible in shaded

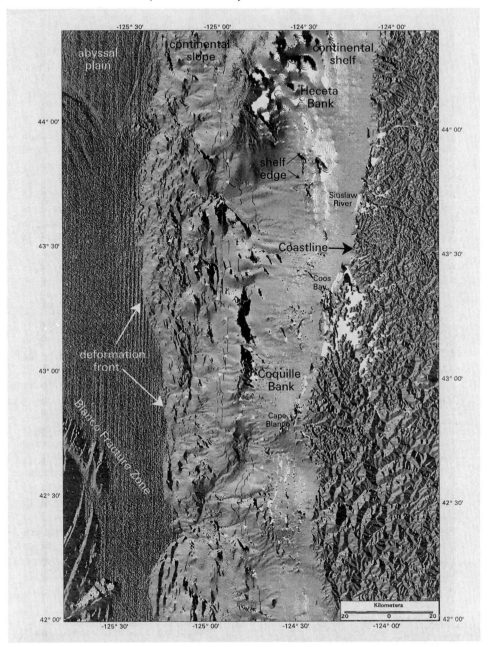

Figure 17.4 Aspect map of the southern Oregon continental margin. This map is made by assigning grey levels to the azimuth of the slope. Black = 0°, grading to white = 360°. The plot emphasises subtle features not easily seen in shaded relief or slope plots. Compare with a shaded plot of the same scene in Figure 17.9.

of tectonic features in sufficient detail that the techniques of tectonic geomorphology can be applied in submarine settings. In the following sections, some examples are detailed from varied tectonic settings to illustrate the present capabilities available for tectonic research.

17.4.1 Thrust Faults and Accretionary Wedges

Our view of convergent margins has largely been shaped by 2-D seismic reflection profiles, however multibeam bathymetry and data integration have revolutionised our view of continental margins and the processes and styles of deformation along active margins. A good example of this is the Cascadia subduction zone. Although the Cascadia convergent margin is frequently cited as a type example of a seaward-vergent accretionary wedge, analysis of bathymetry data has shown that this characterisation is only applicable to a small part of the northern Oregon margin. The Cascadia margin is better characterised by significant along strike variability in structural style and wedge morphology (Figure 17.5). The northern Oregon and Washington accretionary wedge is a broad landward-vergent thrust system with widely spaced folds, and a décollement that steps down to near basement levels, with virtually all incoming sediment being frontally accreted (MacKay, 1995; Goldfinger, 1994). This low-tapering wedge is composed primarily of the Pleistocene Astoria and Nitinat Fans, which have been accreting outboard of a narrow, older Cenozoic accretionary complex. In contrast, the southern Oregon margin is characterised by a steep narrow chaotic continental slope outboard of the outer arc high and forearc basin. Between these distinct provinces is a limited transitional region characterised by the seaward vergent accretionary wedge for which the Oregon margin has become known.

The growth of individual structures can be analysed with GIS techniques. Figure 17.6 shows a channel on the continental shelf with distinct backscatter patterns visible on the channel floor. The dark areas are mud, the light areas are sand, confirmed by submersible dives. Not apparent from the sidescan data, is an active anticline that underlies the area and is transverse to the channel. The anticline is not apparent in the imagery because it is subtle, and was truncated by the latest Pleistocene lowstand of the sea. Effects of its growth can be seen in the ponding of mud on the landward side. As the fold grew, the gradient of the channel was reversed, resulting in ponding. The mud is Holocene in age, so we can infer that this anticline, which began its growth in the Miocene, is currently growing. From the growth rate we can infer an approximate slip-rate for the underlying reverse fault, and its seismic potential (Yeats *et al.*, 1998).

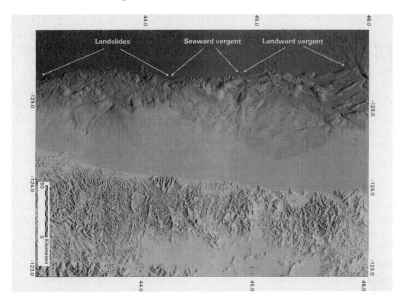

Figure 17.5 Onshore-offshore shaded relief plot of the Oregon continental margin. This compilation revealed that the margin showed a highly variable structural style. The lower continental slope is separable into three morphologic domains, indicated by the arrows. "Vergence" refers to the direction the upper block moves relative to the lower block at the plate boundary fault.

Figure 17.6 AMS 150 sidescan sonar mosaic of part of Stonewall Bank, Oregon. The channel crossing diagonally was previously cut to grade during Pleistocene lowered se-level. Subsequent uplift of the anticline (shown by the fold axis) has locally reversed this gradient, causing ponding of mud (black).

17.4.2 Strike-Slip Faults

Strike-slip faults are notoriously difficult to interpret with seismic reflection data alone, as block motion is usually normal or oblique to the plane of section. This results in tectonic interpretations that emphasise, and sometimes mistake, the compressional or extensional aspects of the structure, which is fundamentally strike-slip. Interpretation of strike-slip faults both in section and in map view is essential to interpretation of the structure, and offers the possibility of calculating or estimating the slip-rate of the fault, an elusive quantity in any tectonic setting.

Some examples of well-studied strike-slip faults are found on the abyssal plain and continental slope offshore Oregon, USA. In migrated multichannel seismic reflection profiles, these faults offset the 3-4-km-thick abyssal plain sedimentary section, as well as the oceanic basement of the incoming Juan de Fuca plate (Goldfinger *et al.*, 1992, 1997). SeaMARC 1A sidescan imagery shows that the Wecoma fault offsets a late Pleistocene channel and an older slump scar on the Astoria submarine fan (Figure 17.7). Horizontal displacements are about 120 and 350 m, respectively (Appelgate *et al.*, 1992; Goldfinger *et al.*, 1992). Prefaulting sedimentary units are offset horizontally 5-6 km. Faulting began at about 600 ± 50 ka, resulting in a slip rate of 7-10 mm/yr. The offset late Pleistocene channel is blocked 18 km to the north by a slump, the age of which is estimated to be 10-24 ka based on core data. Because the fault offset records post-slump motion, a slip rate of 5-12 mm/yr can be calculated from these data, comparable to the 7-10 mm/yr based on offset abyssal plain sediments (Goldfinger *et al.*, 1992).

The Oregon faults, generated by offset of the descending slab, show offset of the continental slope, and are easily imaged and interpreted with integration of sidescan sonar and bathymetric data, but could not be interpreted without such integration. The integration of a variety of data made key interpretation of these faults possible. Recently, spectacular images of strike slip faults along the central California margin have been made from data collected with a SIMRAD EM-300 system in a joint effort by USGS, and Monterey Bay Aquarium Research Institute (e.g., Eittreim *et al.*, 1998). The new generation of high-resolution multibeam systems will allow major advances in our understanding of seafloor morphology and tectonic interpretation.

17.4.3 Normal Faults

Listric normal faulting is a common feature of passive margins, where fault movement contributes to crustal thinning and margin subsidence. Extension and normal faulting are also a fairly common phenomenon on convergent margins throughout the world. GIS analysis of extensional faulting can reveal surprising tectonic relationships. McNeill *et al.* (in press) present an unusual case of extension, the implications of which were revealed by data integration in a GIS. Their analysis of multichannel seismic reflection profiles revealed that listric normal faulting is widespread on the northern Oregon and Washington continental shelf and upper slope, suggesting E-W extension in this region. Fault activity began in the late Miocene, and in some cases has continued into the Holocene. Most listric faults sole out into a sub-horizontal décollement coincident with the

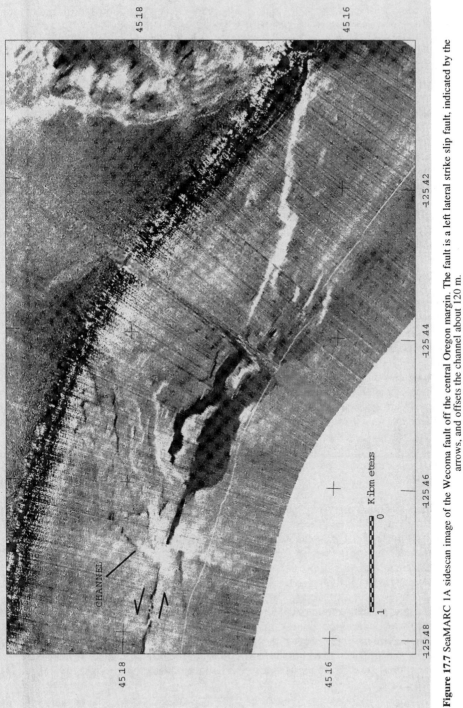

Figure 17.7 SeaMARC 1A sidescan image of the Wecoma fault off the central Oregon margin. The fault is a left lateral strike slip fault, indicated by the arrows, and offsets the channel about 120 m.

upper contact of an Eocene to middle Miocene mélange and broken formation (MBF), known onshore as the Hoh rock assemblage. Other faults penetrate and offset the top of the MBF. The areal distribution of extensional faulting on the shelf and upper slope is similar to the subsurface distribution of the MBF. Evidence onshore and on the continental shelf suggests that the MBF is overpressured and mobile. For listric faults that become sub-horizontal at depth, elevated pore pressures may be sufficient to reduce effective stress and to allow downslope movement of the overlying stratigraphic section along a low-angle (0.1°-2.5°) detachment coincident with the upper MBF contact. Mobilisation, extension, and unconstrained westward movement of the MBF may also contribute to brittle extension of the overlying sediments. Quaternary extension of the shelf and upper slope is contemporaneous with active accretion and thrust faulting on the lower slope, suggesting that the shelf and upper slope are decoupled from subduction-related compression.

Extension and thinning of the MBF are supported by the increased width and low taper angle of the Washington and northern Oregon margins relative to those of the central and southern Oregon margins. Mapped fold trends appear to wrap around a feature on the upper slope of the central Washington margin which is coincident with a convex-seaward "protrusion" or "bulge" in the shelf edge. This shelf edge and fold trends represents the expression of downslope movement of the upper slope. Current extension of the continental shelf and upper slope is contemporaneous with accretion and thrust faulting on the lower slope of the accretionary wedge (Figure 17.8). In addition, extensional faulting appears to be contemporaneous with mapped fold structures on the continental shelf (Goldfinger *et al.*, 1997). McNeill *et al.* (1997) have reinterpreted many of the folds in the vicinity of the normal faults as rollover folds, drape structures, and folds driven by downslope spreading of the continental slope. These structures were misinterpreted as purely convergence-related structures without the integrated data set used for the later study.

17.4.4 Submarine Slides

Submarine slides naturally lend themselves to analysis using GIS techniques (e.g., McAdoo, 1999). Analysis of the age, speed, size, and tsunami implications of slides at many scales have been reported in the literature (see review by Hampton *et al.*, 1996). Regional sidescan sonar mapping and surficial analysis revealed the presence of some of the largest submarine slides on earth off the Hawaiian Islands. The presence of these slides had been speculated long before the acquisition of the sonar data, based on the presence of numerous large cliffs (pali) on Hawaii, and Oahu (Moore, 1964). Moore (1964) observed that these cliffs bore a strong resemblance to headwall scarps of landslides. More than 20 years would pass before the Exclusive Economic Zone GLORIA sidescan sonar surveys of the 1980's would confirm the presence of these giant slides (Moore *et al.*, 1989; 1994; Normark *et al.*, 1993). Morphologic analysis of the region around Hawaii revealed that the runout zones from these slides extended over 140 km, partly uphill on the

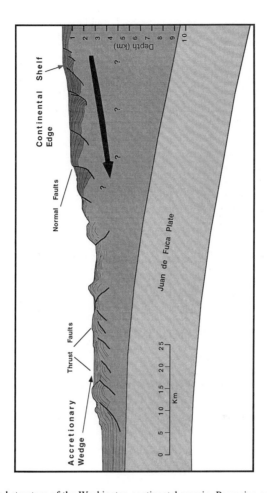

Figure 17.8 Interpreted structure of the Washington continental margin. Pervasive normal faulting on the upper slope, apparent in seismic reflection profiles, is transporting the upper slope Neogene sedimentary section downslope, where it collides with accreting sediments rafted in on the incoming Juan de Fuca plate. The mass downslope transport was deduced from an analysis of fold trends and individual structures together with bathymetry in a GIS.

flanks of the Hawaiian swell, attesting to the catastrophic nature of the slide events. Many of the large cliffs in the Hawaiian Islands have been reinterpreted as headwall scarps from the giant slides, as first speculated by Moore (1964).

Comparable to the Hawaiian slides in scale, three large submarine slides have recently been documented off the southern Cascadia continental margin using Sea Beam bathymetry and multichannel seismic reflection records (Goldfinger *et al.*, in press). The area enclosed by the three arcuate slide scarps is approximately 8,000 km², and involves an estimated 12,000-16,000 km³ of the accretionary wedge (Figure 17.9). The bathymetric scarps correlate with listric detachment faults

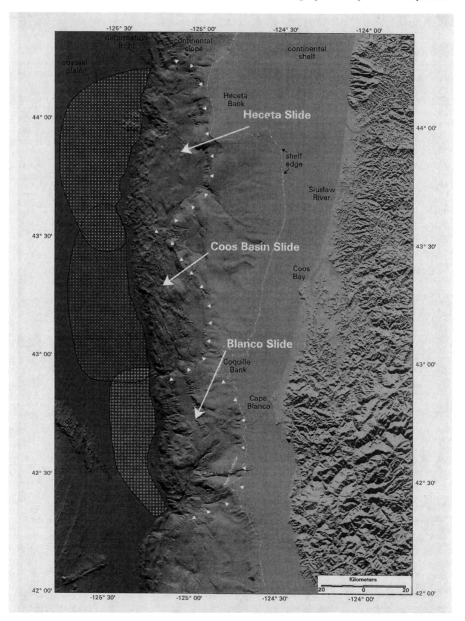

Figure 17.9 Mega-landslides of the southern Oregon margin. Topographic analysis of the southern margin revealed anomalously chaotic surface topography, particularly in the northern part of the view. This led to discovery of buried debris aprons (shaded polygons) and the interpretation of mega-landslides (Goldfinger *et al.*, in press).

identified on reflection profiles that show large vertical separation and bathymetric relief. The slide areas were recognised through analysis of the surface morphology of the continental slope. Reflection profiles on the adjacent abyssal plain image buried debris packages extending 20-35 km seaward of the base of the continental slope. This series of slides travelled 25-70 km onto the abyssal plain in at least three probably catastrophic events, which were probably triggered by subduction earthquakes. The lack of internal structure in the slide packages, and the considerable distance they travelled suggests catastrophic events.

17.4.5 Vertical Tectonics and Subsurface Deformation

The increasing availability of high-resolution bathymetric data has made it possible to address the vertical tectonics of continental margins directly. Previously, vertical deformation has been indirectly derived from low-resolution micro-paleontological analysis of the depths at which various sampled formation were deposited, and what uplift or subsidence had occurred subsequently.

Drowned sea cliffs resulting from episodes of lowered sea level during the Pleistocene offer a powerful means of deducing the vertical (and sometimes horizontal) tectonic movements of a continental shelf over the past 18,000 years. During the last glacial maximum, sea level was 120 m lower than present, and in many areas a sea-cliff and wave-cut erosion platform were cut at this level (Figure 17.1). By mapping the present depth and configuration of the former sea-cliff, we can form a picture of how the continental margin has been deformed since 18 ka.

Goldfinger *et al.* (in prep.) investigated the low-stand shoreline along the Oregon margin with sidescan sonar and a submersible. Submarine traverses found Pholad (intertidal clam) borings, oxidised cross-stratified sands, and shallow-water fossil debris at the base of beach cliffs. Former shorelines were originally horizontal and close to sea level. Now they are warped and faulted, locally to depths of 200-300 m, indicating that the shorelines reflect vertical tectonics rather than eustatic sea level change. Two of the major submarine banks in Oregon, Heceta and Coquille, are strongly tilted to the south (margin parallel), and deformed by faults and folds, while the third (Nehalem) appears relatively undeformed. Although the subsiding banks have undergone ~1 km of uplift since the Miocene, approximately 130 m of subsidence has occurred during the Pleistocene. This could reflect one of many vertical fluctuations, or a significant change in margin tectonics. Subsidence of the banks may be due to gravitational collapse or tectonic erosion of the margin, despite the presence of an accreting prism of sediments to the west.

Similar drowned shorelines are commonly observed, and offer the potential of tectonic analysis along many continental shelves. Recent swath bathymetry has revealed similar shorelines off southern California (Gardner *et al.*, 1997) and off Chile and Peru (C. Goldfinger and A. Mix, unpublished data, 1997).

Similar analyses can be performed on subsurface data to determine regional or local deformation patterns. Plate 17.10 shows an example of integration and analysis of subsurface data in the context of regional deformation patterns along a subduction zone. The structure contour map of a formerly planar subsurface unconformity clearly outlines the deformation of this surface. Variations in vertical tectonic uplift rates of the unconformity on the shelf in a margin-parallel direction show agreement with coastal geodetic and geomorphic data, suggesting a connection between short-term elastic deformation, and much longer-term permanent deformation. Broad uplifts and downwarps may result from elastic/inelastic shortening in the upper plate, coupling variations on the subduction interface, or both. This information contributes to the assessment of the position of the locked zone on the plate boundary and of potential segment boundaries to rupture along the subduction zone, both of critical importance to understanding earthquake hazards.

17.5 SUMMARY AND CONCLUSION

GISs and visualisation packages naturally lend themselves to application in marine sciences, for which most data is remotely sensed, and is commonly presented as imagery. Tectonic geomorphology in the marine environment is now made possible by the rapid development of software and hardware that allow the integration, visualisation, and analysis of large and disparate data sets from a wide variety of data collection devices. With proper data, detailed analysis of active tectonic structures can be done in the submarine environment as well, or in some cases better than can be done on land. On land, urbanisation, natural land cover, and high erosion rates hide or obscure subtle cues from active structures. At sea, these problems are minimised, though data collection is more difficult. Future software advances will bring high-speed visualisation and GIS to a more integrated state, allowing the simultaneous interpretation of more layers of data than presently feasible.

17.6 REFERENCES

Appelgate, B., Goldfinger, C., Kulm, L.D., MacKay, M., Fox, C.G., Embley, R.W. and Meis, P.J., 1992, A left lateral strike slip fault seaward of the central Oregon convergent margin. *Tectonics*, **11**, pp. 465-477.
Blondel, P. and Murton, B.J., 1997, Handbook of Seafloor Sonar Imagery (New York: John Wiley & Sons).
Chadwick, W.W. Jr. and Embley, R.W., 1995, Relationships between hydrothermal vents, intrusions, and ridge structure: High resolution bathymetric surveys using Mesotech sonar, Juan de Fuca Ridge. *Eos, Transactions of the American Geophysical Union*, **76**, pp. 410.
Eittreim, S.L., Stevenson, A.J., Maher, N. and Greene, G.H., 1998, Southern San Gregorio fault and Sur platform structures revealed. *Eos, Transactions of the American Geophysical Union, Fall Supplement*, **79**, p. F825.
Fox, C.G., Murphy, K.M. and Embley, R.W., 1988, Automated display and statistical analysis of interpreted deep-sea bottom photographs. *Marine Geology*, **78**, pp. 199-216.
Gardner, J.V., Dartnell, P., Mayer, L.A. and Clarke, J.-H., 1997, The physiography of the Santa Monica continental margin from multibeam mapping. *Eos, Transactions of the American Geophysical Union, Fall Supplement*, **78**, p. F350.

Goldfinger, C., 1994, *Active Deformation of the Cascadia Forearc: Implications for Great Earthquake Potential in Oregon and Washington*, Ph.D. Thesis (Corvallis, Oregon: Oregon State University).

Goldfinger, C. and McNeill, L.C., 1997, Case study of GIS data integration and visualization in submarine tectonic investigations: Cascadia subduction zone. *Marine Geodesy*, **20**, pp. 267-289.

Goldfinger, C., Kulm, L.D., Yeats, R.S., Appelgate, B., MacKay, M. and Moore, G.F., 1992, Transverse structural trends along the Oregon convergent margin: implications for Cascadia earthquake potential. *Geology*, **20**, pp. 141-144.

Goldfinger, C., Kulm, L.D., Yeats, R.S., McNeill, L.C. and Hummon, C., 1997, Oblique strike-slip faulting of the central Cascadia submarine forearc. *Journal of Geophysical Research*, **102**, pp. 8217-8243.

Goldfinger, C., Kulm, L.D., Yeats, R.S. and McNeill, L.C., in press, Super-scale slumping of the southern Oregon Cascadia margin: tsunamis, tectonic erosion, and extension of the forearc. *Pure and Applied Geophysics Special Volume on Landslides*, edited by Keating, B. and Waythomas, C.

Hampton, M.A., Lee, H.J. and Locat, J., 1996, Submarine landslides. *Reviews of Geophysics*, **34**, pp. 33-59.

Hatcher, G. and Maher, N., 1999, Real-time GIS for marine applications, in this volume, Chapter 10.

Johnson, H.P. and Helferty, M., 1990, The geological interpretation of sidescan sonar. *Reviews of Geophysics*, **28**, pp. 357-380.

MacKay, M.E., 1995, Structural variation and landward vergence at the toe of the Oregon accretionary prism. *Tectonics*, **14**, pp. 1309-1320.

Mallinson, D., Hine, A., Naar, D., Hafen, M., Schock, S., Smith, S., Gelfenbaum, G., Wilson, D. and Lavoie, D., 1997, Seafloor mapping using the Autonomous Underwater Vehicle (AUV) Ocean Explorer. *Eos, Transactions of the American Geophysical Union, Fall Supplement*, **78**, p. F350.

McAdoo, B., 1999, Mapping submarine landslides, in this volume, Chapter 14.

McNeill, L.C., Piper, K.A., Goldfinger, C., Kulm, L.D. and Yeats, R.S., 1997, Listric normal faulting on the Cascadia continental shelf. *Journal of Geophysical Research*, **102**, pp. 12,123-12,138.

McNeill, L.C., Goldfinger, C., Kulm, L.D. and Yeats, R. S.., in press, Tectonics of the Neogene Cascadia forearc basin: Investigations of a deformed Late Miocene unconformity. *American Association of Petroleum Geologists Bulletin*.

Mitchell, C. E., Vincent, P., Weldon II, R. J., and Richards, M. A., 1994, Present-day vertical deformation of the Cascadia margin, Pacific northwest, U.S.A. *Journal of Geophysical Research*, **99**, pp. 12,257-12,277.

Moore, J.G., 1964, Giant submarine landslides on the Hawaiian Ridge. *U.S. Geol. Survey Prof. Paper 501-D*, pp. D95-D98.

Moore, J.G., Normark, W.R. and Holcomb, R.T., 1994, Giant Hawaiian landslides. *Annual Review of Earth and Planetary Sciences*, **22**, pp. 119-144.

Moore, J.G., Clague, D.A., Holcomb, R.T., Lipman, P.W., Normark, W.R. and Torresan, M.E., 1989, Prodigious submarine landslides on the Hawaiian Ridge. *Journal of Geophysical Research*, **94**, pp. 17,465-17,484.

Normark, W.R., Moore, J.G. and Torresan, M.E., 1993, Giant volcano-related landslides and the development of the Hawaiian Islands. In *Submarine Landslides: Selected Studies in the U.S. Exclusive Economic Zone. U.S. Geological Survey Bulletin*, **2002**, pp. 184-196.

Wright, D.J., 1996, Rumblings on the ocean floor: GIS supports deep-sea research. *Geo Info Systems*, **6**, pp. 22-29.

Yeats, R.S., Kulm, L.D., Goldfinger, C. and McNeill, L.C., 1998, Stonewall anticline: An active fold on the Oregon continental shelf. *Bulletin of the Geological Society of America*, **110**, pp. 572-587.

CHAPTER EIGHTEEN

Managing Marine and Coastal Data: A National Oceanographic Data Centre Perspective on GIS

Bronwyn Cahill, Theresa Kennedy and Órla Ní Cheileachair

18.1 INTRODUCTION

The Irish Marine Data Centre (MDC) is an integral component of the Marine Institute, which was established under the Marine Institute Act of 1991. Its mission is defined by statute as follows:

"... To undertake, to co-ordinate and to assist in marine research and development and to provide such services related to marine research and development, that in the opinion of the Marine Institute will promote economic development and create employment and protect the marine environment."

Marine Institute Act, 1991

Data and information management play a vital role in supporting the sustainable development of marine resources. Key to the success of marine research and development is the speed and ease with which users can identify, locate, access, exchange and use marine data and information. Systematic management of reliable marine data and information is a prerequisite to supporting and broadening our understanding of the marine resource. Furthermore,

"... It shall be the general duty of the Marine Institute to collect, maintain and disseminate information relating to marine matters."

Marine Institute Act, 1991

The focus, therefore, of the Marine Institute's mission for the MDC is a mandate to develop a data and information service which cross cuts the business, programmatic and statutory requirements of the marine research and development community and which endeavours to strengthen the overall quality, service and performance of marine data and information management.

Since its inception, the MDC routinely maintains a comprehensive inventory of marine data sources. These sources include universities, research institutes, government bodies, a variety of public and private enterprises and can be very diverse in the types of data they produce, their data policy and their geographical location. The Marine Institute is also one of the largest data collection institutes within Ireland and the MDC itself is a source of national bathymetry, physical, chemical and biological oceanographic data.

Although it's primary function is one of a National Oceanographic Data Centre (NODC), the MDC also act as data management partners in several multi-national marine projects funded within the European Union's Marine Science and Technology (MAST) and Environment and Climate programmes. These projects typically involve several European Institutes and are inherently multi-disciplinary in nature. Through this EU work, the MDC regularly maintains contact and data transfer facilities with a total of 90 institutes from 16 different European countries. These projects also involve the generation and synthesis of very large volumes of data and culminate in the assimilation and publication of this data in a widely usable form, for example data products on CD-ROM.

18.2 INVOLVEMENT WITH GIS

The MDC's first involvement with GIS came about through the recognition of its potential for visualising geospatial marine information, not only in a scientific context, but also as a tool for management and policy making. It also became clear throughout the mid-1990s that the use of GIS in the marine science community was becoming widespread (see Wright, 1999), and that there was a demand for the provision of data sets that could be easily incorporated into GIS projects. In our role as an NODC, our basic function is to endeavour to satisfy the demand for information, provide it in an accessible and desirable format and keep abreast of current technological trends and innovations. The move of our community in to the GIS arena was therefore our impetus to investigate how we might benefit from its implementation in-house and how we might service the needs of marine GIS scientists in Ireland.

The use of GIS functionality was piloted first in several internal and collaborative projects without adhering to a particular GIS. The objective of this was to evaluate the potential future of the technology within an NODC framework and assess its effectiveness as a tool for operational data management. Also under review was its potential in the context of product-driven data management projects and its scope for electronic publication of data. One of the first opportunities arose as a result of the need to track the progress of research cruises for multi-national data collection campaigns and resulted in the design and development of the worldwide web (WWW) Cruise Data Tracking System (CDTS).

18.2.1 Cruise Data Tracking System

The original concept of the CDTS arose from a need to manage, update and disseminate research cruise information and data from national and multi-national projects. Typically, any collection of information involving the deployment of a research vessel is very expensive with operating costs in the order of several thousand pounds per day, and campaigns of only a few days can generate hundreds of megabytes of data. Research vessel activities therefore, represent a very large investment, and return on this investment can be judged in the subsequent management and re-use of the data. The research vessel activity from two EU MAST programmes, European North Atlantic Margin II (ENAM II) and the Canary

Islands Azores Gibraltar Regional Seas (CANIGO), can illustrate this point further. Over a period of two years, approximately 60 different research vessel campaigns were carried out amounting to almost one year at sea. Over 1,500 deployments were made, and 500 scientists from 90 research institutes in 16 European countries were involved. The area covered by these campaigns amounts to over 2,000,000 km² and hundreds of miles of cruise track, translating into a data collection cost of over IEP£2.5m. Managing this voluminous information, and making it available to its users, was a considerable undertaking.

The MDC proposed to implement a geographically referenced database of cruise activity, administered centrally at the MDC, which would be accessible over the WWW. Scientists required up-to-date information on cruise activities, research objectives and the availability of data, whilst the EU funding body required a means by which the status and productivity of research cruises could be assessed. As data managers, the MDC required a system whereby the exchange and re-use of the data would be promoted and maximised. Moreover, the MDC stood to benefit from an efficient means of tracking the status of data collection and processing throughout the term of the project.

Version 1.0 of the CDTS allows the user to query by project, parameter, geographic location, vessel and co-ordinating institute. It displays considerable detail pertaining to an individual cruise and the data collected and is consistent with Cruise Summary Reporting required by International Council for the Exploration of the Seas (ICES). Some of the features of the CDTS are outlined in the Figure 18.1.

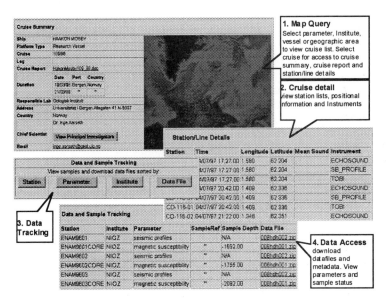

Figure 18.1 Cruise Data Tracking System.

Each user of the CDTS has the ability to access information on the status of research cruises for their particular project, and where desirable, download the data resulting from the cruise. This has promoted an "awareness" between the different disciplinary groups of the research projects, of complementary work and on-going research in their study area, which was previously less acute. The CDTS therefore may promote data integration and collaboration, as well as provide dynamic information on project activities.

While its GIS functionality is rudimentary in terms of spatial querying and display, it is clear from user statistics that it provides a useful resource. It has also proved its effectiveness for the MDC in terms of tracking the progress of individual project tasks and deliverables and the distribution and use of project data within the life of the project. Given the success of Version 1.0 of the CDTS, a potential future investment is the integration of its database with a fully functional GIS WWW interface.

18.2.2 Product Driven Data Management and GIS

Our multi-national data management projects are essentially product-driven activities. The main objectives of project data management, as specified by the funding bodies, are to guarantee the longevity of project data, aid dissemination of the project data and results and maximise the re-use of the data. The traditional method of achieving these objectives is to publish a "product", usually on CD-ROM, to hold the data and, in some cases, to provide custom-made software to interrogate or extract the data. Increasingly, the need to visualise spatial aspects of large multi-disciplinary project data has become an essential requirement driving the development of data management products. The ENAM II project is a good example of this development. The project is multi-disciplinary and involves 14 partners from eight European countries. Its scientific objective is to quantify and model large-scale sedimentary processes and material fluxes in the north Atlantic and to assess their relation to the variability of oceanic and cryospheric processes (Mienert *et al.*, 1996). The project involves the collection of geological, geophysical and geochemical information from the Atlantic Margin extending from northern Norway to southern France between 1996 and 1999.

ENAM II data are inherently geospatial in nature. Where many biological oceanographic projects concentrate on geographically limited areas over long periods of time, ENAM II is heavily dominated by spatial comparisons in three dimensions. Spatial correlations between different collection methodologies are of paramount importance. For example, the proximity of sediment cores to seismic profiles and the intersection of side-scan sonar data with bottom sampling and shallow profiling. Data types are represented in 2-D (e.g., sidescan sonar) and 2.5- to 3-D (e.g., seismic profiles and sediment cores). The project involves the integration of numbers with imagery, direct observations and subjective interpretations and all of these parameters exist within a complex environmental framework. The ENAM II data product therefore is required to manage and present the data, maintaining the scientific relationships between data types and providing a geospatial framework through which to integrate the data. It was clear, through a detailed analysis of user requirements however, that in-depth data analysis was not a requirement, and that a major strength of the product should be the visualisation of spatial relationships between the different types of data and an intuitive way to readily locate, access, and export the information of interest.

Development of the product centred on the design of a relational database with a GIS front-end. Since no prior knowledge of GIS was presumed for the target audience, an intuitive interface was designed to feature simple icons and pre-defined geographical queries. For example, the ability to display certain layer of information such as bathymetry, was enabled through simple click boxes (see Figure 18.2), and pre-defined queries for displaying cruise tracks by data collection mode were enabled in a similar way. The user is essentially treated to a visual, geographical overview of the project when entering the product, and has the power to manipulate the information shown on the overview map, or access the data directly – through the map interface or using traditional browser tables. The complexity of the data model allows for advanced querying of the database, and export of the results to external analysis software. This was a priority feature identified by the users. A screen grab of the product prototype is shown in Figure 18.2 to give an overview of the interface and an illustration of some of the features.

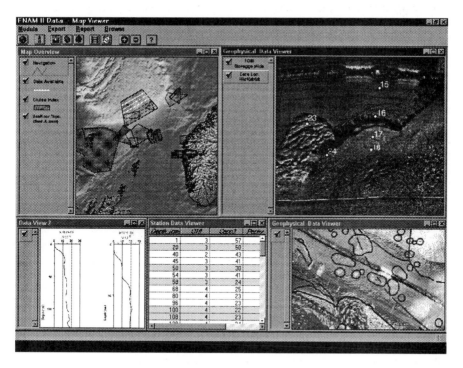

Figure 18.2 ENAM data product.

18.3 FUTURE INTEGRATION OF GIS AT THE MDC

The application of GIS at the MDC to date has been driven by two factors. On the one hand, its potential as a powerful tool to improve the quality of service as an NODC has been demonstrated through the development of the CDTS. On the other, the need to publish data electronically has driven developments of user-oriented products such as the ENAM data product. We are therefore experiencing a parallel development of the technology and will continue to move toward customising GISs for particular needs and objectives as well as integration of the technology as part of our operational infrastructure.

The long-term scope of our data and information management brief extends beyond classical oceanographic data and will demand sophisticated information products, which cater for the needs of coastal zone managers and sustainable economic development. Statistics on tourism, fisheries and socio-economics have a prominent position in our National Marine Policy (Marine Institute, 1998) and consolidate our need for a powerful visualisation and analysis infrastructure.

Furthermore, our function to support GIS development will be dependent upon the availability of large data sets to support innovative research and the timely provision of data to users in a format which is readily usable and comprehensively documented. Implicit in this, is the need for rigorously designed data models for scientific relational databases.

18.4 REFERENCES

Mienert, J., 1996, European North Atlantic Margin (ENAM II): Quantification and modelling of large-scale sedimentary processes and fluxes. In *Technical Annex, MAS3-CT95-0003* (Dublin: Irish Marine Data Centre), http://www.marine.ied/datacentre/projects/ENAM.

Marine Institute, 1998, *A Marine Research, Technology, Development and Innovation Strategy for Ireland. A National Team Approach* (Dublin: Irish Marine Data Centre).

Wright, D.J., Down to the sea in ships: The emergence of marine GIS, in this volume, Chapter 1.

CHAPTER NINETEEN

Significance of Coastal and Marine Data within the Context of the United States National Spatial Data Infrastructure

Millington Lockwood and Cindy Fowler

19.1 INTRODUCTION

Interest in coastal and ocean spatial data in the United States has motivated national policy since the establishment of the Coast Survey in 1807. Except for limited periods when national security was a concern, the U.S. has always had a policy of open access to spatial data – historically in the form of maps and charts – and more recently in the form of digital data. It is felt that this information, regardless of the form, should be regarded as a public asset and is necessary to support commerce, health, safety and environmental protection. There is a significant element within the spatial data community in the U.S. that has an interest in coastal and ocean data. This community of users has looked to the National Oceanic and Atmospheric Administration (NOAA) to be the suppliers of these data. In response, NOAA has made a commitment to support the data needs of the coastal community within the context of the overall national mandate to create a National Spatial Data Infrastructure (NSDI).

NOAA has traditionally conducted coastal and ocean mapping to support its overall mission of promoting safe navigation. With the passage of the Coastal Zone Management Act (CZMA) in 1972, there was an additional need to support the new mandate of coastal stewardship. This new mission, combined with user demands driven by new mapping technologies, such as geographic information systems (GIS), has caused NOAA to give considerable attention to the management of geospatial data. Though still evolving, NOAA's major effort for the NSDI will be to develop data, metadata, standards, and clearinghouses to support shoreline, bathymetry and marine cadastral framework data[1] for use in coastal and ocean GIS. Other useful non-framework data sets will be of next priority, such as marine transportation, imagery, marine biology, and other supporting geospatial data types.

---

[1] A framework data set is a data layer upon which subsequent mapping may be developed. It provides the geospatial foundation to which an organization can add additional detailed information (Federal Geographic Data Committee, 1995). Shoreline, bathymetry and marine cadastral are most certainly framework datasets. Depending on perspective, one could argue that coastal imagery, marine navigation, tidal benchmarks and benthic habitats could also be considered framework datasets.

19.2 BACKGROUND

19.2.1 Background of U. S. Policy for Geospatial Data

The National Partnership for Reinventing Government, formerly known as the National Performance Review (NPR), included a call for a consistent, reliable means to share geographic data among all users, nationwide and to reduce wasteful duplication of effort among all levels of government with respect to these data. In April 1994, President Clinton signed Executive Order 12906, calling for the establishment of the NSDI (Clinton, 1994). This Executive Order strengthened the earlier Office of Management and Budget (OMB) Circular A-16 and created a co-ordination role for the Federal Geographic Data Committee (FGDC) in the development of the NSDI.

The NSDI, as defined by the Executive Order, is identified as the technologies, policies, and people necessary to promote sharing of geospatial data throughout all levels of government, the private and non-profit sectors, and the academic community. As requested by the NPR, the NSDI works to develop incentives to do more with less within government. The NSDI strives to make the best use of taxpayer dollars and provide the spatial data to support problem solving on a state, regional, tribal, and local community level. Federal agencies were given nine months to begin documenting new data and one year to develop a plan for documenting existing spatial data and making them accessible via National Geospatial Data Clearinghouses.

Although an Executive Order exists for the NSDI, there is no enforcement department monitoring compliance. Federal agencies are given some leeway in determining which data might contribute to the NSDI. OMB Circular A-16 uses language such as "the most expeditious and economical manner possible with available resources" or "practical and economical to do so" (OMB, 1990). Individual agencies must understand the big picture of why the sharing of geospatial data can help support their individual mandates. Because no additional resources are available to federal agencies for participating in the NSDI, implementation requires creative arrangements. One example of such innovation is the development of strategic partners outside of the individual agency, such as state councils, other federal agencies, universities, private industry, and others, to work together to develop spatial data. Another example is leveraging data within the organisation to support multiple mandates. In response, NOAA has made a commitment to support the data needs of the coastal and ocean community within the context of the overall national mandate to create the NSDI.

19.2.2 Background of Coastal and Ocean Mapping within NOAA

The Coast Survey has been mapping the U.S. coastal waters since the 1800s. The basic authority for the "Survey of the Coast" comes from the Organic Act of February 10, 1807 (2 Stat 413). It "authorised and requested to cause a survey to be taken of the coasts of the U.S. in which shall be designated the islands and shoals...for completing an accurate chart of every part of the coast". In 1970, under Presidential Reorganization Plan No. 4 (84 Stat 2900), NOAA was formed and coastal mapping and charting activities for the coastal and adjacent ocean areas of the U.S., U.S. Possessions and Territories, the Great Lakes, and other inland navigable areas were consolidated into NOAA's Coast and Geodetic Survey, currently known as the Office of Coast Survey.

In 1972, the passage of the CZMA authorised the Coastal Zone Management Program, a voluntary partnership between the Federal Government and U.S. coastal states and territories. To date, 27 coastal states and five U.S. territories have developed coastal zone management programs. Until the passage of the CZMA, coastal mapping activities were predominately focused on supporting safe navigation of merchant ships and naval fleets. The needs of the navigator were fairly straightforward and had been well known for many decades. The result of the passage of this act was an expanded need for coastal mapping that would use the information for the wise development and protection of the coastal zone. NOAA now had two mandates requiring coastal and ocean data: safe navigation and coastal stewardship. To support these mandates, NOAA is charged with mapping 95,000 linear miles of oceanic and Great Lakes coastline and 3.4 million square nautical miles of ocean area to support these missions.

Another driving force in coastal and ocean mapping has been the impact of new technologies. The introduction of modern mapping tools such as GIS, raster navigational charts (RNC), electronic chart display and information system (ECDIS), and the Global Positioning System (GPS) have forced the agency to depart from the traditional paper chart to the development of digital spatial data products to meet user needs. In addition, advanced technologies in the collection of data, such as side scan sonar, airborne laser, GPS-controlled photogrammetry and swath bathymetry, have changed the processes by which data are acquired and stored. These highly accurate data streams improve the quality of coastal and marine mapping, but require considerable resources to utilise. NOAA is working to incorporate advanced technologies in coastal and ocean mapping to meet the changing user needs. An important aspect of this change is the development of institutional co-operation with respect to spatial data found in the development of the NSDI.

19.3 COASTAL AND MARINE COMPONENTS OF THE NSDI

Chaired by the Department of the Interior, the FGDC consists of 14 federal agencies and departments that are involved in surveying, mapping, and related spatial data activities. There are numerous working groups (e.g., Standards, Clearinghouse, and Framework) and thematic subcommittees (e.g., Cadastral, Geodetic, and Water) within the FGDC structure that intersect coastal and marine data (Figure 19.1). Generally, the Bathymetric Subcommittee, chaired by NOAA, is the primary group involved with fostering the NSDI with respect to coastal and marine data. The Subcommittee, which held its first meeting in April 1993, has concentrated on bathymetry and shoreline data standards development. In addition to shoreline and bathymetry, committee members within NOAA have

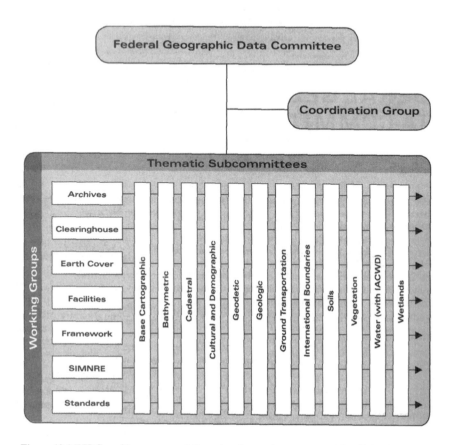

Figure 19.1 FGDC working groups and thematic subcommittee structure (after Federal Geographic Data Committee, 1997).

identified maritime boundaries as the other framework data set held by NOAA that is uniquely needed by the coastal and ocean mapping community. The subcommittee will also be working to bring other non-framework data sets, such as marine navigation, tidal benchmarks, tide-controlled imagery, fisheries, marine mammals, and corals into the NSDI. The subcommittee has been working within NOAA, more specifically in the National Ocean Service[2] (NOS) office in which these data predominately reside, to foster the development of the NSDI with respect to coastal and marine data. In addition, NOAA has responsibility for the development of the geodetic control framework. This critical element is crucial to all mapping activities, but is not unique to the coastal zone, and therefore will not be discussed here in detail.

---

[2] NOS is the office within NOAA whose mission requires the development of these marine framework data sets.

The FGDC has determined that coastal data warrant special attention by the creation of a separate subcommittee, but many of the data sets important to coastal mapping do intersect other FGDC data themes. There is considerable room for interpretation on what data element falls into what theme. For example, the shoreline is a unique coastal feature that could also be considered to fall into the jurisdiction of the Water Subcommittee. "Where does the river bank end and the shoreline begin?" is the question that must be answered for this component. Generally, it is considered the Head-of-Tide, but wide-scale location of this feature can be somewhat problematic. Bathymetry is often considered negative elevation and therefore could be part of the elevation theme. Like elevation, bathymetric data share the component of a vertical datum, but uniquely have the attribute of a tidal datum. Maritime navigation features could be considered within the jurisdiction of the Transportation Subcommittee. Can a navigation channel be considered an ocean-based highway? In addition, many of the maritime boundaries, such as the territorial sea or Exclusive Economic Zone (EEZ), could be considered under the auspices of the Cadastral Subcommittee. Is not a benthic habitat data layer just a continuation of the land-based vegetation theme? The FGDC has identified coastal data as having unique elements that warrant the attention of their own subcommittee, but considerable overlap exists with their land-based counterparts. Moreover, as expected, the thematic subcommittee structure follows federal agency responsibilities. NOAA, with its many years of experience mapping the coastline and the oceans, must naturally lead the effort to make these data part of the NSDI.

19.4 WHY ARE COASTAL AND MARINE DATA UNIQUE?

The land-water interface of the coastal zone makes it one of the most challenging areas to map and understand. The Great Lakes, open water, wetlands, mudflats, buoys, channels, intracoastal waterways, islands, bays, low waterline, high water line, estuaries, levees, mangroves, coral reefs, beaches, dunes, piers, groins, 3, 9, 12, and 200-nautical mile lines are just a few of the features that must be depicted when mapping the coastal and ocean zone (Figure 19.2).

Knowing and understanding the condition of the tide is critical for accurate coastal mapping and is often overlooked. Knowing the tidal datum, the base elevation defined by a certain tide phase, must be known when mapping in areas affected by tidal fluctuations.

A few of the many terms used to reference tidal datums are mean higher high water (MHHW), mean high water (MHW), mean low water (MLW), mean lower low water (MLLW), and mean sea level (MSL). These boundaries can be critical for determining the limits between international, national, state, and private rights. Moreover, individual states vary in which tidal datum is used for the demarcation between private and state-owned tidelands. Tidal influences in this zone make mapping extremely complex and only accurate for a specific state of tide. Tidal ranges between low and high water can vary from a virtually zero to over 50 feet in areas such as the Bay of Fundy, Canada. Tidal arrival times can also vary greatly spatially from one location to another, greatly complicating coastline mapping.

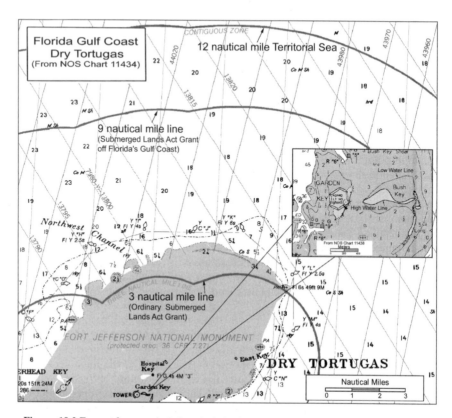

Figure. 19.2 Excerpt from nautical chart depicting low water line, high water line, 3, 9, and 12-nautical mile boundaries, and bathymetric data.

Tide is generally defined as the rise and fall in sea level with respect to the land. Tide is produced by the gravitational forces of the moon and sun, but additional non-astronomical factors such as coastline configuration, depth of the water, and topography are all factors influencing the interval and the arrival time of the tides. To achieve the maximum accuracy in tidal prediction, a series of continuous observations at one location for 18.6 years is required. Within this number of years all significant astronomical modifications of tides will have occurred. For practical reasons, NOAA maintains a tidal prediction control network of 140 permanent tide gauges and supplemental secondary temporary gauges, where needed. Tidal scientists use the control network data and other known ancillary information to predict tidal ranges and times. As technology advances and user needs demanded more accurate mapping of coastal features, the need for accurate tidal predictions and real-time tidal data has become an increasingly important mapping element.

In addition to tidal influences, other elements complicate mapping in the coastal and ocean region. Hydrologic forces on the landscape in the coastal zone cause many parts of this region to be extremely dynamic. Depending on the region, tidal influences exacerbated by extreme weather can make an accurate coastal survey quickly obsolete. In addition to a constantly changing environment, mapping in the ocean is problematic because many measurements must be taken indirectly through the water column. The tidal phase, clarity, or wave activity of the water column can greatly influence this measurement. In addition, the ocean is a multi-dimensional feature and many of the current spatial data models were developed for two-dimensional space. The addition of the water column (i.e., the third dimension) and the changing nature of this environment (i.e., time, the fourth dimension) require new data structures that will adequately model this feature within a GIS. The dynamic natural forces at work in the coastal and marine environment, the movement of water along the surface of the earth, the nature of sampling in and through the water column, and the need for multi-dimensional data structures are elements that make these coastal and marine data sets unique from their land-based counterparts.

19.5 FRAMEWORK DATA ELEMENTS

19.5.1 Shoreline

The nature of the shoreline is such that it is a transient boundary with changes occurring from both human-induced and natural forces. Despite its complexity, the shoreline is one of the data elements most widely used by the coastal community. Some form of shoreline delineation is used in almost every example of coastal and ocean resource mapping. The International Union of Geological Sciences (IUGS) has identified shoreline as one of 27 global geo-indicators (Berger and Iams, 1996). In addition, an examination of geographic data needs by each of the FGDC thematic subcommittees revealed that eight of the 11 subcommittees have a critical interest in shoreline data (Federal Geographic Data Committee, 1996).

The term "shoreline" is defined as the intersection of the land with the water surface (Ellis, 1978). Generally, this definition is sufficient for those not involved in coastal mapping or boundary determination. The problems with shoreline data geometry readily become apparent, however, when co-ordinates are assigned to this line, as they are in a GIS or boundary survey. Moreover, the uncertainty surrounding the nomenclature becomes apparent when data are shared between disciplines. As discussed in Section 19.4, tidally influenced shoreline cannot be easily captured as a fixed line; it can only be represented as a condition at the water's edge during a particular instant of the tidal cycle (Shalowitz, 1962).

NOAA has been mapping the shoreline since the early 1800s. The mapping component of this effort that includes the shoreline is the topographic sheet, commonly referred to within the mapping community as the T-sheet. These coastal survey maps are special use planimetric or topographic maps ranging in scale from 1:5,000 to 1:40,000. They accurately define the shoreline and other natural and man-made features and are carefully controlled for tidal fluxes. Before aerial photography was available, traditional plane table and alidade survey techniques were applied to construct the T-sheet. After 1927, NOAA began using aerial photography and the science of photogrammetry to delineate the shoreline. NOAA developed specialised techniques to acquire tide-controlled photography. Using tide gauges to monitor water level, NOAA photography is acquired within specified time intervals of low tide. Photogrammetric techniques are then used to manually delineate the high and low water line. In the 1990s, technological advances in digital processing provided for digital versions of the data that had previously only been available on the paper T-sheet. Although new satellite technologies are being explored, operationally, photogrammetric techniques are still used to map all of the shoreline within NOAA.

Approximately 13,000 T-sheets dated pre-1990, are housed in NOAA or the National Archives, and in many cases, are the most accurate record of the state of the coast at a particular point in time. To date, NOAA has mapped approximately two-thirds of the nation's coast. The missing third is predominately in Alaska, where weather conditions make aerial mapping difficult. The historic T-sheets provide a valuable resource for shoreline change analysis. For most coastal states, this is a necessary component of determining building setback lines. The contemporary shoreline is needed for accurate nautical charts as well as in modern mapping systems such as GIS and ECDIS.

These shoreline data are extremely valuable, but there are many technical issues that must be resolved before these data may be utilised in computer systems. The bulk of the T-sheet archive is only available in a hardcopy format (i.e., paper, linen, negative) and must be converted to a digital format. This is a multi-step process which involves scanning the hardcopy resource, georeferencing the resulting raster scan, vectorising the georeferenced file, and finally creating the topologically structured and attributed shoreline data layer (Figures 19.3 and 19.4). The importance of metadata is all too evident when the technician attempts to georeference the raster T-sheet and finds that little metadata is available describing the horizontal datum or projection. Many of the historic documents require research into the survey methods of the period. This is no small undertaking for 13,000 historic documents.

The T-sheet archive provides the largest consistent source of shoreline data and will be a major component of the NSDI, but other versions of shoreline data are equally important elements. These complexities of mapping the shoreline, such as its ephemeral nature, effects of tide, and widely varying nomenclature, compound the everyday data problems that include knowledge of the projection, datum, and scale. This confusion over terms and the determination of which line is "meant" by shoreline creates uncertainty in coastal management and boundary decisions. In an effort to clarify the situation, individuals within the FGDC, NOAA and, other interested agencies have been working to develop a Shoreline Standard, which is currently in the public review stage of the process. Within the NSDI

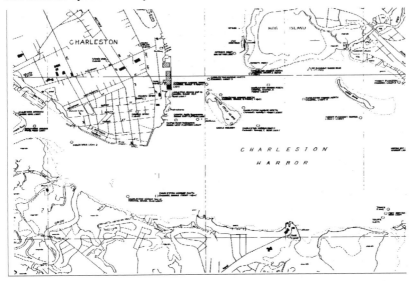

Figure. 19.3 Raster scan of paper topographic sheet (T-sheet).

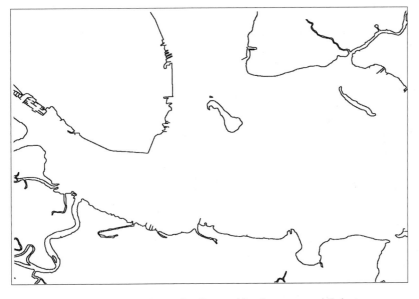

Figure. 19.4 Vector shoreline resulting from processed T-sheet.

structure, the shoreline data element falls under the auspices of the Bathymetric Subcommittee. A shoreline standard will enhance data sharing by providing common shoreline definitions and metadata structure for the acquisition and management of shoreline data. In addition, the creation of a standard will assist in minimising duplication within the government agencies collecting shoreline data.

19.5.2 Bathymetry

In the NSDI structure, bathymetric data are a part of the Elevation Data Framework element and refer to a spatially referenced vertical position above or below a datum surface. Generally, bathymetry differs from elevation because the surface is covered with water. As defined by the FGDC, bathymetric framework data will consist of soundings (point data) and a gridded surface model of the bottom. Currently, bathymetric data are referenced to a local tidal datum, but ideally they could also be contained in a vertical reference system based on the global ellipsoid model.

A coastal GIS is required to perform analyses and modelling across the land-water interface and will require the merging of topographic and bathymetric data into one common reference system. Merging data in this zone requires a greater knowledge of the data than normally needed to combine data in a strictly land-based area (Currie and Walker, 1997). The user must be cognisant of the vertical datums, such as the chart datum in the water, and MSL on land. The simplest transformation from one vertical datum to another requires only adding or subtracting the differences between the two datums. This process is sufficient if the area of interest is relatively small and the vertical accuracy requirements are minimal. Because tidal ranges may vary widely, however, a uniform shift may not be possible. The coastal GIS user must be mindful when combining topographic data across the land-water interface. Metadata are critical in interpreting the data. Because much of the existing bathymetry for the U.S. can be classified as historic data, complete metadata records do not exist. The user may be required to examine the corresponding chart, supporting documents, date of survey, and other ancillary information to accurately interpret the reference systems used.

The largest single source of bathymetric point data is from NOAA hydrographic surveys[3]. Surveys are conducted to support a variety of activities: nautical charting, port and harbour maintenance (dredging), coastal engineering (beach erosion and replenishment studies), coastal zone management, and offshore resource development. Most surveys are conducted to identify water depth but may include additional information such as the nature of the sea floor material, revisions to the shoreline, and cultural features seaward of the low water line.

Up until 1930, the state-of-the-art technology for hydrographic surveys was a knotted lead line dropped from the boat and dead reckoning to navigate. In the 1930s, echosounders (instruments that can measure depth by bouncing sound waves off the bottom) were used. Many of these older technologies were amazingly accurate, but were only able to produce single point data. Bottom depths were interpolated from widely spaced point data and many submarine canyons and peaks were missed.

With the advent of swath bathymetric surveys, uncertainty concerning the ocean floor was greatly diminished. Multibeam sonar systems provided fan-shaped coverage of the seafloor by recording the time required for the acoustic signal to travel from the transmitter (transducer) to the bottom and back to the receiver. Multibeam sonars are generally attached to a vessel and the coverage area on the seafloor is dependent on the depth of the water, typically two to four times the water depth. Although promising new aircraft-based sensors, such as the Laser

---

[3] Data are available from NOAA's National Geophysical Data Center in Boulder, Colorado, USA.

Infrared Digital Airborne Radar (LIDAR) may also prove useful for shallow water bathymetry, they have not been widely adopted by the primary bathymetric mapping agencies. In addition, satellite technologies such as the Geosat satellite accurately measure the distance to the ocean's surface and use an understanding of the differences in the earth's gravity field to estimate bathymetry on a global scale. This broad perspective is useful for a general understanding of the ocean floor in those areas not yet surveyed by ships. The NSDI solution for bathymetry must take into consideration many forms and scales of bathymetric data.

As previously stated, NOAA is responsible for mapping 3.4 million nautical square miles of the ocean bottom, and, until relatively recently, the agency focus has been on safe navigation with respect to mapping. During the years from 1981 to 1993, NOAA was actively involved in mapping the newly-annexed EEZ for a determination of the value of this region. Both the expense of working in the marine environment and the accuracy requirement for bathymetric data necessary to ensure safe navigation are considerable. Modern swath systems produce enormous amounts of data that must be managed. Because of fiscal constraints and the extensive area involved, much of the U.S. seafloor has not yet been mapped with modern surveys. In addition to not having adequate modern survey coverage, many of the data for an area are not currently available in digital form. Efforts have been ongoing within NOAA to convert archived data into digital products. Currently, new NOAA hydrographic surveys are focused only on the 40 priority ports and only in those areas 3 fathoms (18 feet) or deeper, so historic data are still of great value.

Many of the data acquired by NOAA for charting purposes are useful to the broader mapping community. Because of potential liability, however, NOAA has been extremely cautious about selecting data that will be released or portrayed on a nautical chart. In the collection process, potentially less than 1 percent of the soundings may be transferred to the nautical chart. The chart production uses a number of rule-based decisions to determine which soundings are included, but generally, decisions are biased toward minimum bottom depths. This is the proper decision for navigation, but not for those interested in resource management. Items such as submarine canyons may be of considerable interest to scientists studying aquatic species or geology, but may be simplified and not included on the chart to facilitate cartographic representation. In addition to NOAA, other agencies are routinely acquiring bathymetric data that are valuable to resource management. Examples of other non-classified bathymetric sources lie within the U.S. Army Corps of Engineers, U.S. Geological Survey (USGS), state resource management agencies, and university research programs. Other classified sources may become available as Cold War mapping becomes declassified. Currently, there is no easy way for the user to locate, assemble, or evaluate these multiple data sources. As a fundamental tenet of the NSDI, the bathymetric framework must be all-inclusive. NOAA, the lead agency with respect to bathymetric data within the NSDI, is taking the lead on making these data sets available.

To support the development of the bathymetric framework, the Bathymetric Subcommittee has been working on a data standard to describe these data. The standard is closely modelled after the International Hydrographic Organisation (IHO) S-44 standard (IHO, 1996; Bathymetric Subcommittee, 1998). Generally, where possible, FGDC standards work to complement IHO standards. Among other attributes unique to bathymetry, the standard includes an accuracy classification field[4] similar to the geodetic control classification system. This accuracy assessment field allows the end user to evaluate the data for appropriate use. The IHO and the FGDC standards both recognise that a diverse user community needs hydrographic survey data and accommodations to the standard have been made for non-navigation users.

In addition, NOAA has been working to identify a structure in which bathymetric soundings can be stored, updated, and made available to public. The size and complexity of this process is requiring considerable time and will take substantial agency resources. Because the majority of the data are not digitally archived, work must be done to rescue and subject these data to quality control before distribution. Some work has been accomplished to this end (see footnote 3) but a considerable amount is yet to be accomplished. The intended use of the bathymetric clearinghouse is to support internal NOAA functions, but in the spirit of the NSDI, this database will be a mechanism for access to bathymetric data for all users.

In addition to data access, as stated previously, the complexity of mapping topography in the coastal zone requires considerable expertise in interpreting these data. In the recent past, NOAA and the USGS jointly mapped this region. The outcome was a product called the Topo/Bathy Map Series. The resulting paper map product resolved the discrepancies of vertical and tidal datums and provided an excellent base map for subsequent mapping in the coastal zone. Due to budget constraints within NOAA, this map series was discontinued in 1993. However, with the increasing interest in coastal mapping (other than for navigational purposes), and the future availability of data sets (such as bathymetry, shoreline, and USGS data), a modern day digital equivalent could be constructed.

19.5.3 Marine Cadastral Boundaries

As described by the NSDI, the Cadastral Theme is concerned with property interests, or the geographic extent of past, current, and future rights and interests in property (Federal Geographic Data Committee, 1997). The Cadastral Thematic Subcommittee is led by the Department of the Interior, specifically the Bureau of Land Management (BLM). BLM, with the assistance of others, have developed a Cadastral Data Content Standard to describe these data.

The standard makes references to marine boundaries but currently, little work has been done to describe and fully document those items needed for offshore standards. Marine boundaries have many similarities with their terrestrial counterparts, but do possess a few unique characteristics that must be included in the NSDI. Currently, NOAA and MMS are involved in developing standards to better describe offshore boundaries.

---

[4] For example, in the IHO standard, the horizontal and vertical accuracy of soundings is depth dependent. For example, an Order 1 survey at a depth of 30 metres, the horizontal accuracy standard is 6.5 metres and the vertical accuracy is 0.63 metres. One hundred percent ensonification may be required in selected areas. These standards apply to nearly all harbours and approach channels. There are less stringent standards for Order 2 and 3 surveys, which are applicable to deeper areas. There is also a more stringent level called a Special Order standard.

As specified in the Articles of the Law of the Sea (LOS) Convention, marine boundaries are those legal boundaries that extend offshore. The responsibility for delimitation fall under the auspices of the Department of State's Office of Ocean Affairs, the Department of Commerce's NOAA and the Department of the Interior's Minerals Management Service (MMS) (United Nations, 1983). Other agencies such as the Department of Justice and the Department of Defense (Army, Navy) are also involved with the legal description of marine boundaries through the Ad Hoc Committee on the U.S. Baseline[5]. The boundaries can generally be divided into the two categories of international and national. The international boundaries include such items as the Contiguous Zone, 12-mile Territorial Sea, and the EEZ. The national boundaries include the State-Seaward Boundary, the Revenue Sharing Boundary (Section 8(g) of the Outer Continental Shelf Lands Act), and state lateral boundaries. In addition, for the NSDI framework, marine boundaries will include marine protected areas such as National Marine Sanctuaries, other offshore-restricted zones, and federal lease blocks.

These boundaries share a common element with their land-based counterparts in that in order to map a boundary, one must adequately interpret the legal description of the law in a spatial context. Unlike their land-based counterparts, however, the marine geography is further complicated by the inability to include any physical boundary markers, such as benchmarks, stakes, fences, or hedgerows that are used on land.

Similar to the terrestrial side, the mapping of the intent of the legal description of a boundary can often be problematic. The legal text describing the rights associated with a piece of geography may be the only legally defensible information about that area. In other cases, as specified by the LOS for the EEZ boundary, the only accepted depiction of the line may be as plotted on the official largest scale nautical chart (Figure 19.2). In other cases, where there is a lack of understanding of mapping principles, the legal description may not adequately describe the geography or it may be extremely complicated to develop a mapping solution.

One example of this is illustrated in the legal description of the Channel Islands National Marine Sanctuary. In this case, the Channel Islands National Park of the U.S. National Park Service (NPS) has exclusive jurisdiction over the islands (land) and shared jurisdiction with NOAA and the State of California's Fish and Game Department from mean high tide out to one nautical mile (water). NOAA has total jurisdiction from one to six nautical miles. Within the one nautical mile buffer, NPS manages that area of the water above the depth of 1 cm while NOAA manages the area below a depth of 1 cm. Not only is the term "mean high tide" not an actual tidal datum, but also the inclusion of a vertical (1 cm along the surface) jurisdiction requires that the representative cadastral boundary be 3-dimensional. This example is fairly complex to map and even more complex to manage the institutional responsibilities.

Other examples of mapping ambiguities include listing co-ordinates without reference to a vertical or horizontal datum, referencing other ambulatory features such as the "wash of the waves at high tide," "following the 100 fathom isobath,"

---

[5] The Ad Hoc Committee on the U.S. Baseline, chaired by the Department of State was established in 1970 within the Law of the Sea Task Force. The Committee provides the forum for discussions on all issues relating to the delimitation of the U.S. coastline.

or using language such as "200 nautical miles from the baseline." In the case of this last example, is the correct boundary 200 miles over the surface of the earth, a chord projected through the earth, or some other interpretation of the appropriate algorithm? Many legal descriptions do not give complete information to accurately map the boundary and require the user to be both cartographer and detective.

Offshore maritime boundaries are referenced by their distance seaward from a baseline. A normal baseline is comprised of a series of points along the line of ordinary low water along that portion of the coast that is in direct contact with the open sea and is interpreted by the U.S. to be the same as the MLLW line (Thormahlen, 1999). However, because coastlines vary from smooth to deeply indented and are interrupted by bays and river mouths, and because the designation of offshore boundaries may confer significant economic rights, establishment of baselines is often complicated and contentious.

A number of rule-based spatial determinations are used in such cases. These aid in establishing closing lines across the mouths of bays and rivers, and help determine the baseline status of islands or intermittently exposed features fringing the shore (Figure 19.5). The revenues tied to the resulting lines make each baseline point of extreme importance. The financial consequence of baseline decisions is evident in the U.S. Supreme Court case of *United States of America* vs. *State of Alaska*, and the issue of "Dinkum Sands". At issue was whether this small formation just off the northern coast of Alaska qualified as an island with baselines from which a 3-mile Submerged Lands Act Grant would belong to the state. Important oil and gas reserves were discovered nearby and both the Federal Government and the State of Alaska sought the right to grant leases for exploration of the area. After years of legal action the court sided with the Federal Government establishing its right to the $19.6 billion in oil and gas revenues.

A number of factors complicate the mapping of baselines. As with mapping all coastal and ocean features, the tidal influences on the marine boundary element are considerable. The major complication with U.S. baselines is that there are two separate interpretations. Regulatory limits controlled by the Outer Continental Shelf Lands Act are under the responsibility of MMS and the baseline can be fixed for a period of time negotiated between the States and the Federal Government. Maritime boundaries controlled by the LOS are under the responsibility of NOAA and that baseline can correspond with the MMS baseline at the time of agreement but is ambulatory in nature and will shift with changes in the low water line. In other words, one baseline is used for determining offshore revenue sharing and the other is used for determining state, federal, and international jurisdictions. This results in two versions of the baseline used for generating different regulatory limits of which each line can often depend on the placement of the other. Attempts to encode the baselines and the resulting boundaries can often result in confusion for the geospatial data community. Currently, the NSDI has not fully evolved in the area of marine boundaries though progress has been made to date. A Memorandum of Understanding between MMS and NOAA has been in place since 1996 to cooperate on boundaries and generation of digital baselines. A methodology is being developed to create digital versions of boundaries that were previously only available on paper nautical charts; this methodology will be presented to the Ad Hoc Committee on the U.S. Baseline for approval. Moreover, the FGDC funded a project ion 1999 to test the Cadastral Data Content Standard on

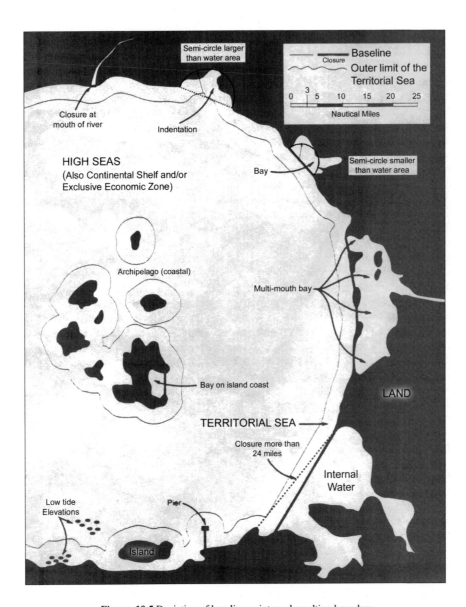

Figure. 19.5 Depiction of baseline points and resulting boundary.

offshore data. This project will use the standard on marine cadastral data elements and, if necessary, make recommendations for changes to better support marine data. Lastly, NOAA has been revising all of the National Marine Sanctuary boundaries in an effort to clarify all of the co-ordinate issues found in the legal descriptions. All of these efforts will support the development of the NSDI cadastral framework with respect to marine data.

19.5.4 Other Coastal Data Sets

Within the development of the NSDI, there is a structure for collecting and serving many types of coastal and marine data. There are considerable data holdings within NOAA and other agencies that are not specifically designated as part of the coastal framework, but could play an important role in the NSDI. Some of these data include marine transportation features, tide-controlled aerial photography, benthic habitat mapping, ocean satellite imagery, and tidal benchmarks, to name a few. Additional work is needed by those involved with coastal data to understand how these data might be best integrated with their land-based counterparts and fully contribute to the NSDI.

Potentially, marine navigation features could be considered within the context of the Transportation Thematic area. Marine transportation data have many similarities to land-based transportation but have the additional requirement of needing to contain a water-levelling component. For example, it is not sufficient to know the height of the bridge above the ground, but the navigation community must know the height of the bridge above the constantly changing water level. Once again, tide becomes a critical factor in documenting coastal data.

Benthic habitat data includes the information collected when the bottom of the seafloor is mapped and includes habitat elements such as seagrass and corals. Currently, there are a number of national and even international initiatives to map these highly productive habitats. Within the NSDI, benthic habitat could be considered a continuation of land cover and should be developed to coexist with their land-based complement. These are just a few additional examples of coastal and marine data that need to be considered as the NSDI is developed offshore.

19.6 METADATA, CLEARINGHOUSE AND STANDARDS

Another critical component in the development of the NSDI is the ability to find, understand and access coastal and marine data. Like their land-based counterpart, coastal and marine data must be documented, must adhere to national standards and, be made readily available via distributed means. In most of these structural components, coastal and marine data share the same problems and solutions as their land-based counterparts. Standards must be developed for each data element and integration activities must evolve such that there is a seamless transition from land to water. In addition, coastal metadata elements must be modified to include a description concerning the state of the tide. Moreover, where appropriate, protocols must be developed to assist state, local and private organisations in the development of framework coastal data. These protocol manuals should be referenced in the associated data standards.

Some progress has been made in all of these areas, but the NSDI development for offshore data has generally lagged behind its land-based counterpart. As previously discussed, standard activities are underway for shoreline, bathymetry and, marine cadastral. In addition, many of the coastal data sets are being documented with FGDC-compliant metadata and being served on FGDC-compliant clearinghouses. There is still considerable progress to be made in all of these areas to properly support the NSDI with respect to coastal and marine data.

19.7 CONCLUSION

Coastal and marine data have many unique requirements that warrant special consideration within the NSDI. Currently, there is considerable effort to increase the scope of the NSDI to look beyond the shoreline and include appropriate framework data, such as shoreline, bathymetry, and the marine cadastre. Although progress has been made, considerable work is still needed to accomplish the goal of a seamless coastal framework bridging the land-water interface. In keeping with its long tradition of coastal mapping, NOAA, along with many partners, is committed to supporting the NSDI with respect to coastal and marine data.

19.8 ACKNOWLEDGEMENTS

The authors would like to offer thanks to Lee Thormahlen, Curt Loy, Maureen Kenny, Dave Pendleton, Eric Treml, Jerry Mills, and Richard Naito for their technical review and Lauren Parker for her editorial comments in the development of this paper. A special thanks goes to Nancy Cofer-Shabica for both technical and graphics assistance.

19.9 REFERENCES

Berger, A.R. and Iams, W.J., 1996, *Geoindicators: Assessing Rapid Environmental Changes in Earth Systems* (Rotterdam: A. A. Balkema).

Clinton, W., *Executive Order 12906*, April 13, 1994. Federal Register, Vol. 59, Number 720, pp.17671-17674.

Currie P. and J. Walker, 1997, *Eleventh Annual Symposium on Geographic Information Systems* (Vancouver, British Columbia, Canada), pp. 124-127.

Ellis, E., 1978, *U.S. Department of the Interior, Geologic Survey and U.S. Department of Commerce, National Oceanic and Atmospheric Administration Coastal Mapping Handbook* (Washington, D.C.: U.S. Government Printing Office).

Federal Geographic Data Committee, 1995. *Development of a National Digital Geospatial Data Framework* (Washington, D.C.: U.S. Government Printing Office).

Federal Geographic Data Committee, 1997, *Framework Introduction and Guide*. Federal Geographic Data Committee (Washington, D.C.: U.S. Government Printing Office).

Federal Geographic Data Committee Bathymetric Subcommittee, 1998, *Bathymetric Data Standard* (draft), working document.

Gore, A., 1994, *Reinventing Government* (Washington, D.C.: U.S. Government Printing Office).

Hicks, S., 1989, *Tide and Current Glossary*, U.S. Department of Commerce. National Oceanic and Atmospheric Administration. National Ocean Service. Rockville, Maryland (Washington, D.C.: U.S. Government Printing Office).

International Hydrographic Organisation, 1996, *IHO Standards for Hydrographic Surveys*, 4th Edition (draft), IHO Special Publication 44 (S-44).

National Oceanic and Atmospheric Administration, 1976, *Our Restless Tides* (Washington, D.C.: U.S. Government Printing Office).

National Oceanic and Atmospheric Administration, National Ocean Service web site, http://www.nos.noaa.gov/.

Office of Management and Budget, 1990, *Circular No. A-16*, October 19, 1990.

Shalowitz, A., 1962, *Shore and Sea Boundaries* (Washington, D.C.: U.S. Government Printing Office).

Thormahlen, L., 1999, *Boundary Development on the Outer Continental Shelf*, U.S. Department of the Interior, Minerals Management Service, Mapping and Boundary Branch, Technical Series Publication, MMS 99-0006.

United Nations, 1983, *The Law of the Sea: The United Nations Convention on the Law of the Sea with Index and Final Act of the Third United Nations Conference on the Law of the Sea* (New York: United Nations).

CHAPTER TWENTY

GIS Applications to Maritime Boundary Delimitation

Harold Palmer and Lorin Pruett

20.1 INTRODUCTION: HISTORICAL DEVELOPMENT OF INTERNATIONAL MARITIME BOUNDARIES

Instructors in introductory oceanography or earth science courses often include a statement that "seventy-two per cent of the planet is covered by ocean waters". Although the planet is named for the remaining 28 per cent, it is possible to view the globe from a point over the central Pacific Ocean and see nothing but water and a few relatively small islands. For millennia, nations have drawn lines on the continents or staked claim to islands and shoals to define territories over which they claim sovereignty. These boundaries had (and have) clear cultural and natural features, which serve as markers to their location. Attempts to perform similar delimitation on bodies of water have often been frustrated by the absence of fixed references and precision tools for determining the boundaries and the means to locate them. Other than islands, shoals or coastal features upon which a line of sight might be established, the location of a maritime boundary often remained ephemeral, conjectural, disputed or, more commonly, ignored.

The question of who owns the oceans has been asked for many centuries. Naval power was more often than not the deciding factor in the rights of commerce, fisheries and expeditions of discovery. According to Ross (1980) the Pope (Alexander VI, a Spaniard) issued a Papal Bull in 1493, which divided the oceans between the two major maritime powers of the time: Spain and Portugal. They were given exclusive trading rights to the Indies, but enforcement lasted only as long as they had the naval power to deter competitors. In 1609, a Dutch lawyer, Hugo Grotius, published the concept paper "Mare Liberum" which declared that nations are essentially free to use the sea for any purpose which does not interfere with another nation's maritime activity. Some limit to jurisdiction in waters proximal to coastal states was needed. Another Dutchman, van Bijnkershoek, as the President of the Dutch Supreme Court in the early 1700s, decreed that a state could claim a "territorial sea" which was as wide as the effective range of a coastal cannon. This was generally accepted as three miles and although this "buffer" was never adopted by international agreement, it remained a working concept until shortly after World War II.

In 1945, U.S. President Harry Truman, in what is now termed the "Truman Proclamations", issued limited claims of ownership for the seafloor and overlying waters of the U.S. continental shelf. Then followed a series of claims by other nations, which ultimately compelled the United Nations to convene a meeting in Geneva (1958) to address international aspects of the "Law of the Sea". The outcome of this and subsequent meetings became the United Nations Convention on the Law of the Sea (UNCLOS) under which most nations acknowledge and accept the 300 plus Articles which define and describe legal maritime activities and the issues of jurisdiction. In 1994 the required number of coastal states had signed and ratified the UNCLOS and it entered into force for those States party to the Convention. The U.S. has signed, but not ratified, the UNCLOS. However, under a Presidential Proclamation issued by President Reagan in 1983, the U.S. has declared that it recognises the UNCLOS Articles and will abide by them. This includes jurisdictional aspects in the context of maritime activity and claims, and thus the nature of maritime boundaries becomes a significant issue in rights of nations engaged in marine commerce, research and resource management. Ross (1980) provides an excellent summary of the history of Law of the Sea developments through 1978. Perhaps the best overall summary of the UNCLOS can be found in the *Dispatch Supplement* published by the U.S. Department of State (*Dispatch*, 1995). Further background on Law of the Sea details, especially specific coastal state's claims, will be found in Prescott (1985), Charney and Alexander (1993), and in the continuing series of publications offered by "New Directions in the Law of the Sea" and the U.N. "Law of the Sea Bulletin".

20.2 MARITIME BOUNDARIES AND ZONES ESTABLISHED BY THE UNCLOS

Various Articles in the UNCLOS specify the manner in which maritime boundaries are to be drawn. Some of the boundaries affect the seabed, some the water column and some affect both. The fundamental reference for most maritime boundaries is the *baseline*. Baseline construction can incorporate the "normal" baseline (shoreline), a series of straight baselines, or a combination of both.

Most shoreline baselines are referenced to a low water datum associated with low tide. There are at least six options available in the definition of a low water datum, and it is essential that this be indicated for the coastal state(s) in question since the extent of offshore zones will be determined by the baseline selected. Straight baselines are used to enclose waters which, due to their close proximity to the mainland, are effectively inland waters. Properly constructed straight baselines do not significantly extend the Territorial Sea beyond that which would be derived from the normal baseline. The UNCLOS has established straight baseline construction formulae, which are designed to preclude excessive or unreasonable claims for legal jurisdiction of bays, estuaries, inland seas and other indentations of the coast. Examples of bay-closing constructions appear in Figure 20.1.

In reality, stipulations in the UNCLOS, which admit straight baseline construction, are somewhat unclear for coasts with "deep indentations" or "a fringe of islands". While Article 10 (see Figure 20.1) furnishes an explicit method for acceptable baselines enclosing bays, Article 7 is sufficiently vague to permit a broad range of interpretation which has led to abuse of UNCLOS intent. A coastal nation may claim several maritime zones (see Figure 20.2) in relation to the baseline and in accordance with UNCLOS guidelines. A description and the maximum extent of the zones follow.

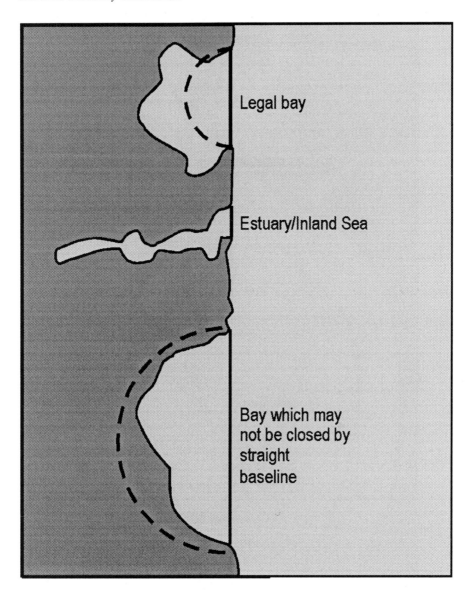

Figure 20.1 Bay-closing constructions; semi-circle test conforms to UNCLOS Article 10.

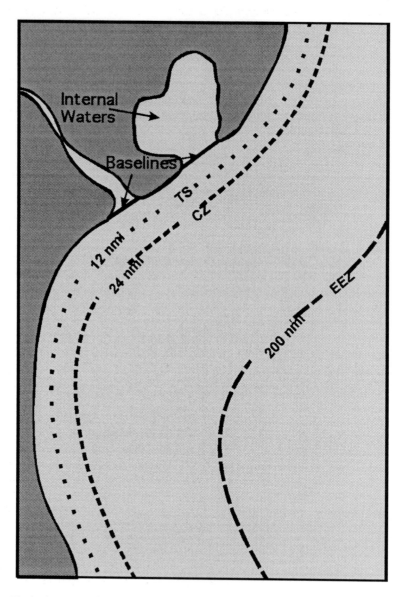

Figure 20.2 Primary maritime boundaries. TS = Territorial Sea, CZ = Contiguous Zone, EEZ = Exclusive Economic Zone, nmi = nautical miles. Continental Shelf boundaries may lie seaward of the EEZ limit.

Waters lying on the landward side of the baseline (normal or straight) are the Internal Waters over which the coastal state has complete sovereignty. Seaward of the baselines, to a maximum distance of 12 nmi (nautical miles: a nautical mile equals 6,076 feet or 1,852 meters or 1 minute of latitude), lies the *Territorial Sea*.

A coastal state may claim a *Contiguous Zone* affording an additional 12 nmi of jurisdiction beyond the outer limit of the Territorial Sea. This zone defines a maritime region in which the coastal state can exercise control over immigration and customs, monitor and exercise pollution abatement, and perform interdiction of illegal activity. The Contiguous Zone does not, by itself, provide rights over economic resources..

A coastal state may also claim a *Fisheries Zone*, up to 200 nmi from the baseline, in order to control exploitation of the living resources by both its own fleet and the vessels of other nations. This zone is also measured from the outer limit of the Territorial Sea and may overlap a Contiguous Zone.

In addition to, or instead of, the Fisheries Zone, many nations claim an *Exclusive Economic Zone* (EEZ) which is generally 200 nmi wide (again measured from the baseline). The EEZ is an area in which a coastal state has sovereign rights for the purposes of managing both living and non-living natural resources below, on, or above the seabed. It can also authorise or deny scientific research, protect and preserve the environmental quality, and exercise other rights and duties as granted by the UNCLOS (Dispatch, 1995).

As a result of jurisdictional license provided under the UNCLOS Articles, nations controlling insular territory can establish all of the boundaries and Zones just described which then surround these outposts. Similarly, small insular nations are entitled to apply the same maritime limits as the largest coastal state. Thus, the tiny Pacific Ocean nation of Tuvalu (with a land area of 26 square miles) is justified in establishing an EEZ that encompasses 225,000 square miles of ocean space!

Deep-water drilling and mining technologies are permitting ever deeper exploration and exploitation of natural resources, and it has become clear that many regions of the sea bed beyond a 200 nmi EEZ may well contain economic deposits of minerals. This situation has not escaped the attention of the U.N. Under Article 76 of the UNCLOS, it has established criteria for defining bounds to the *Continental Shelf*. At least 33 coastal states may claim seabed jurisdiction beyond their EEZs, but to do so they must collect and analyse sea bed data, prepare justification defending their choice(s) and ultimately present a claim to the U.N. Commission of the Limits of the Continental Shelf.

Because the Continental Shelf delimitation involves geospatial analyses well beyond those required for other boundary resolutions, a GIS approach is essential in the organisation and handling of both enormous quantities and disparate formats of data. Australia has recently completed extensive surveys in preparation for submitting their claims to the Commission (Borissova *et al.*, in press). Because their survey and synthesis efforts spanned several years, they needed a stable GIS product, which was commercially widespread and popular, robust, and portable. Their analyses using GIS linked bathymetric profiles and individual soundings with precise navigation positioning.

One element in Continental Shelf delimitation is locating and marking the "foot of the continental slope", the point of maximum change in bottom gradient along sounding profiles crossing the base of the slope. Another is the point where the sediment thickness vs. the distance to the foot of the slope is 1% (a ratio of 0.01). Yet another is the precise location of the 2,500-m isobath. The outer bound of the Continental Shelf Zone may be established at a distance of up to 350 nmi from the baseline, 60 nmi from the foot of the slope, or 100 nmi seaward of the 2,500-m isobath and combinations of these criteria are permitted. In fact, an even broader zone may be sought if a state can show a ridge of continental material extends further into the deep sea, but these are special cases to be argued separately. Within the Continental Shelf the state can exercise control of seabed resource exploration and extraction. The main GIS application supporting such boundary definition is again a buffering exercise utilising the most advantageous criteria, which yield the maximum enclosed area.

Finally, the water column in the region beyond the outer limit of a claimed Exclusive Economic Zone is considered *International Waters* or the *High Seas*. In such regions the UNCLOS reaffirms the rights of all states to operate on, over and within the High Seas. However, the issue of claims to sea bed mineral resources in High Seas regions falls under the purview of the U.N.'s International Seabed Authority (ISBA). It has designated these regions "The Area" and at least 7 nations have claimed mining rights to specific tracts of the seafloor through boundary delimitation registered with the ISBA.

These are the fundamental maritime limits. Other categories such as joint development zones, archipelagic zones and specific joint fisheries zones have regional significance, but constitute a relatively small number of maritime boundaries. The booklet previously cited (*Dispatch*, 1995) provides a clear description of these and other boundaries and zones which are incorporated in the UNCLOS.

20.3 COMPLICATIONS OF THE REAL WORLD

In the simplest situation, ignoring potential Continental Shelf claims, a coastal state may extend its boundaries seaward until 200 nmi from the baseline is reached. The marine environment beyond this limit falls under the High Seas regime. In a few areas, such as some South American and African states, such constructions work well (Figure 20.3). However, the majority of coastal countries encounter someone else's maritime claim in the process. In such cases, there are several options open to the parties: they may negotiate a mutually satisfactory boundary through their respective government agencies; they can petition an international body for a decision; they may claim a tacit "median" or "equidistant" line between their coasts or baselines until agreement is reached; or they may declare the overlapping area(s) to be "disputed".

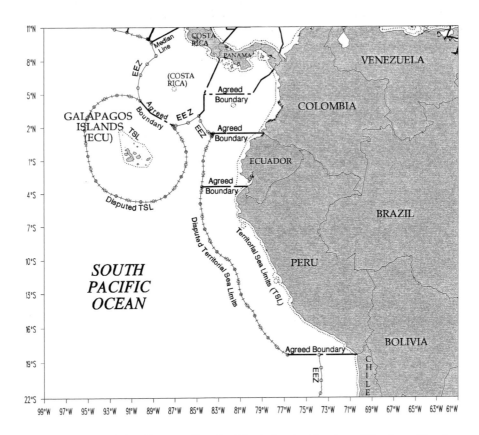

Figure 20.3 Boundaries off western Central and South America. Ecuador and Peru have claimed a 200 nmi Territorial Sea Limit (TSL) which far exceeds the stated UNCLOS limit of 12 nmi. Note the expanse of claims afforded by small islands. Mercator projection.

The "happy ending" scenario may be achieved through country to country dialog and resolution; one of may examples is the case between Italy and Greece in the eastern Mediterranean Sea. In 1980, both countries agreed to define their respective jurisdictions as being separated by a line running roughly north-south along an equidistant line. Overlapping claims to potential exploration drilling sites for petroleum resources led Libya and Malta to seek adjudication from the International Court of Justice (ICJ) with regard to their Continental Shelf boundary. The resulting 1985 judgement left neither country completely happy, but it was accepted by both parties. The ICJ placed the boundary line closer to Malta than to Libya primarily due to the length of their opposing shorelines. Egypt, on the other hand, simply claims an EEZ to "the median line" pending future agreements with neighbouring countries (Figure 20.4).

Even when two countries agree to a boundary along a median line, the solution to delimitation may remain complicated. Problems with adequate scale and accuracy of reference maps, agreeable baseline claims, the weight given to minor landmasses (rocks, shoals and even islands) and historical appeals to past provincial claims and former real or imagined rights have all entered into negotiations. Many of the semi-enclosed bodies of water such as the Mediterranean, Baltic, South China and other Seas have coastal states in such proximity that a maritime claim, especially an EEZ and in some cases even a Territorial Sea, will create overlaps with other country's claims. The hypothetical median lines that separate jurisdictions in the eastern Mediterranean Sea appear in Figure 20.4.

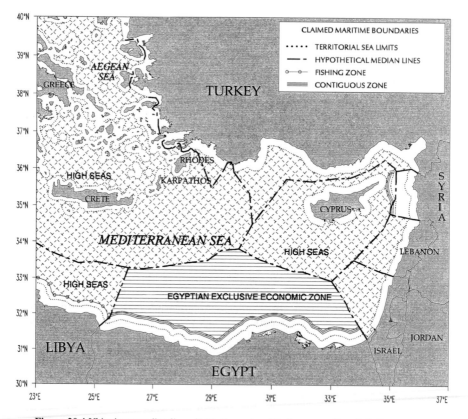

Figure 20.4 Ubiquitous median lines divide the eastern Mediterranean Sea. Only Egypt asserts an EEZ, but it is everywhere constrained by median lines and thus cannot extend to the 200 nmi width afforded by the UNCLOS. Mercator projection.

A number of researchers have addressed the question of median line construction with various applications well suited to a GIS approach. Christensen (1998) has revived the concept of "waterlines" which employ lines drawn parallel to the edges of a shoreline of a country. They are then extended until they coalesce along points of intersection with similar lines of a neighbouring nation which thus define an equidistant line. In another approach, Tobler (1997) advances the "equidominance" concept in which a circle drawn from a point on the water encompasses equal proportions (land areas) of adjacent coastal states. A line connecting such points defines the plane along which both (or multiple) countries have equal terrestrial representation. In both cases, the construction of successive buffers moving away from a shoreline or from a point can be supported by a number of GIS programs. Finally, Carrera (1989) discusses several modules of the DELMAR computer program which are designed to generate various maritime boundaries. One such module generates median lines between two countries on the surface of a geodetic reference ellipsoid (Carrera, 1989).

20.4 DISPUTED AREAS: FROM AMICABLE DISCUSSION TO FLASH POINTS

In a number of the world's enclosed or marginal seas, sovereignty claims have created overlapping regions in which the perceived jurisdictions of one state are at odds with others. Perhaps the best examples of the inherent complexity are the claims in the South China Sea. The maritime claims situation is presented in Figure 20.5. In one area alone, the tiny Spratly Island Group, seven Southeast Asian nations claim some or all of these coral reefs and islands. Claimants have built structures on several of the islets and installed troops and/or fishermen. Shots have been exchanged and ensuing naval engagements have occurred. Research cruises and seismic surveys for oil and gas by conflicting countries have been performed. And, in attempts to resolve the situation, countless conferences, workshops, and commission agendas have been convened but have thus far failed to resolve the maritime jurisdictions of this area. On the other hand, disputed areas may be acknowledged in a more friendly fashion like the ongoing discussions addressing the overlapping fisheries area in the Baltic, which both Finland and Sweden are hoping to resolve. The tools and techniques of GIS can be utilised in the presentation and analysis of these otherwise confusing overlapping boundaries. As shown in Figure 20.5, various symbology can be employed to designate lines claimed by states, and shading, patterns or colour for disputed areas may be used. Attribution of both line and area features allow slicing of the data set in a number of ways; Figure 20.5 shows just one slice, highlighting overlapping claims.

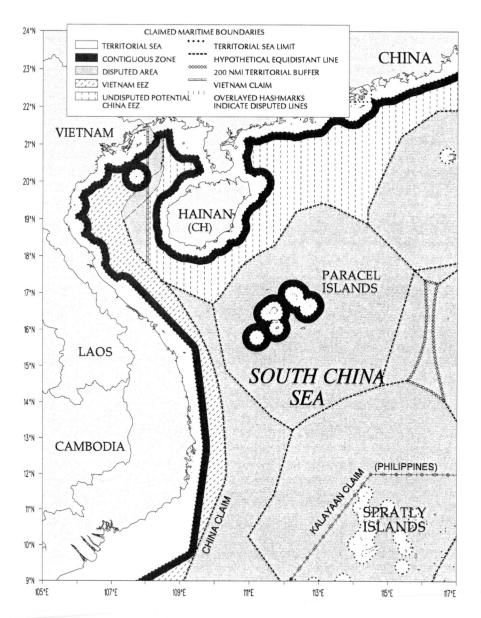

Figure 20.5 Portions of the South China Sea may have as many as seven claimants for the same parcel of water. The bewildering mix of lines and encompassed overlapping zones requires a GIS approach to track and display geospatial relationships of both valid claims and contested areas. Mercator projection.

20.5 APPLICATIONS OF GIS TOOLS AND TECHNIQUES TO BOUNDARY DELIMITATION

The cataloguing and display of maritime boundaries might be seen, at first glance, to be a straightforward exercise in routine GIS applications. After all, we are dealing with the basics: points, lines and polygons. For all but the Continental Shelf, boundary elements are based on two-dimensional construction on the surface of the sea. For the Continental Shelf, construction is based on three-dimensional data (discussed later). All of the boundaries, however, have complexities that may not be obvious to the casual user. These arise from both the interpretations of a claim or an agreement and the accurate portrayal of the interpretations without oversimplification. Also, the relationship of the claim at hand must be evaluated with respect to other claims and agreements. Several factors that contribute to complications in claim portrayal are: the variable application of UNCLOS Articles by individual coastal states; disputed claims to landmasses that make up the natural baselines; incomplete resolution of overlapping sea claims between countries; and, methodologies for resolving claim disputes.

For over ten years MRJ Technology Solutions has developed and maintained a Global Maritime Boundaries Database (GMBD) and we have experienced all of the problems noted above, and more. Our database was originally compiled regionally for a variety of research activities, used a World Data Bank II (WDBII) coastline base map and contained basic essential attribution for line features. For specific cases of overlay analysis, polygon features were created and attributed. This approach served the needs at the time: general desktop for in-house research and analysis in a GIS setting.

Eventually, with requirements expanding to include research as well as accelerated delivery times, we deemed it prudent to expand the data set to represent all basic maritime boundaries of the world with World Vector Shoreline 1:250K (WVS) as the coastline base map. As discussed earlier, this was not a trivial task and was subject to the same source accuracy limitations discussed earlier. As a result, and until better source data are available, we consider the GMBD to be accurate to within 1/3 nmi for straight baselines and agreed boundaries at the waypoints. All other boundaries generated from the baseline are deemed accurate to approximately one nmi.

Construction of GMBD boundary lines are initiated from the baseline. For coastal states that do not claim any straight baselines, the WVS coastline is used. Due to the small scale of the WVS as compared to the larger scale of charts (i.e. 1:25,000 to 1:100,000) from which listed boundary co-ordinates may be derived, a review is made of nautical charts and other sources to ensure islands, reefs, shoals and rocks are represented in WVS. MRJ Technology Solutions is currently pursuing larger scale coastlines to serve as a base map. The raw data for all agreed boundaries, baselines, and cited datums are maintained and can be easily used with a larger scale coastline in the future. Boundaries buffered from the coastal baseline may then be re-buffered. With the proliferation of Electronic Charting Display Systems (ECDIS), large-scale accurate portrayal of maritime boundaries is becoming increasingly important.

Countries with straight baselines, by proposing data more accurate than WVS, present an additional problem. Again, due to scale differences between WVS and baseline source maps with scales as large as 1:25K, co-ordinates of straight baselines do not match up with the WVS coastline. Straight baselines can end offshore, inland, but almost never exactly on the WVS coastline. Our approach has been simple: err on the side of caution. With maritime boundaries we take the furthest seaward baseline, whether straight or coastline, to generate buffered Territorial Sea Limits and other buffered claims. This results in portraying a country's claim slightly outside, rather than slightly inside, their jurisdiction.

An in-house Arc/INFO Arc Macro Language (AML) performs the buffering of the baselines and initial attribute documentation. Small blocks of the baseline are fed to the AML, which projects, buffers, densifies, re-projects and provides initial attribution of the claim. An Azimuthal Equidistant projection with automatically generated line(s) of true scale based on the block latitude extent is used in the AML. The boundary is projected back to geographic co-ordinates for storage in the GMBD where additional attribution is included. Further, buffered boundary lines often indicate overlapping claims to areas. However, buffered boundaries are not blindly added to the GMBD, but rather are compared to the literature to determine whether overriding criteria may come in to play. Often country claims to boundaries are made with the preface that they are extended to the median line, where potential overlaps with neighbouring coastal states may occur.

Boundaries that arise from multilateral agreements provide resolution in areas of overlapping claims and are generally published as co-ordinate lists. However, several problems arise in the portrayal of even these boundaries. While our goal is to convert all data to a geodetic standard datum (WGS84), this clearly is not possible at the present time. In most cases, not all of the necessary source information is published in widely distributed media. Information often lacking includes the following: source chart; chart datum, projection and lines of true scale; co-ordinates; and the nature of the line (geodesic, rhumb, and great circle). Sometimes two lists are published, both missing necessary information but each using the originating country's own source maps. At this time we enter and store all of the raw co-ordinate data and indicate whether or not the data has been converted to WGS84. This attribute serves as a flag for further research and updating if additional source data become available.

The GMBD maritime boundaries described thus far have as their positional origin either a co-ordinate list or buffer of the baseline. Yet, there exists another form of maritime boundary: the median line. As mentioned, there are many reasons to employ a median line: they may be useful when opposing claims are "to the median line". They may be useful when opposing claims threaten to or do overlap; or, they may be useful when no claim has been made but potentially may be made in the future. The GMBD utilises median lines in all of these cases. One of the major reasons for doing so is based on the need to demonstrate the extent of maritime claims. Another reason is to understand resource potential of the marine environment. Also, in the case of unclaimed potential zones, it is often desirable to know which coastal states will most likely have an interest in activities within a specific area in the future.

Lines in a database are sufficient for depicting the locations of boundaries. However, to perform overlay analysis of the data, they must also be stored as areas. In fact, much of the attribute information of maritime boundaries reflects the nature of areas, not lines. For example, a claim to a 200 nmi EEZ is an area claim. The lines that enclose the area may, or may not, be at 200 nmi from the coast when other countries' claims and agreed boundaries are taken into account. Further, the rights of a claim pertain to the area, not the line.

Storing maritime claim areas as polygons in a GIS has its own drawbacks, since many overlap or are transected by other claims. For example, a country's own claims may overlap such as an EEZ measured from the baseline which overlaps an existing TSL, CZ and/or EEZ. Additionally, in semi-enclosed seas with overlapping claims such as the South China Sea, the number of lines is overwhelming. These areas could be stored as polygons but entry and upkeep of the graphical features is expensive and prone to errors of oversight. This is why we have chosen to use the Arc/INFO Regions data model for storage of these features in the GMDB. This model allows not only storage of overlapping areas, but includes tools for analysing regions. For instance, one simple example in ArcView is the placement of the cursor at a location in the sea to identify all countries that claim that particular site.

20.6 SOURCES OF ERROR

Inasmuch as the coastal state's baselines become the reference line from which buffers and median lines are created, the precise geodetic location of these lines is crucial to the construction of the subsequent maritime zones. These co-ordinates must be referenced to a geodetic datum for accurate representation. However, in practice, necessary information is generally lacking in publications citing maritime claims.

Together, the appearance of a boundary line on a chart or map and an accompanying text describing how that line was constructed may easily convey a perception of geospatial exactitude. The assumption that the lines are errorless or that any errors are so small that they are practically irrelevant is a common one. Yet in some cases (such as resource jurisdiction) even a small discrepancy of a few tens of meters may involve actual trespass and result in costly legal confrontations.

According to Vanicek (1998) there are three different origins of error in boundary delimitation. First are errors in measurement of boundary construction; these may be both random and systematic and cannot be eliminated without careful review of mensuration procedures. Secondly, definitions and interpretations of the UNCLOS text may be misunderstood or misrepresented. Finally, transformations of co-ordinates (positions) between two or more geospatial co-ordinate systems may introduce errors.

The first instance is of special concern to the application of GIS techniques. If the boundaries are defined only on the basis of graphical presentations (points and lines on a map) the question of scale and projection enter as significant sources of error. Nautical charts are designed specifically for navigation and their coverage is not conducive to boundary delimitation. Consider only the issue of line width; many charts are at a scale of 1:1,500,000 wherein printed lines for coastlines are rarely less than 0.25 mm wide. At this scale, a boundary line of equal weight drawn on the sea surface is 375 meters wide! Cartographic notation of a turning point for baseline construction at this common scale becomes even less rigorous since it is in turn dependent upon two such basepoints. Such errors propagate seaward as the boundary buffers are developed through linear extensions from the baseline. In the world of digital maps, it is easy to ignore scale as the ability to zoom into an area in almost limitless. However, zooming to a scale larger than that of the original data results in misleading perceptions of accuracy.

Projection is another source of error. On the common Mercator projection used by mariners the surface of a three-dimensional earth is projected onto a two-dimensional paper (or electronic screen) chart. Errors in line length and in true distances increase as one moves further from the equator and in one real case of median line construction the geometric equidistance adopted between opposing coastal points departed by four nautical miles (7.2 km) from the geodetic equidistant line! For MRJ Technology Solutions' purposes every attempt is made to use original projection information to accurately reflect co-ordinate positions. However, as previously discussed, requisite information for accurate line reconstruction is seldom available. Therefore, all co-ordinate lists point and line data are maintained in their original form as published to facilitate improved reconstruction as additional source date becomes available. An excellent discussion of the issues related to geodetic datums and the UNCLOS requirements to specify them will be found in Lathrop (1997).

20.7 USER COMMUNITIES AND THEIR NEED FOR MARITIME BOUNDARY INFORMATION

Reflecting on the breadth of topics encompassed by the UNCLOS, nearly every conceivable maritime activity has been addressed in these Articles in one fashion or another. Indeed, it was the growing awareness of conflicting claims to jurisdiction and the desire of coastal states to "stake out" their maritime heritage that led to the Geneva Conventions of 1958. While these were specific to activities such as navigation, fishing, mineral resources and laying of pipelines and cables, the scope grew until, in 1982, a broad and comprehensive international "Law of the Sea" was promulgated.

Since that time, the text of the UNCLOS has continued to evolve to accommodate other interests such as sea bed mining on the High Seas, continental shelf resource jurisdictions, geopolitical resolutions, and issues of cultural heritage (jurisdiction over historic wrecks and sites beyond Territorial Sea waters).

The need for maritime boundary information in the context of permissible offshore activity may be summarised through a brief review of some current protocols affecting work at sea. In all cases, the application of GIS constructions backed by attribute data containing the explanatory information on sources of information provide the requisite boundary information necessary for planning offshore activity that will conform to legal framework of the Law of the Sea.

Fishing: Global capture fisheries which employ roving vessels and fleets have peaked at nearly 100 million tons annually. Most commercial species are either at a point of "maximum sustainable yield" or are demonstrably overfished. Many states have thus established Fishing Zones or EEZ which incorporate buffers similar to other offshore boundaries. Within these areas strict rules are enforced regarding which countries may fish, limits to the number of vessels, the period of fishing, and quotas or maximum catch are enforced.

Hydrocarbon Resources: Oil, gas and gas hydrate deposits are being found in ever deeper water. The limits, which will be drawn under the Continental Shelf provisions, will rely heavily on data presented in GIS formats for review by the U.N. Commission. In addition, the coastal states will employ GIS tools to define lease block boundaries to exploration drilling.

Geopolitical Aspects: We have noted the turmoil associated with overlapping claims. As we have seen in the South China Sea, unilateral (but contested) geopolitical solutions may be drawn by countries simply to deny potential entry and activity by others. GIS will continue to play a major role in facilitating claim boundaries with reference to the UNCLOS stipulations on maritime limits.

Marine Scientific Research (MSR): Elaborate protocols are in place regarding permission and access to coastal state's waters for oceanographic research. This activity is covered in detail by Fenwick (1992) and more recently by Roach (1996). Utilisation of GIS in the planning of cruises and in defining a sampling and survey plan facilitates the permit request process as well as furnishing the scientists with positioning information for the data/sample collection phase and post-cruise analyses. Included in MSR is the growing activity in marine archaeology. In this field, questions of ownership and jurisdiction are paramount in determining legal aspects of search, excavation and artefact recovery. O'Keefe (1996) has summarised the current status of the protection of objects constituting a state's underwater cultural heritage, and this issue is presently under review by the U.N. as yet another facet of the Law of the Sea.

Other offshore activities and/or endeavours which can incorporate the power of GIS tools and techniques include waste disposal options, electronic chart production, defence, maritime law, submarine telecommunications routing, and marine recreation. In every case, the issue of maritime boundary location becomes an integral part of establishing limits in the seas. As the poet Robert Frost wrote, "good fences make good neighbours". While one may argue that a marine "fence" represented by a maritime boundary may be disputed, the application of GIS technology coupled with precise geodetic positioning offers the best hope for resolution of these watery claims.

20.8 ACKNOWLEDGEMENTS

Laura Crenshaw, Suenette Hunsberger, and Tracie Penman, MRJ Technology Solutions GMBD Analysis Group prepared the graphics for this chapter. Kurt Christensen, MRJ Technology Solutions, contributed to the construction of median line generation program. Ron Macnab, Canadian Geologic Survey and Irina Borissova, Australian Geological Survey Organisation kindly shared references and information on the development of Continental Shelf delimitation for their respective countries.

20.9 REFERENCES

Borissova, I., Symonds, P.A. and Gallagher, R., in press, Law of the Sea: Defining the outer limit of Australia's marine jurisdiction. *GIS Asia Pacific.*

Carrera, G., 1989, DELMAR, A computer program library for the delimitation international maritime boundaries, *International Centre for Ocean Development.*

Charney, J.I. and Alexander, L.M., 1993, *International Maritime Boundaries,* Vols 1-3, American Society of International Law (Dordrecht: Martinus Nuhoff Publishers).

Christensen, A.H.J., 1998, A global method for the delineation of maritime boundaries. *The Hydrographic Journal,* **89,** pp. 3-6.

Dispatch, 1995, *Law of the Sea Convention: Letters of Transmittal and Submittal and Commentary,* U.S. Department of State, Office of Public Communication, Bureau of Public Affairs, Vol. 6, Supplement No. 1, edited by Macdonald, C. (Washington: U.S. Government Printing Office).

Fenwick, J., 1992, *International Profiles on Marine Scientific Research Jurisdiction, and Research Histories for the World's Coastal States* (Woods Hole, Massachusetts: Woods Hole Oceanographic Institution Sea Grant Program).

Lathrop, C., 1997, The technical aspects of international maritime boundary delimitation, depiction and recovery. *Ocean Development and International Law,* **28,** pp. 167-197.

O'Keefe, P.J., 1996, Protection of the underwater cultural heritage: developments at UNESCO. *The International Journal of Nautical Archaeology,* **25,** pp. 169-176.

Prescott, J.R.V., 1985, *The Maritime Political Boundaries of the World* (London: Methuen and Co.).

Roach, J.A., 1996, Marine scientific research and the new law of the sea. *Ocean Development and International Law,* **27,** pp. 59-72

Ross, D.A., 1980, *Opportunities and Uses of the Ocean* (New York: Springer-Verlag).

Tobler, W.R., 1997, The equidominance line: A new geopolitical concept. *Applied Geographic Studies,* **1,** pp. 7-11.

Vanicek, P., 1998, On the errors in the delimitation of marine spaces. *International Hydrographic Review,* **75,** pp. 59-64.

CHAPTER TWENTY-ONE

Information Quality Considerations for Coastal Data

Nancy von Meyer, Kenneth E. Foote and Donald J. Huebner

21.1 INTRODUCTION

Information quality is an indication of the usability of a data set based on a composite of factors. The composite of factors includes relative accuracy and precision and the procedures for collecting, maintaining, and distributing a data set. Information quality is important to the coastal community because of the dynamic nature of coasts and shorelines. A constantly changing environment can have varying information covering a wide range of conditions. Knowing which data set covers which conditions and how well those conditions are described is paramount to coastal analysis. Beyond data accuracy and precision, it is important to know how a data set is collected, how often it is updated, and how it has been distributed to determine its usability in a given situation. This chapter examines accuracy, precision and error factors that may impact information quality for coastal managers.

21.2 BACKGROUND

Information quality assessment is a concern of the digital age. The ability to analyse large spatial data sets, to compare spatial data over time, and to reformat and transmit data easily is attributable to computers and geographic information systems (GIS). However, until recently people involved in developing and using digital data paid little attention to the problems caused by error, inaccuracy, and imprecision in spatial data sets. Certainly there was an awareness that all data suffers from inaccuracy and imprecision, but the effects on the application of digital data to problems and solutions were not considered in detail. Major introductions to the field such as C. Dana Tomlin's *Geographic Information Systems and Cartographic Modeling* (1990), Jeffrey Star and John Estes's *Geographic Information Systems: An Introduction* (1990), and Keith Clarke's *Analytical and Computer Cartography* (1990) barely mention the issue. This situation has changed substantially in recent years. It is recognised that error, inaccuracy, and imprecision can "make or break" many GIS projects. That is, errors left unchecked can make the results of any digital analysis nearly worthless.

The irony is that the problem of error devolves from one of the greatest strengths of GIS. GIS gain much of their power from being able to collate and cross-reference many types of data by location. They are particularly useful because they can integrate many discrete data sets within a single system. Unfortunately, every time a new data set is imported, the GIS also inherits its errors. These may combine and mix with the errors already in the database in unpredictable ways. The key points are that even though error can disrupt GIS analyses, there are ways to maintain high information quality through careful planning and methods for estimating the effects of error and there are questions a data consumer can ask to ascertain information quality.

21.3 DEFINITIONS

Accuracy and precision play an important role in characterising information quality. It is important to distinguish between accuracy and precision. Recognise also that accuracy and precision apply to spatial and attribute information.

Accuracy is the degree to which information recorded in a data set matches the true or real-world value. Using the analogy of a target, accuracy is a measure of how close the information in the data set is to the bull's eye. Accuracy indicates how close the value at hand is to the true value.

Precision is the repeatability or exactness with which information is measured or described. Returning to the target analogy, precise information would be clustered around a single point, but that point may or may not be the bull's eye or may or may not be the true value. Precise spatial data may be measured to a fraction of a unit. Precise attribute information may specify the characteristics of features in great detail. Precise data, no matter how carefully measured, may be inaccurate.

Precision is a measure of the method, while accuracy is a measure of the result and how carefully it is captured. Neither precision nor accuracy account for errors in recording information, missing information, or omitted information.

At first blush it may seem that all coastal and shoreline information needs to be as accurate as possible. But consider a trend analysis related to vegetation stands in dunes. If a method yields a consistent and repeatable measure that is consistently repeatable over time such that a true trend can be detected, then the accuracy may be less important. In this case the method is precise. On the other hand if it is necessary to get an absolute count on the numbers, variety and volume of vegetation, then accurate measurements are needed. That is, a method that will provide the correct indication of the real world value.

Be aware also that GIS practitioners are not always consistent in their use of these terms. Sometimes the terms are used almost interchangeably. This should be guarded against.

21.4 INFLUENCES ON INFORMATION QUALITY

There are many factors that may affect information quality. Some are obvious, but others can be difficult to discern. Few of these will be automatically identified by the GIS itself. It is the user's responsibility to prevent them. Particular care should be devoted to checking for errors because GIS are quite capable of lulling the user into a false sense of accuracy and precision unwarranted by the data available. For example, smooth changes in boundaries, contour lines, and the stepped changes of chloropleth maps are "elegant misrepresentations" of reality. In fact, these features are often "vague, gradual, or fuzzy" (Burrough, 1998). There is an inherent imprecision in cartography that begins with the projection process and its necessary distortion of some of the data (Koeln *et al.*, 1994), an imprecision that may continue throughout the GIS process. Recognition of error, and importantly what level of error is tolerable and affordable, must be acknowledged and accounted for by GIS users.

Burrough (1998) divides sources of error into three main categories. These categories are used here with some additions in each category:

1. Obvious sources of error.
2. Natural variations and original measurements.
3. Compounding error.

Generally errors of the first two types are easier to detect than those of the third because errors arising through processing can be quite subtle and may be difficult to identify.

21.4.1 Obvious Sources of Error

21.4.1.1 Age of Data

Data sources may simply be too old to be useful or relevant to current GIS projects. Past collection standards may be unknown, non-existent, or not currently acceptable. For instance, John Wesley Powell's nineteenth century survey data of the Grand Canyon lacks the accuracy of data that can be developed and used today. Additionally, much of the information base may have subsequently changed through erosion, deposition, and other geomorphic processes. Despite the power of GIS, reliance on old data may unknowingly skew, bias, or negate results.

21.4.1.2 Areal Cover

Data on a given area may be completely lacking, or only partial levels of information may be available for use in a GIS project. For example, vegetation or soils maps may be incomplete at borders and transition zones and fail to accurately portray reality. Another example is the lack of remote sensing data in certain parts of the world due to almost continuous cloud cover. Uniform, accurate coverage may not be available and the user must decide what level of generalisation is necessary, or whether further collection of data is required.

21.4.1.3 Source Map Scale

The ability to show detail in a map is determined by its scale. A map with a scale of 1:1,000 can illustrate much finer points of data than a smaller scale map of 1:250,000. Scale restricts type, quantity, and quality of data (Star and Estes, 1990). One must match the appropriate scale to the level of detail required in the project. Enlarging a small scale map does not increase its level of accuracy or detail.

21.4.1.4 Density of Observations

The number of observations within an area is a guide to data reliability and should be known by the map user. An insufficient number of observations may not provide the level of resolution required to adequately perform spatial analysis and determine the patterns GIS projects seek to resolve or define. A case in point: if the contour line interval on a map is 40 feet, resolution below this level is not accurately possible. Lines on a map are a generalisation based on the interval of recorded data, thus the closer the sampling interval, the more accurate the portrayed data.

21.4.1.5 Relevance

Quite often the desired data regarding a site or area may not exist and "surrogate" data may have to be used instead. A valid relationship must exist between the surrogate and the phenomenon it is used to study but, even then, error may creep in because the phenomenon is not being measured directly. An example of the use of surrogate data are habitat studies of the golden-cheeked warblers in central Texas. It is very costly (and disturbing to the birds) to inventory these habitats through direct field observation. But the warblers prefer to live in stands of old growth cedar *Juniperus ashei*. These stands can be identified from aerial photographs. The density of *Juniperus ashei* can be used as a surrogate measure of the density of warbler habitat. But, of course, some areas of cedar may be uninhabited or inhibited to a very high density. These areas will be missed when aerial photographs are used to tabulate habitats.

Another example of surrogate data are electronic signals from remote sensing that are used to estimate vegetation cover, soil types, erosion susceptibility, and many other characteristics. The data are being obtained by an indirect method. Sensors on the satellite do not "see" trees, but only certain digital signatures typical of trees and vegetation. Sometimes these signatures are recorded by satellites even when trees and vegetation are not present (false positives) or not recorded when trees and vegetation are present (false negatives). Due to cost of gathering on site information, surrogate data are often substituted and the user must understand variations may occur and although assumptions may be valid, they may not necessarily be accurate.

21.4.1.6 Format

Methods of formatting digital information for transmission, storage, and processing may introduce error in the data. Conversion of scale, projection, changing from raster to vector format and pixel resolution are examples of possible areas for format error. Expediency and cost often require data reformation to the "lowest common denominator" for transmission and use by multiple GIS. Multiple conversions from one format to another may create a ratchet effect similar to making copies of copies on a photo copy machine. Additionally, international standards for cartographic data transmission, storage and retrieval are not fully implemented.

21.4.1.7 Accessibility

Accessibility to data is not equal. What is open and readily available in one country may be restricted, classified, or unobtainable in another. Prior to the break-up of the former Soviet Union, a common highway map that is taken for granted in this country was considered classified information and unobtainable to most people. Military restrictions, inter-agency rivalry, privacy laws, and economic factors may restrict data availability or the level of accuracy in the data. Some data such as archaeological sites must have restricted access to protect historically valuable assets.

21.4.1.8 Cost

Extensive and reliable data are often quite expensive to obtain or convert. Initiating new collection of data may be too expensive for the benefits gained in a particular GIS project and project managers must balance their desire for accuracy the cost of the information. True accuracy is expensive and may be unaffordable.

21.4.2 Natural Variations and Original Measurements

Many of the natural variations and original measurement issues revolve around accuracy and precision. These variations concern both spatial and attribute data. Although these error sources may not be as obvious, careful checking will reveal their influence on the project data.

21.4.2.1 Positional Accuracy

Positional accuracy is a measurement of the variance of map features and the true position of the attribute (Antenucci *et al.*, 1991). The ability to determine accuracy depends on the type of phenomenon being used or observed. Mapmakers can accurately place well-defined objects and features such as roads, buildings, and discrete topographical units on maps and in digital systems, whereas less discrete boundaries such as vegetation or soil type may reflect the estimates of the cartographer. Climate, biomes, relief, soil type, drainage and other features lack sharp delineation in nature and are subject to interpretation. Faulty or biased field work, map digitising errors and conversion, and scanning errors can all result in inaccurate information for these data.

There is also a danger in false accuracy. That is reading locational information from data sets to levels of accuracy beyond which they were created. This is a very great danger in computer systems that allow users to pan and zoom at will to an infinite number of scales. Accuracy and precision are tied to the original data set and do not change even if the user zooms in and out. Zooming in and out can however mislead the user into believing, falsely, that the accuracy and precision have improved.

21.4.2.2 Content Accuracy

GIS data sets should be correct and free from bias. Qualitative accuracy refers to the correct labelling and presence of specific features. For example, a pine forest may be incorrectly labelled as a spruce forest, thereby introducing error that may not be known or noticeable to the data user. Certain features may be omitted from the spatial database through oversight, or by design.

Other errors in quantitative accuracy may occur from faulty instrument calibration used to measure specific features such as altitude, soil or water pH, or atmospheric gases. Mistakes made in the field or laboratory may be undetectable in the GIS project unless the user has conflicting or corroborating information available. Information about how observations were made and the instrumentation that was used is valuable in assessing content accuracy. For example, if it is important to know the mean water level at a particular tide cycle, then it is important how the determination of time and the tide cycle were established.

21.4.2.3 Attribute Accuracy

Attribute data linked to location may also be inaccurate or imprecise. Inaccuracies may result from methods of collection or misunderstandings of the content and definition of attribute information.

21.4.2.4 Attribute Completeness

This is a measure of the extensiveness of an attribute data set. For example, a complete description of a shoreline property might include the address, the current owner, the assessed value, the date of purchase, a legal description, and any zoning or restrictive covenant limitations on use. An incomplete description might include just the current owner and the assessed value.

21.4.2.5 Precision of Attribute Information

GIS depend upon the abstraction and classification of real-world phenomena. The users determine what amount of information is used and how it is classified into appropriate categories. Sometimes users may use inappropriate categories or misclassify information. For example, classifying major storm events by date and name would probably be an ineffective way to study the effect of storm strength on property damage. Failing to classify storm events by strength and duration would limit the effectiveness of a GIS designed to predict property damage and emergency government requirements. Even if the correct categories are employed, data may be mis-classified. A study of drainage systems may involve classifying streams and rivers by "order", that is where a particular drainage channel fits within the overall tributary network. Individual channels may be mis-classified if tributaries are miscounted. Yet some studies might not require such an accurate categorisation of stream order at all. All they may need is the location and names of all stream and rivers, regardless of order.

21.4.2.6 Logical Accuracy and Precision

Information stored in a database can be employed illogically. For example, permission might be given to build a residential subdivision on a floodplain unless the user compares the proposed subdivision plan with floodplain maps. Then again, building may be possible on some portions of a floodplain but the user will not know unless variations in flood potential have also been recorded and are used in the comparison. The point is that information stored in a GIS database must be used and compared carefully if it is to yield useful results. GIS systems are typically unable to warn the user if inappropriate comparisons are being made or if data are being used incorrectly. Some rules for use can be incorporated in GIS designed as "expert systems", but developers still need to make sure that the rules employed match the characteristics of the real-world phenomena they are modelling.

21.4.2.7 Variation in Data

Variations in data may be due to measurement error introduced by faulty observation, biased observers, or by mis-calibrated or inappropriate equipment. For example, one cannot expect sub-meter accuracy with a hand-held, non-differential GPS receiver. Likewise, an incorrectly calibrated dissolved oxygen meter would produce incorrect values of oxygen concentration in a stream.

There may also be a natural variation in data being collected, a variation that may not be detected during collection. As an example, salinity in Texas bays and estuaries varies during the year and is dependent upon freshwater influx and evaporation. If one was not aware of this natural variation, incorrect assumptions and decisions could be made, and significant error introduced into the GIS project. In any case if the errors do not lead to unexpected results their detection may be extremely difficult.

21.4.2.8 Numerical Errors

Different computers may not have the same capability to perform complex mathematical operations and may produce significantly different results for the same problem. Burrough (1998) cites an example in number squaring that produced 1,200% difference. Computer processing errors occur in rounding off operations and are subject to the inherent limits of number manipulation by the processor. Another source of error may from faulty processors, such as the recent mathematical problem identified in Intel's Pentium chip. In certain calculations, the chip would yield the wrong answer.

A major challenge is the accurate conversion of existing to maps to digital form (Muehrcke, 1986). Because computers must manipulate data in a digital format, numerical errors in processing can lead to inaccurate results. In any case numerical processing errors are extremely difficult to detect, and perhaps assume a sophistication not present in most GIS workers or project managers.

21.4.2.9 Topological Analysis

Logic errors may cause incorrect manipulation of data and topological analyses (Star and Estes, 1990). One must recognise that data is not uniform and is subject to variation. Overlaying multiple layers of maps can result in problems such as slivers, overshoots, and dangles. Variation in accuracy between different map layers may be obscured during processing leading to the creation of "virtual data which may be difficult to detect from real data" (Sample, 1994).

21.4.2.10 Classification and Generalisation Problems

For the human mind to comprehend vast amounts of data it must be classified, and in some cases generalised, to be understandable. According to Burrough (1986) about seven divisions of data are ideal and may be retained in human short term memory. Defining class intervals is another problem area. For instance, defining a cause of death in males between 18-25 years old would probably be significantly different in a class interval of 18-40 years old. Data are displayed and manipulated most accurately in small multiples. Defining a reasonable multiple and asking the question "compared to what" is critical (Tufte, 1990). Classification and generalisation of attributes used in GIS are subject to interpolation error and may introduce irregularities in the data that are hard to detect.

Figure 21.1 illustrates the boundary between tidal inflow and fresh water The boundary between is a transition zone, but the appearance of that zone varies depending on the accuracy of the original collection and the level of generalisation that was applied to sample points.

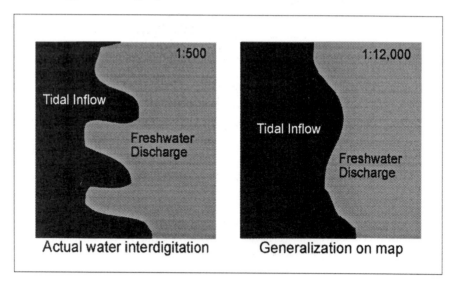

Figure 21.1 The relationship of accuracy and level of map generalization. Many natural phenomena shade into one another along a continuous or interdigitated boundary. When these boundaries are generalized and drawn on a map of a different scale, accuracy of position may be sacrificed.

21.4.2.11 Processing Errors

Processing errors can occur during any phase of data manipulation. This includes computing original measurements, digitising and geocoding, overlay and boundary intersections, and errors from rasterizing a vector map.

21.4.3 Compounding Error

The discussion thus far has focused on the information quality factors that may be present in single sets of data. GIS usually depend on comparisons of many sets of data. This schematic diagram shows how a variety of discrete data sets may have to be combined and compared to solve a resource analysis problem. It is unlikely that the information contained in each layer is of equal accuracy and precision. Errors may also have been made compiling the information. If this is the case, the solution to the GIS problem may itself be inaccurate, imprecise, or erroneous.

The point is that inaccuracy, imprecision, and error may be compounded in GIS that employ many data sources. There are two ways in which this compounding may occur.

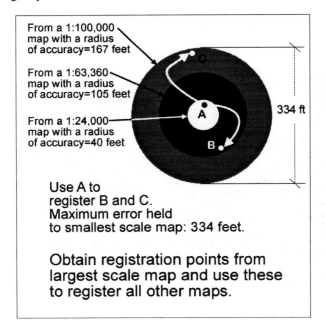

Figure 21.2 Positional accuracy and error propagation resulting from the registration of maps of different scales. Because positional accuracy is related to map scale, it is best to register all maps to the one with the largest scale. In this case, A is the largest scale map (1:24,000) and it is used to register points from maps B (1:63,360) and C (1:100,000). If instead map A were used to register map B, and map B then used to register map C, the inaccuracy inherent in each map would propagate to the next. This would mean that points might vary as much as 624 feet from their true positions, rather than by the 334 feet obtained using the suggested procedure.

21.4.3.1 Propagation

Propagation occurs when one error leads to another. For example, if a map registration point has been mis-digitized in one coverage and is then used to register a second coverage, the second coverage will propagate the first mistake. In this way, a single error may lead to others and spread until it corrupts data throughout the entire GIS project. To avoid this problem use the largest scale map to register points. Figure 21.2 illustrates a propagation error. Often propagation occurs in an additive fashion, as when maps of different accuracy are collated.

21.4.3.2 Cascading

Cascading means that erroneous, imprecise, and inaccurate information will skew a GIS solution when information is combined selectively into new layers and coverages. In a sense, cascading occurs when errors are allowed to propagate unchecked from layer to layer repeatedly.

The effects of cascading can be very difficult to predict. Figure 21.3 illustrates cascading errors. They may be additive or multiplicative and can vary depending on how information is combined, that is from situation to situation. Because cascading can have such unpredictable effects, it is important to test for its influence on a given GIS solution. This is done by calibrating a GIS database using techniques such as sensitivity analysis. Sensitivity analysis allows the users to gauge how and how much error will effect solutions. It is also important to realise that propagation and cascading may affect horizontal, vertical, attribute, conceptual, and logical accuracy and precision.

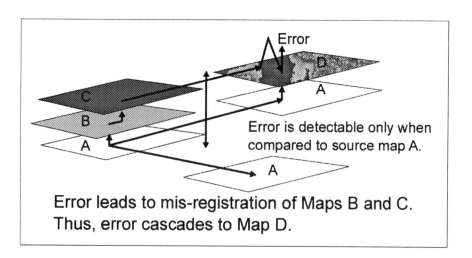

Figure 21.3 Cascading error in solving a GIS problem. In the case, errors introduced different stages of analysis skew the data to the point where the solution is no longer accurate.

21.5 CAUTIONS

GIS users are not always aware of the difficult problems caused by error, inaccuracy, and imprecision. They are often fall prey to false precision and false accuracy, that is, they report their findings to a level of precision or accuracy that is impossible to achieve with their source materials.

If locations in a GIS data set are measured within a hundred feet of their true position, it makes no sense to report predicted locations in a solution to a tenth of foot. That is, just because computers can store numeric figures with many decimal places does not mean that all those decimal places are "significant." It is important for GIS solutions to reported honestly and only to the level of accuracy and precision they can support. This means in practice that GIS solutions are often best reported as ranges or ranking, or presented within statistical confidence intervals.

Given these issues, it is easy to understand the dangers of using undocumented data in a GIS project. Unless the user has a clear idea of the accuracy and precision of a data set, mixing this data into a GIS can be very risky. Data that have been prepared carefully may be disrupted by mistakes someone else made. This brings up three important issues.

21.5.1 Ask for a Data Quality Report

Many major governmental and commercial data producers work to well-established standards of accuracy and precision that are available publicly in printed or digital form. These reports will report on exactly how maps and data sets were compiled (e.g., Table 21.1). Obtain and study these reports carefully.

Data quality reports take other forms when obtained from local and state government agencies or from private suppliers. The following report is employed by the Texas Natural Resources Information System to document data sets listed in its catalogue. The Federal Geographic Data Committee (FGDC) Metadata Standards for Geospatial Data is the best guide for capturing this information. The following data quality report is not in the FGDC format, but it contains the type of information that is in the FGDC standard.

Table 21.1 Sample data quality report.

Texas Water Development Board – Bays and Estuaries **Ambient Water Quality Monitoring** **Quality Report**	
What:	Salinity, Conductivity, Water Temperature, pH, Dissolved Oxygen
Where:	Selected stations in bays of Texas Gulf of Mexico estuaries, see map and text.
When:	Most records late 1986 through August, 1989. Some Stations 1987 through present. Some stations May 1990 through present. Other stations present with limited records. Collection interval hourly, 90 minute (most), or every two hours
How:	Program uses Hydrolab Datasondes: self-contained, battery powered, programmable instruments armed with sensors to measure above parameters
Why:	The purpose of this high frequency water quality monitoring effort was primarily to support calibration of estuary circulation and salinity simulation models and for development of statistical inflow-salinity relationships for the estuaries.
Data Characteristics	

Hydrolab Datasonde I's were deployed in the fall of 1986 in a number of Texas bays. After September 1989 some new sites were established and old sites abandoned. Map shows approximate locations of sondes. Locations were determined in part by the need for salinity data near the heads and mouths of major estuaries for purposes of salinity modeling. Locations were also determined by ease of access and availability of anchoring structures, in compromise with ideal locations.
At each location, the Datasonde was suspended to approximately mid-depth from a fixed structure. The sonde was housed in a heavy iron slotted pipe, to prevent damage from physical buffeting and to deter the curious. The protective pipe, but not the sonde or probes, was painted with anti-foulant. When fouling of the case did occur, the case was cleaned at monthly servicing. During the month-long deployment, probe surfaces became fouled to varying degrees, dependent somewhat on salinity.
Datasondes were calibrated prior to deployment according to manufacturers instructions. To check instrument accuracy, independent measures of water quality parameters were made at installation and again at removal of the instrument.
Temporal coverage is not complete. There are data gaps. Not all data files have all the parameters. In most cases in which inspection showed probe readings to be unreliable the probe records were deleted. Users should still be cautious. The instruments were deployed for one-month periods. During this time, biological fouling occurs to greater or lesser extents dependent on temperature, salinity, nutrients in the water. This fouling degrades accuracy of readings, particularly dissolved oxygen. Data files have text headers including disclaimers.

21.5.2 Prepare a Data Quality Report

Data will not be valuable to others unless there is an accompanying data quality report. Even if data are not going to be shared, the data quality report is important for data maintenance and data tracking.

21.5.3 Ask Data Quality Report Questions

If there is no information quality for a data set, then ask questions about undocumented data before you use it.

- *What is the age of the data?*
- *Where did it come from?*
- *In what medium was it originally produced?*
- *What is the areal coverage of the data?*
- *To what map scale was the data digitised?*
- *What projection, coordinate system, and datum were used in maps?*
- *What was the density of observations used for its compilation?*
- *How accurate are positional and attribute features?*
- *Does the data seem logical and consistent?*
- *Do cartographic representations look "clean?"*
- *Is the data relevant to the project at hand?*
- *In what format is the data kept?*
- *How was the data checked?*
- *Why was the data compiled?*
- *What is the reliability of the provider?*

21.6 MEASURE AND MANAGE QUALITY

A number of methods are employed to measure and manage error in GIS and attribute data sets. One of the most important procedures is to calibrate or sample the data set against known information. In this way it is possible to understand strengths, weakness and fitness for use of data sets. Understanding information quality is a key component in understanding the usability of data.

21.7 REFERENCES

Antenucci, J.C., Brown, K., Croswell, P.L., Kevany, M. and Archer, H. 1991, *Geographic Information Systems: A guide to the technology* (New York: Chapman and Hall).

Burrough, P.A. and Rachael A. McDonnell, 1998, *Principles of Geographical Information Systems for Land Resources Assessment* (Oxfordshire: Oxford).

Clarke, K. 1990, *Analytical and Computer Cartography* (Englewood Cliffs, New Jersey: Prentice Hall).

Koeln, G.T., Cowardin, L.M., and Strong, L.L. 1994, Geographic information systems. In *Research and Management Techniques for Wildlife and Habitat*, edited by Bookhout, T.A. (Bethesda, Maryland: The Wildlife Society), p. 540.

Muehrcke, P.C. 1986, *Map Use: Reading, Analysis, and Interpretation* (Madison Wisconsin: J.P. Publications).

Sample, V.A., editor, 1994, *Remote Sensing and GIS in Ecosystem Management* (Washington, D.C: Island Press).

Star, J. and Estes, J. 1990, *Geographic Information Systems: An Introduction* (Englewood Cliffs, New Jersey: Prentice Hall).

Tomlin, C. Dana, 1990, *Geographic Information Systems and Cartographic Modeling: An Introduction* (Englewood Cliffs, New Jersey: Prentice Hall).

Tufte, E.R. 1990, *Envisioning Information* (Cheshire, Connecticut: Graphics Press).

21.8 ADDITIONAL READING

Bolstad, P.V. and Gessler, P., 1990, Positional uncertainty in manually digitised map data. *International Journal of Geographical Information Systems*, **4**, pp. 399-412.

Goodchild, M. and Gopal, S., editors, 1989. *Accuracy of Spatial Databases*, (Bristol: Taylor & Francis).

King, J.L. and Kraemer, K.L., 1985. *The Dynamics of Computing* (New York: Columbia University Press).

Openshaw, S., Charlton, M. and Carver, S., 1991. Error propagation: A Monte Carlo simulation. In *Handling Geographic Information: Methodology and Potential Applications*, edited by Masser, I. and Blakemore, M. (New York: John Wiley & Sons), pp. 102-114.

Scott, L.M., 1994, Identification of GIS attribute error using exploratory data analysis. *The Professional Geographer*, **46**, pp. 378-386.

CHAPTER TWENTY-TWO

Epilogue

Darius J. Bartlett and Dawn J. Wright

22.1 INTRODUCTION

As we write this chapter, a mere year and a half remains of the 20th century. The closing of a century (and a millennium) is a natural point at which to take stock of past achievements, assess current practices, and make plans for the future. The authors and editors of this book have attempted to provide just such an assessment with regard to research and recent developments in marine and coastal geographical information systems.

For practitioners in these fields, these are exciting times. Born in the 1960s (though conceived out of theories, concepts, technologies and visions laid in earlier times), the earliest GISs were designed to help solve land-based problems. Over the ensuing decades they have thrived and multiplied in that environment, growing ever more powerful and sophisticated in the process, so that in the late 1990s, their use for most terrestrial applications is so widely accepted as to frequently go unremarked.

We have seen earlier in this book, that GISs started their migration seawards in the mid-1970s (Bartlett, 1999). For much of the subsequent decade, the effective limit for most GIS applications lay at the coast. Here, there were challenges enough: the conceptual and technical challenges of representing a highly dynamic, multidimensional, fuzzy-bounded environment in a digital framework; and the institutional challenges (some would say these were much the harder to resolve) of reconciling multiple demands and frequently conflicting interests, both human and non-human, at the shore. Gradually, technical innovations, patient research, and political developments in equal measure addressed these different concerns, and prevailed to the point where coastal GIS is now starting to come of age.

With emerging success at the shore, it was only a matter of time before the first incursions of GIS to the deep ocean realm occurred. Wright (1999) has charted the key stages of this important evolutionary progression. Some of the challenges encountered in coastal and littoral environments are also met in the deep ocean, where they combine with others unique to the benthic and, particularly, the abyssal domains. Probably the most taxing problems in these latter categories arise from the sheer lack of knowledge regarding much of what happens below the surface of the world's oceans. With few exceptions, human engagements with marine environments have literally only scratched the surface, and we enter the 21st century still knowing remarkably and embarrassingly little about the two-thirds of the surface (and what happens below the surface) of this planet we live on. Thus, for deep ocean GIS, one of the biggest and most pressing challenges is simply that of acquiring more and more reliable data to work with.

We should also consider the changing milieu in which this evolution has occurred. In the early days of GIS, particularly in the 1960s and 1970s, most of the pioneering research and development in GIS arose within the academic sector. By the start of the 1980s, the commercialisation of mainstream GIS was well underway and, for the next two decades, much GIS development work was vendor-led. For specialist applications however (and this would include both the coastal and the marine sectors) it is still more usual, and perhaps more appropriate, for the core research to remain located within the academic sector. History shows that in disciplines that are undergoing rapid expansion and evolution (a situation which applies to marine and coastal science, as well as to GIS itself), many of the greatest advances arise from the work of radicals and visionaries, operating in academic environments that are comparatively free from the conservatism imposed by the commercial constraints of market viability. It is no accident that most of the chapters of the current volume dealing with technical and conceptual aspects of marine and coastal GIS are written by authors working in universities or in government-funded research settings.

Nevertheless, after many years of focus on terrestrial (environmental and socio-economic) applications, there are signs that the commercial sector is increasingly paying heed to the specialist needs of marine and coastal GIS users. Within the industry, many of the leading GIS developers and vendors are involved in collaborative research with marine scientists, to the certain benefit of both sides of the partnership. Recent industry-wide moves towards open, extendable GIS will certainly play an important role in this process, and it is likely that the future will see an increasing range and quantity of sector-specific extensions being developed for and by specialist end-user communities. These will "plug into" and add to a core set of generic system capabilities provided by the GIS vendors. Such developments are already evident in the growing desktop GIS sector: for example in the expanding suite of coastal, marine and other extensions available or under development for the ArcView and MapInfo systems respectively.

As has been indicated repeatedly in the various contributions to this book, much research remains to be done into improving, extending and optimising the capabilities of spatial information systems in the marine and coastal realms. Essential areas ripe for investigation include:

22.1.1 Strategies and Techniques for Marine and Coastal Data Collection

Data are the raw materials that fuel GIS use. Without adequate data, even the most sophisticated analytical techniques are rendered useless. The current paucity of reliable, relevant and usable data for the marine and coastal domain has been a recurring thread in the chapters of this book. Arising from this need, there is a clear imperative to expand significantly the range of data collecting technologies available to the scientist and manager. Even around our shores, routine collection of relevant environmental and other data is far from established, and huge gaps occur in the sampling density of such data, in both the spatial and the temporal dimensions. Offshore, the situation is even worse. Recent generations of satellite remote sensors have greatly helped routine monitoring and data collection for the ocean surface, but current sensors are unable to penetrate the water column. At present, remote sensing of the ocean bed, or the water columns overlying the seafloor, is essentially restricted to the use of sonar and occasionally moored buoys or drifters bearing data loggers.

22.1.2 Marine and Coastal Data Standards

The need for data standards is now recognised in all GIS application areas, and no less with regard to marine and coastal data. It may even be argued that the need for clear and unambiguous standards for data definition, quality control and data exchange is all the more important in marine and coastal environments, given the multiplicity of users and interest groups acquiring and using these data and also, frequently, the international dimension of the applications concerned. Although, as was indicated above, the global marine and coastal database is still limited in both size and content, the growing number of national and international research initiatives planned or under way make it certain that sizes of data holdings, and the flow of data and information between institutions, will grow rapidly and substantially. This mobility of data carries with it many benefits, but also a number of possible hazards: among the latter, the most pressing and potentially serious is the risk of data quality loss and creeping error or uncertainty in the results obtained from using such data; plus the concomitant danger that the wrong data will be applied inadvertently to the wrong situation. It is essential that the whole issue of creating and applying sound, internationally agreed data standards is investigated thoroughly, so as to maximise benefits and minimise any concomitant disadvantages.

22.1.3 Automating and Optimising Shoreline Definition and Maritime Boundaries

As was indicated in Chapter 21 (Palmer and Pruett, 1999), there have been numerous recent advances in the definition of shorelines and maritime boundaries. Nevertheless, much research remains to be done in this area. In particular, while WGS-84 provides a global horizontal datum for shoreline determination, the development of a global vertical datum is an area of ongoing investigation and discussion. Additionally, since shorelines are constantly undergoing change due to the forces of nature (and these changes can frequently be very rapid, or even catastrophic, in response to extreme weather events and other intense disturbances), new and more automated collection techniques are being explored to locate and extract rapidly the land/sea interface from multi-spectral and other imagery. Thus, key areas of research and development that need support from the international community include:

- the definition and adoption of a global vertical datum to improve the accuracy of future shoreline determination; and
- the development and adoption of an objective process for the automated or semi-automated extraction of shoreline from imagery to provide timely updates for shoreline changes.

Improvements in the definition of maritime boundaries are contingent upon the decisions of the international hydrographic community, the United Nations, unilateral and bilateral agreements of and between coastal states, and adjudication of disputes by authoritative bodies. Palmer and Pruett (1999) provides a summary of an approach for creating a comprehensive source of global maritime boundaries. The utility of a global maritime boundary database (GMBD) is a function of the currency of information. As subsequent claims, counter-claims, disputes and adjudicated boundaries are announced, they will be incorporated in the GMBD. This is a challenging task, since the dynamic nature of geo-political declarations ensures the need for frequent updates: this, too, poses many important questions worthy of research, including the development and testing of consensus-building techniques, for ensuring wider acceptance of international decisions on boundary definition; and investigation of the role of GIS as a tool for communicating such decisions to governments, other maritime interests, and the wider public.

22.1.4 Working with Three-dimensional (3-D) Data

The present generation of GIS evolved largely out of a translation of conventional cartographic metaphors into a digital environment. As a result, many of the inherent limitations of traditional mapping are embedded within current geoinformation systems. Foremost among these is the continuing emphasis on working in two, or (in the case of surface modelling) two and a half, spatial dimensions (Raper, 1999). Much geography, particularly in the coastal and marine realms, requires all three spatial dimensions to be considered equally, whether simply for visualisation, or else for more demanding spatial analyses. In fact visualisation and analysis are two separate, but complementary, aspects of working in three spatial dimensions: the former is concerned with the graphic presentation of data to the viewer in a volumetric (3-D) form, either in hard copy or on a computer display; while the latter involves actual manipulation and processing of the data in order to derive information or further data as output, which may or may not then be visualised.

While GIS supports two-dimensional spatial analysis very well, it does not provide easy-to-use 3-D visualisation or volumetric analysis, as is found with more specific software such as Fledermaus, Dynamic Graphics, IBM Visualization Data Explorer or Spyglass. Some attempts have been made to extend capability into the third dimension, most notably with Intergraph's Voxel Analyst, but so far such developments, while welcome, are still limited in the functions they provide. For example, a 3-D Analyst extension is now available for ESRI's ArcView, but despite the name this tool is suitable only for 2.5-D (surface) modelling and viewing, since the underlying data structure is not truly 3-D. Likewise, current database management systems (DBMSs) do not directly provide this capability either, although the prevailing view from industry is that it is probably easier and more appropriate to provide data to scientific visualisation applications from a DBMS than from a system such as ArcView.

22.1.5 Time Series Data

Most current GISs have only rudimentary support for analysis of the time series data needed in many marine and coastal studies. While considerable advances have been made in recent times, this weakness in GIS is still rooted in imperfect conceptual understandings of temporal data generally and, more specifically, of the spatial dynamics of many marine and coastal environments. Where temporality has been incorporated into marine or coastal GIS, it has mostly arisen through extension of the DBMS underpinning the software, where support for maintaining and viewing data as time series is better understood, primarily because of heavy use of time series by banks and stock exchanges who have lobbied DBMS companies for this functionality. The current fix for most GIS specialists is to provide links from the GIS to special purpose programs that provide time series analysis (e.g., Fourier analysis, power spectra, stochastic simulation, etc.). This is not an ideal solution, however, and the quest for truly temporal GIS remains one of the "holy grails" of much current GIS research. Any benefits accruing from such research would be felt across almost the entire span of the marine and coastal science, and would have important roles in many other application domains.

22.1.6 Computational Models and Experimental Flow

GIS have excellent capabilities for connecting maps to computational models or for linking programs running on one platform to those running on another. Linkages have been successfully made between GIS and a wide variety of process models drawn from the terrestrial domain: in the natural environment, these include groundwater contamination models, climate models, soil loss equations, surface hydrological models and others; while in the socio-economic domain they include modelling spheres of influence and sales pitches, modelling epidemics and flows of commodities, etc. In contrast, many of the techniques involved in coupling marine and coastal models to GIS are still poorly investigated or understood, and thus the benefits and synergy that can arise from bringing these different tools together are rarely seen.

Even where such coupling of models and GIS can be achieved, linking and scheduling these to run as a logical sequence for scientific investigations is no easy task. Just running a single model from within (or alongside) a GIS can be non-trivial. Running a series of computational (numerical) experiments with a GIS is even more problematic. Thus, an important direction for future research would appear to lie in developing and providing templates and computational tools for extracting data or importing results to, from, and between multiple process models and GIS. At present, such operations are best done via scientific notebook systems (e.g., Cuny *et al.*, 1997; Skidmore *et al.*, 1998) specifically designed to support scientific collaboration and simultaneous computational experimentation (i.e., several scientists working on the same data set, at the same time via a collaboration, even though they may be physically separated by great distances). It would be a significant advantage for marine and coastal scientists alike, to be able to schedule and conduct such experiments within a unified and integrated spatial information system environment.

22.1.7 Human Factors

While some of the research issues outlined above are technological in focus, it should be emphasised that, in common with recent concerns in many other areas of information system application, a growing degree of attention is now being devoted to the human and institutional contexts within which coastal/marine GIS use takes place. The past two decades have seen a huge, rapid uptake of GIS in academic, commercial, governmental, scientific, and other sectors, for a range of marine and coastal applications, and this awareness is translating into a gradually-expanding user base for the technology. However, using GIS, and using GIS *well*, are not necessarily one and the same thing. There is a need to ensure that end-users remain aware of both the strengths and the limitations of their tools, and this in turn requires that training of end-users keeps pace with advances in concepts and technology. It is important, also, to remember that GIS should always be a means to an end, and should never become the end in itself.

There is, therefore, great need for research into these human aspects: how can the technology be fine-tuned and adjusted to best serve the working environments and information needs of its marine or coastal users? And what are the training requirements of system end users, and how can these needs be best met? Specific areas where such human-oriented research appears to be most needed include the design and development of decision support systems to assist coastal or marine sector managers. What decisions require to be made, and what data and rules apply in the making of such decisions? How can these rules be translated into expert or other systems to provide necessary technological support, while still keeping sufficient control in the hands of the human operator? The design of appropriate user interfaces (e.g., Su, 1999), including, interfaces that will allow efficient use of GIS on the bridge of a pitching ship in a storm; within the confines of a submersible on the deep ocean floor; or on hand-held computers being operated by scuba divers; etc.; and the use of GIS as a teaching, training and awareness-building tool for promoting sustainable ocean use, integrated coastal zone management and other long- and short-term strategies, and for assisting in community participation in (especially coastal) planning and decision-making.

22.2 CONCLUSION

We live in a very rapidly changing world. Space exploration and travel, now taken almost for granted, are the product of a mere 30 years; the Internet which, as indicated in the Preface, was instrumental in helping this book see the light of day, is only some 20 years old; and the cell phones that are so ubiquitous in modern society are even younger. GISs are still also comparatively recent innovations and, against a plethora of potential user communities and application areas, continue to evolve and develop with sometimes bewildering rapidity.

The nature of science in general and, specifically, marine and coastal science, has also undergone radical change and evolution in the past three decades. In common with many other branches of, particularly the environmental sciences, marine and coastal science has largely rejected the reductionism and the hard-engineering-dominated ethos of "command and control" that so characterised much of the science applications in the first half of the 20th century. Gone is the idea of the seas as a sink, unlimited in its capacity to soak up the pollutants and waste products of human society. Gone, too, is the notion of the ocean as a boundless source of fish and other resources, without need for control or rationalisation of exploitation. And gone is the concept of armouring the shore and waging war between humanity and the raging ocean. Instead, in both the marine and the coastal domains, the new ethos preaches an holistic message, with sustainability of resource use, and the needs or rights of nature and of future human generations embodied firmly within its texts.

In the closing years of the 20th century, many concerns have been expressed by scientists, politicians and the public alike, regarding the health of the world's oceans and coastal zones. There are also indications that these warnings are being heard and heeded by decision-makers at all levels from the international to the strictly local. Whether in the natural sciences or the humanities, whether pitched at the conceptual, the technical or the application end, or, as is increasingly being demanded and attempted, in the integration of data and applications from all of these interest groups and professions, the role of GIS in this new thinking is clear. Applying GISs to marine and coastal environments presents taxing, but particularly satisfying challenges to end users and system developers alike. The chapters of this book demonstrate eloquently the many advances that have been achieved in recent years and the steps currently underway to address remaining issues. There is much research and development yet to be done, and we, the editors and authors of this volume, very much hope that the lead and examples demonstrated in these pages will encourage other scientists, investigators and end-users to join the ranks of the marine and coastal GIS communities.

22.3 REFERENCES

Bartlett, D.J., 1999. Working on the frontiers of science: applying GIS to the coastal zone, in this volume, Chapter 2.

Cuny, J.E., Dunn, R., Hackstadt, S.T., Harrop, C., Hersey, H., Malony A.D., and Toomey, D.R., 1997, Building domain-specific environments for computational science: A case study in seismic tomography. *International Journal of Supercomputer Applications*, **11**, pp. 1-21, http://www.cs.uoregon.edu/~harrop/papers/ETPSC96/.

Raper, J., 1999. 2.5- and 3-D GIS for coastal geomorphology, in this volume, Chapter 9.

Palmer, H. and Pruett, L., 1999, GIS applications to maritime boundary delimitation, in this volume, Chapter 21.

Skidmore, J., Sottile, M., Cuny, J. and Malony, A., 1998, A prototype notebook-based environment for computational tools. In Proceedings, Supercomputing 98, Orlando, Florida (Piscataway, New Jersey: IEEE), pp. 1-13, http://www.csi.uoregon/edu/nacse/vine/pub/sc98.html.

Su, Y., 1999, A user-friendly marine GIS for multi-dimensional visualisation, in this volume, Chapter 16.

Wright, D.J., 1999. Down to the sea in ships: The emergence of marine GIS, in this volume, Chapter 1.

Index